Nanocellulose

Scrivener Publishing
100 Cummings Center, Suite 541J
Beverly, MA 01915-6106

Publishers at Scrivener
Martin Scrivener (martin@scrivenerpublishing.com)
Phillip Carmical (pcarmical@scrivenerpublishing.com)

Nanocellulose

A Biopolymer for Biomedical Applications

Edited by
Mainak Mukhopadhyay
Department of Bioscience, JIS University, West Bengal, India
and
Debalina Bhattacharya
Department of Microbiology, Maulana Azad College, Kolkata, India

Scrivener Publishing

WILEY

This edition first published 2024 by John Wiley & Sons, Inc., 111 River Street, Hoboken, NJ 07030, USA and Scrivener Publishing LLC, 100 Cummings Center, Suite 541J, Beverly, MA 01915, USA

For more information about Scrivener publications please visit www.scrivenerpublishing.com.

Wiley Global Headquarters
111 River Street, Hoboken, NJ 07030, USA

For details of our global editorial offices, customer services, and more information about Wiley products visit us at www.wiley.com.

Library of Congress Cataloging-in-Publication Data

978-1-394-17237-5

Cover image: Pixabay.Com
Cover design by Russell Richardson

Set in size of 11pt and Minion Pro by Manila Typesetting Company, Makati, Philippines

Printed in the USA

10 9 8 7 6 5 4 3 2 1

Contents

Preface xvii

1 Nanocellulose: A Cutting Edge Biopolymer - An Overview 1
Bidisha Saha and Mainak Mukhopadhyay
1.1 Introduction 2
1.2 Nanocellulose: A Brief Overview 3
1.2.1 Structure and Source of Cellulose 3
1.2.2 Nomenclature and Types of Nanocellulose 5
1.3 Extraction of Nanocellulose 6
1.3.1 Mechanical Extraction 6
1.3.2 Chemical Extraction 9
1.4 Surface Modification and Functionalization for Nanocelluloses 10
1.4.1 Esterification 11
1.4.2 Silylation 11
1.4.3 Amidation 12
1.4.4 Sulfonation 12
1.5 Applications of Nanocellulose in Polymeric Composites 13
1.5.1 Nanocellulose-Reinforced Polymer Nanocomposites 13
1.5.1.1 Dispersion of Nanocellulose in the Matrix 14
1.5.1.2 Mechanical Performance of Resulting Nanocomposite 14
1.5.2 Nanocellulose-Based Hydrogels 16
1.5.3 Nanocellulose-Based Filaments 17
1.5.4 Nanocellulose for Biomedical Application 18
1.6 Summary and Prospects 19
1.6.1 Advantages of Nanocellulose-Reinforced Polymer Composites 19
1.6.2 Current Challenges 20
1.6.2.1 Functionalization 20
1.6.2.2 Mass Production 20

1.6.3 Conclusion and Future Developments 21
References 21

2 Cellulose Nanofibers (CNF) and Nanocrystals (CNC): Pre-Treatment, Preparation, and Characterization 27
Jiawei Chen, Yu Chen and Tanushree Ghosh
2.1 Introduction 28
2.2 Cellulose Nanofibers (CNF) 29
2.2.1 Pre-Treatment 29
2.2.1.1 Retting Process 30
2.2.1.2 Desalting Process 30
2.2.1.3 Dewaxing Process 31
2.2.1.4 Alkali Treatment 31
2.2.1.5 Ionic Liquid Extraction 31
2.2.1.6 Bleaching Treatment 32
2.2.1.7 Acid Treatment 32
2.2.1.8 Microwave-Assisted Hydrolysis 33
2.2.1.9 TEMPO-Oxidation 33
2.2.2 Extraction 34
2.2.2.1 Homogenization 34
2.2.2.2 Microfluidization 35
2.2.2.3 Ball Milling 35
2.2.2.4 Grinding 35
2.2.2.5 Blending 36
2.2.2.6 Sonication 36
2.2.2.7 Steam Explosion 36
2.2.2.8 Enzymatic Hydrolysis 36
2.2.3 Post-Treatment 37
2.2.3.1 Dialysis 37
2.2.3.2 Centrifugation 37
2.2.3.3 Drying 37
2.3 Cellulose Nanocrystals (CNC) 38
2.3.1 Pre-Treatment 39
2.3.1.1 Alkali Treatment 39
2.3.1.2 Bleaching 40
2.3.1.3 Other Pre-Treatments 40
2.3.2 Extraction 40
2.3.2.1 Acid Hydrolysis 40
2.3.2.2 Enzymatic Hydrolysis 41
2.3.2.3 TEMPO Oxidation 42

2.3.2.4 Mechanical Process 42
2.3.3 Post-Treatment 42
2.4 Characterization 43
2.4.1 Morphology 43
2.4.1.1 Microscopy 43
2.4.2 Physical Characterization 43
2.4.2.1 XRD 45
2.4.2.2 Zeta Potential 46
2.4.2.3 TGA 46
2.4.3 Chemical Characterization 47
2.4.3.1 FTIR 47
2.4.3.2 XPS 48
2.4.3.3 Elemental Analysis 48
2.4.4 Mechanical Characterization 48
2.5 Conclusion 49
References 55

3 Synthesis and Characterization of Bacterial Nanocellulose 63
Moushali Sil, Kamolika Mukhopadhyay, Durba Roy, Sampriti Kundu and Debalina Bhattacharya
3.1 Introduction 64
3.1.1 Production of Bacterial Nanocellulose 65
3.1.2 Classification of Bcs Operon 66
3.2 Structure and Functions of Proteins Involved in Bcs Operon 70
3.3 Diverse Nature of Bcs Operon 72
3.4 Regulation of Bacterial Nanocellulose Biosynthesis 73
3.5 Genetic Manipulation of BNC Producing Strains to Increase Yield 74
3.6 Factors for BNC Production 77
3.7 Characterization of Bacterial Nanocellulose 80
3.7.1 X-Ray Diffraction Spectroscopy (XRD) 81
3.7.2 Nuclear Magnetic Resonance (NMR) 81
3.7.3 Fourier Transform Infrared (FTIR) 81
3.7.4 Scanning Electron Microscopy (SEM) 81
3.7.5 Transmission Electron Microscopy (TEM) 82
3.7.6 Atomic Force Microscopy (AFM) 82
3.7.7 Electrophoretic Light Scattering (ELS) 82
3.7.8 Thermogravimetric Analysis (TGA) 82
3.7.9 Differential Scanning Calorimetry (DSC) 83

3.7.10 Particle Size Analysis 83
3.8 Conclusion 83
References 84

4 Process and Applications of Electrospinning 89
Anushka Lather and Shubhangi Paliwal
4.1 Introduction 89
4.2 A Brief History of Electrospinning 90
4.3 Setup for the Experiment 92
4.4 Principle of the Process 93
4.5 Factors Affecting the Process of Electrospinning 93
4.5.1 Operating Parameters 94
4.5.1.1 Applied Voltage 94
4.5.1.2 Flow Rate 94
4.5.1.3 The Separation between the Metallic Needle's Tip and the Collector 94
4.5.2 Material Parameters 94
4.5.2.1 Polymer Concentration 95
4.5.2.2 Viscosity 95
4.5.2.3 Surface Tension 95
4.5.2.4 Conductivity 95
4.5.3 Properties of Solvents 96
4.5.3.1 Vapor Pressure 96
4.5.3.2 Allowability 96
4.5.4 Environmental Aspects 96
4.6 Electrospinning Variations 97
4.6.1 Needle-Based Electrospinning 97
4.6.1.1 Single Nozzle 97
4.6.1.2 Coaxial Electrospinning 98
4.6.1.3 Tri-Axial Electrospinning 99
4.6.1.4 Multichannel Electrospinning 99
4.6.2 Needleless Electrospinning 100
4.6.3 Melt Electrospinning 100
4.6.4 Emulsion Electrospinning 100
4.6.5 Solution Electrospinning 101
4.7 Applications of Electrospun Fibers 101
4.7.1 Biomedical Applications 101
4.7.1.1 Drug Delivery Systems 104
4.7.1.2 Tissue Engineering 104
4.7.1.3 Wound Healing 104

4.8 Summary, Conclusion and Future Prospects 105
References 105

5 Development of Nanocellulose-Based Nanocomposites and Its Properties 113
Sunita Adak
5.1 Introduction 114
5.1.1 Nanocomposite 114
5.1.2 Nanofillers 116
5.2 Nanocellulose 117
5.2.1 Types and Nomenclature 118
5.2.2 Sources and Preparation 119
5.2.2.1 Top-Down Process 120
5.2.2.2 Bottom-Up Process 123
5.3 Nanocellulose-Based Nanocomposites 125
5.3.1 Nanocellulose Properties 125
5.3.2 Nanocomposite Fabrication 126
5.3.2.1 Solution Casting 126
5.3.2.2 Melt Mixing 128
5.3.2.3 *In-Situ* Polymerization 129
5.3.2.4 Electrospinning 130
5.3.2.5 Other Fabrication Techniques 130
5.4 Properties of Nanocellulose-Based Nanocomposites 131
5.4.1 Mechanical Property 131
5.4.2 Thermal Property 132
5.4.3 Barrier Property 132
5.4.4 Optical Property 133
5.4.5 Other Properties 133
5.5 Conclusion 134
References 134

6 Surface Functionalization Process: Its Advantages and Disadvantages 141
Pranabi Maji and Dibyarupa Pal
6.1 Chemical Approach 141
6.1.1 Acetylation/Esterification 142
6.1.2 Polymer Grafting 143
6.1.3 Silylation 144
6.1.4 Acylation/Alkanoylation 144
6.1.5 Sulfonation 145
6.1.6 TEMPO-Mediated Oxidation 145

6.2 Enzymatic Approach 146
6.2.1 Direct Enzymatic Approach 146
6.2.1.1 Phosphorylation via Hexokinases 146
6.2.1.2 Cellulose Acylation/Esterification by Hydrolases 147
6.2.1.3 Esterification via Lipases 147
6.2.2 Indirect Enzymatic Approach 148
6.2.2.1 Surface Modification via Xyloglucan Endotransglucosylase 148
6.2.2.2 Grafting of Active Molecules 148
6.3 Physical Techniques 148
6.3.1 Plasma Treatment 148
6.3.2 Flame Treatment 154
6.3.3 Corona Treatment 154
6.3.4 Ion Beam Treatment 154
6.4 Conclusion 154
References 155

7 Applications of Nanocellulose in Tissue Engineering and Tissue Grafting 159
Sneha Dey, Paramjeet Singh, Sampriti Sarkar and Avik Acharya Chowdhury
7.1 Introduction 160
7.2 Nanocellulose and Its Properties 161
7.2.1 Physical Properties 162
7.2.2 Surface Chemistry 163
7.2.3 Biological Characteristics 164
7.2.3.1 Biological and Hematological Compatibility 164
7.2.3.2 *In Vivo* Biodegradability 165
7.3 Classification of Nanocellulose and Their Synthesis 166
7.3.1 Cellulose Nanocrystals (CNCs) 166
7.3.2 Cellulose Nanofibrils (CNFs) 167
7.3.3 Bacterial Cellulose (BC) 168
7.4 Composites Based on Nanocellulose 169
7.4.1 Nanocellulose and Graphene Composite 169
7.4.2 Nanocellulose and Polyvinyl Alcohol Composite 170
7.4.3 Nanocellulose and Chitosan Composite 171
7.5 Applications of Nanocellulose in Tissue Engineering and Tissue Grafting 172
7.5.1 Nanocellulose for Engineering Skin and Tissues 172
7.5.2 Nanocellulose for Repairing Blood Vessels 173

7.5.3 Nanocellulose in Neural Tissue Engineering 175
7.5.4 Nanocellulose in Bone Tissue Engineering 176
7.6 Limitations of Nanocellulose 177
7.6.1 Production Cost 177
7.6.2 Mechanical Properties 178
7.6.3 Scalability 179
7.6.4 Compatibility with Cells 179
7.7 Future Prospects 180
7.8 Conclusion 181
References 181

8 Application of Nanocellulose for Wound Dressings 193
B.A. Aderibigbe
8.1 Introduction 194
8.2 Types of Wounds 195
8.3 Nanocellulose-Based Wound Dressings 196
8.3.1 Hydrogels 196
8.3.2 Sponges 199
8.3.3 Nanofibers 201
8.3.4 Hydrocolloids 203
8.3.5 Films/Membranes 204
8.3.6 3D-Printed Wound Dressings 206
8.3.7 Composites 208
8.3.8 Other Scaffolds 209
8.4 Commercialized Nanocellulose-Based Wound Dressings 210
8.5 Future Perspective and Conclusion 211
Acknowledgments 212
References 212

9 Use of Nanocellulose for Drug Carriers for Drug Delivery Applications 221
Rubai Ahmed, Swarnali Das, Jhimli Banerjee, Sovan Samanta, Sandeep Kumar Dash and Arindam Pramanik
Abbreviations 222
9.1 Introduction 223
9.2 Strategies for Drug Delivery 224
9.2.1 Active Targeting Drug Delivery 224
9.2.2 Passive Targeting Drug Delivery 225
9.3 Application of Various Biomaterials for Drug Delivery 226
9.3.1 Synthetic Biomaterials 226

9.3.2 Natural Biomaterials 227
9.4 Different Methods Involved in Nanocellulose Synthesis 229
9.4.1 Mechanical Synthesis 229
9.4.2 Chemical Synthesis 230
9.4.3 Biological Synthesis 231
9.5 Types of Nanocellulose 232
9.5.1 Cellulose Nanofibers (CNFs) 232
9.5.2 Cellulose Nanocrystals (CNCs) 233
9.5.3 Microfibrillated Cellulose (MFC) 234
9.5.4 Bacterial Nanocellulose (BNC) 234
9.6 Different Methods Used for Drug Loading/Encapsulation of Drug in Nanocellulose 235
9.6.1 1D Formulation Strategies 236
9.6.2 2D Formulation Strategies 236
9.6.3 3D Formulation Strategies 238
9.7 Importance of Nanocellulose as Drug Carrier 239
9.7.1 Biodegradability and Biocompatibility 239
9.7.2 Toxicological Concern 239
9.8 Drug Delivery via Different Nanocelluloses 241
9.8.1 Nanocellulose Aerogels 241
9.8.2 pH-Responsive Hydrogels 242
9.8.3 Injectable Hydrogels—Implants 243
9.8.4 Magnetic Nanocellulose 243
9.9 Nanocellulose-Based Drug Delivery in Pathological Disorder 243
9.9.1 Inflammation 244
9.9.2 Cancer 244
9.9.3 Bone-Related Disorder 247
9.9.4 Pulmonary Disorder 247
9.9.5 Wound Healing 247
9.9.6 Optical-Related Disease 248
9.10 Commercialized and Research-Associated Nanocellulose Products Reported in Clinical Trials 249
9.11 Future Perspective 252
9.12 Conclusion 253
Acknowledgments 253
References 253

10 Preparation of Antibacterial Nanocomposite Materials Using Nanocellulose **267**
Moumita Bishai
10.1 Introduction 268
10.2 Nanocellulose 269
10.3 Functionalization of Nanocellulose 272
10.4 Nanocellulose-Based Nanocomposite 273
10.5 Properties of the Nanocomposites 273
10.6 Nanocellulose-Based Nanocomposite Processing 274
10.6.1 Solution Casting 274
10.6.2 Melt Intercalation 275
10.6.3 Impregnation 275
10.6.4 *In Situ* Polymerization 276
10.6.5 Coating 276
10.6.6 Freeze Drying 277
10.6.7 3D Printing 277
10.6.8 Electrospinning 277
10.6.9 *In Situ* Composites 278
10.6.10 Other Methods 278
10.7 Antibacterial Properties of the Nanocomposite Material 279
10.8 Conclusions and Future Trends 279
References 293

11 Application of Nanocellulose for Treatment of Renal Failure **297**
Prapti Chakraborty and Priyanka Ray
11.1 Introduction 298
11.2 Renal Disease and Its Implications 299
11.2.1 The Structure and Function of the Kidneys 300
11.2.2 Types of Renal Disease 300
11.2.2.1 Acute Kidney Injury (AKI) 300
11.2.2.2 Chronic Kidney Disease (CKD) 300
11.2.2.3 Glomerulonephritis 300
11.2.2.4 Polycystic Kidney Disease (PKD) 301
11.2.3 Implications of Renal Disease 301
11.2.3.1 Electrolyte Imbalances 301
11.2.3.2 Accumulation of Waste Products 301
11.2.3.3 Hypertension 301
11.2.3.4 Anemia 301
11.2.3.5 Bone Disease 302
11.2.4 Causes of Renal Disease 302
11.2.5 Symptoms of Renal Disease 303

11.2.6 Diagnosis of Renal Disease 303
11.3 Nanocellulose—Concept and Formulation 304
11.3.1 Methods of Extraction 305
11.3.1.1 Mechanical Extraction 305
11.3.1.2 Chemical Extraction 306
11.3.2 Cellulose Nanofibers (CNFs) 306
11.3.3 Cellulose Nanocrystals (CNCs) 307
11.3.4 Bacterial Cellulose (BC) 307
11.3.5 Cellulose Nanofibrils (CNFib) 308
11.3.6 Nanocellulose Aerogels 309
11.4 Kidney Targeted Drug Delivery by Nanocellulose 309
11.5 Conclusion 312
References 312

12 Use of Nanocellulose Hydrogels for Ophthalmic Applications 319
Gouranga Dutta, Debabrata Ghosh Dastidar and Abimanyu Sugumaran
12.1 Introduction 320
12.2 Nanocellulose Overview 321
12.2.1 Structure, Source, and Isolation of Nanocellulose 321
12.2.2 Types and Properties of Nanocellulose 325
12.3 Nanocellulose Compatibility for Ophthalmic Applications 327
12.4 Nanocellulose Hydrogel 328
12.5 Ophthalmic Application of Nanocellulose 330
12.6 Conclusion and Future Aspect 331
References 332

13 Application of Nanocellulose as Implant and Grafting Materials 335
Shubhangi Paliwal and Anushka Lather
13.1 Introduction 336
13.2 Chemical as Well as the Structural Makeup of Cellulose Fibers 337
13.3 Cellulose—Origin, Extraction, and Sources 339
13.4 Isolation, Chemical Transformations, and Purification of Cellulose 341
13.4.1 Nanofibrillar Cellulose (NFC) 341
13.4.2 Cellulose Nanocrystal or CNC 341
13.4.3 Microcrystalline Cellulose (MCC) 344
13.4.4 Nanocellulose—Chemical Alterations 344

13.5 Application of Nanocellulose as Implant and Grafting Materials 345
13.6 Conclusion 352
References 353

14 Use of Nanocellulose for Dental Applications 359
Rupankar Ghosh, Amlan Roychowdhury and Ditipriya Hazra
14.1 Introduction 359
14.1.1 Nanocellulose versus Cellulose 360
14.1.2 Bacterial Nanocellulose Production 361
14.1.3 Plant-Based Nanocellulose Preparation 361
14.2 Application of Nanocellulose in Dental Science 362
14.2.1 BNC-Based Membrane Barrier in Guided Tissue Regeneration 362
14.2.2 Periodontal Disease Treatment 363
14.2.3 Class II Furcation Lesion Treatment 363
14.2.4 Oral Mucosa-Based Surgical Wound Treatment 364
14.2.5 Treating Canker Sore 364
14.2.6 Dental Pulp Tissue Treatment and Soft Tissue Reinforcement 365
14.2.7 Dental Root Canal Treatment 366
14.2.8 Chitosan-Based Hydrogel Reinforced with Oxidized Nanocellulose in the Field of Tooth Engineering 367
14.2.9 Pickering Emulsion-Based Oral Drug Delivery 367
14.3 Conclusion 368
References 369

15 Future Prospects of Bacterial Nanocellulose and Its Composites 373
Anbazhagan Thirumalai, Mahashweta Mitra Ghosh, Agnishwar Girigoswami and Koyeli Girigoswami
Abbreviations 374
15.1 Introduction 374
15.2 Cellulose 375
15.3 Bacterial Nanocellulose (BNC) 376
15.3.1 BC/Natural Polymer Composites 376
15.3.2 BC/Water-Soluble Synthetic Polymer Nanocomposites 377
15.3.3 BC/Thermoplastic Nanocomposites 378

15.3.4 BNC-Based Electroconductive Polymer Nanocomposites 379
15.4 Bacterial Nanocellulose-Based Hybrid Nanocomposites 379
15.4.1 BNC/Metal and Metal Oxide Nanocomposites 379
15.4.2 Bacterial Cellulose Carbon Nanocomposites 380
15.5 Biomedical Applications of BNC 382
15.5.1 Drug Delivery 382
15.5.2 Biosensing 382
15.5.3 Wound Dressing Material 383
15.5.4 Antimicrobial Activity 383
15.5.5 Artificial Skin 383
15.5.6 Tissue Engineering 384
15.5.7 Vascular Tissue Engineering 384
15.5.8 Bone Tissue Engineering 384
15.5.9 Cartilage Tissue Engineering 384
15.5.10 Anticoagulant Character 385
15.6 Conclusion 386
Acknowledgments 386
References 386

Index **391**

Preface

Nanocellulose is a promising nanomaterial with unique qualities including low cost. durability, non-toxicity, accessibility, etc. Basically, cellulose can be classified into two types: nanocrystals and nanofibrils, depending on the way it is extracted from trees, plants, or other cellulose-containing species. Textiles, cosmetics, and food products are just a few of the commercial uses for nanocellulose. It also has a strong potential for use in medicine.

This book presents the most recent findings in scientific research on nanocellulose as a biopolymer and its potential uses in medicine. The most recent research on nanocellulose structure and modification is presented in this book, along with in-depth details on the material's present and potential future uses.

Chapter 1 focuses on the basic structure of nanocellulose their different sources, nomenclatures, and types. Different types of nanocellulose (nanocrystals and nanofibrils), their production, pretreatment, and characteristics have been discussed in Chapter 2. Currently, bacterial nanocellulose has gained more attention than plant cellulose or fungal cellulose.

Chapter 3 explains the synthesis of bacterial nanocellulose from different bacterial species and their characteristics. Nanofibrillated cellulose is produced by a simple technique using a high-voltage electrode known as electrospinning. In Chapter 4, the detailed processes and applications of electrospinning have been summarised. Nanocellulose is also studied for the synthesis of novel nanocomposite materials. Chapter 5 focuses on the development of nanocellulose-based nanocomposites and their properties. Cellulose is insoluble in water, so surface modification is required to maintain the balance between its hydrophobic and hydrophilic constituents.

Chapter 6 discusses the various surface functionalization processes of cellulose and their advantages and disadvantages. Based on various surface functionalizations, nanocellulose is extensively used in several biomedical fields. Chapter 7 emphasizes the application of nanocellulose in the field of tissue engineering and grafting, their advantages, future prospects, etc. As nanocellulose-based biomaterials show excellent biocompatibility, reduced cytotoxicity, and biodegradability, one of their outstanding biomedical applications is wound dressing. Chapter 8 focuses on the application of nanocellulose in wound dressing. Based on various surface functionalizations, nanocellulose is extensively used as a carrier for drug delivery, which is discussed in Chapter 9. Another very promising application of nanocellulose is the synthesis of antibacterial nanocomposites for treating multi-drug-resistant bacteria.

Chapter 10 delves into the synthesis of antibacterial nanocomposites from nanocellulose, while Chapter 11 explores the role of nanocellulose in the treatment of renal failure. Chapter 12 emphasises nanocellulose hydrogel for ophthalmic applications. Chapter 13 focuses on the utilisation of cellulose in the biomedicine sector, mainly for grafting and implants. In Chapter 14, the role of nanocellulose in dental applications is explained. In the last chapter of the book, Chapter 15, the future prospects of bacterial nanocellulose and its composites are outlined.

This book can be used in education, research, and industrial applications. It covers both general and advanced topics, and has been designed to be pedagogical so that even beginners can learn the fundamentals of nanocellulose.

We wish to express our gratitude to everyone involved in this project for their efforts, as well as Wiley and Scrivener Publishing for their cooperation and assistance in the publication of this book.

Mainak Mukhopadhyay and Debalina Bhattacharya
November 2023

1

Nanocellulose: A Cutting Edge Biopolymer - An Overview

Bidisha Saha and Mainak Mukhopadhyay*

Department of Bioscience, JIS University, Kolkata, West Bengal, India

Abstract

As the ever-increasing need for plastics and other polymeric materials, producing them ethically is an essential part of the business. Making it a sustainable and environmentally friendly industry is therefore critical. Filler for either a synthetic matrix or a natural starch matrix, cellulose presents a great chance to lessen the impact of non-biodegradable elements. Due to their excellent mechanical properties, biodegradability, biocompatibility, high specific surface area, and rich hydroxyl groups for extensive chemical modification, nanocelluloses have recently become increasingly popular as naturally derived, bio-based nanometer-sized reinforcement in a wide variety of technological areas. However, for its intrinsic hydrophilicity and difficulties in dispersion inside a hydrophobic matrix, the extraction of nanocellulose from cellulosic biomass and its dispersion in the matrix remain major hurdles. The literature provides a summary of recent advancements in nanocellulose research, including methods of extraction, surface modification, and polymeric composite applications. This chapter demonstrates how the source of the cellulosic materials and the processing factors affect the morphologies and performances of nanocellulose. Although nanocellulose is derived from organisms or plants that initially appear to be fragile, it can be used as a reinforcement material or the primary component to create high-value, cutting-edge products like high-performance nanocomposites, multifunctional hydrogels, conductive filaments, medical dressings, and energy storage materials. The development of high-performance nanocellulose-reinforced polymer composites is presented, along with research priorities and directions.

Keywords: Nanocellulose, natural biopolymer, nanocellulose-based nanocomposites, reinforced polymer composites, polymer, functionalization

**Corresponding author*: m.mukhopadhyay85@gmail.com

Mainak Mukhopadhyay and Debalina Bhattacharya (eds.) Nanocellulose: A Biopolymer for Biomedical Applications, (1–26) © 2024 Scrivener Publishing LLC

1.1 Introduction

The use of biomasses for the preparation of polymer-based composites is expanding quickly in academia as well as industry due to the overuse and exhaustion of petroleum-based feedstocks as well as significant damage to the environment. Cellulose is being used for decades as a renewable energy source since it is the biomass resource with the highest availability. It is the main carbon source for the growing renewable sector that is creating functional composites [1–3]. It can be derived from numerous different sources including plants, fungi, and marine species, and because of its sustainability, non-toxicity, and biodegradable properties, it has a broad spectrum of applications in the creation of paper, coatings, cosmetics, and pharmaceutical industries. Cellulose is composed of glucose units connected by 1,4-glycosidic linkages. Many freely reactive hydroxyl groups are also present on the exterior of cellulose at the C2, C3, and C6 locations, which results in hydrogen bonding between the linear chains [4–6]. Cellulose is a perfect bio-filler for synthetic or natural polymers due to its excellent physicochemical, mechanical, and chemical characteristics. The inherent hydrophilicity, poor solubility, and infusible process ability of cellulose limit its employment in high-value-added domains, despite the fact that it is a good bio-filler with many beneficial features [7, 8].

In recent years, it has become increasingly difficult to obtain cellulose's nanoscale structured components for use in the production of numerous valuable bio-compatible nanocomposite and commercialized cellulose compounds. In general, there are three different varieties of this cellulose-based nanostructured material, often known as nanocellulose: cellulose nanofibrils (CNFs), cellulose nanocrystals (CNCs), and bacterial nanocellulose (BNC) [9, 10]. This classification is based on the raw material, production process, and fiber morphology. However, there are still some significant challenges to the use of nanocellulose. Although there are numerous physical, chemical, and biosynthetic methods for producing nanocellulose, these processes remain dependent on intensive chemical processing, specialized equipment, or unorthodox raw materials. An economically feasible, efficient, and sustainable processing technology for nanocellulose is needed to decrease manufacturing expenses and prices. Polymeric composites with excellent biocompatibility and low toxicity can benefit from using nanocellulose as nanofiller for reinforcement. Though it has a propensity to clump easily because of its hydrophilicity which frequently leads to poor dispersion in the matrix and decreased performance of composites. Additionally, the keratinization that may happen as a result of nanocellulose drying could make it difficult to obtain dry nanoparticles or result in product flaws [11].

The goal of this chapter is to summarize the recent advances in the study of nanocellulose, provide a brief overview to the processes involved in its extraction and modification, and discuss its uses in polymeric composites, hydrogels, and filaments, as well as nanocellulose's potential uses in biomedicine.

1.2 Nanocellulose: A Brief Overview

1.2.1 Structure and Source of Cellulose

A crystalline structure of cellulose known as primary fibrils is created when the newly produced cellobiose units are connected to one another. To create macro-fibrils, or cellulosic fibres, these latter are grouped together to make micro-fibrils. The unique characteristics of cellulose, such as chirality, ease of chemical functionalization, hydrophilicity, insolubility in most aqueous solvents, and infusibility, are due to its intra- and intermolecular chemical groups. Anhydroglucose units (AGUs), which make up the majority of cellulose, are repeated in the 4C1-chain structure with each monomer unit corkscrewed at an angle of 180 degrees with respect to its neighbours [12]. Evidently, cellulose is produced from wood and cotton, with respective polymerization degrees of 10,000 AGUs and 15,000 units, respectively. The degree of polymerization and the polymeric chain size imparts significant result on cellulose's characteristics. Structured (crystalline) and unstructured (amorphous) regions both combine to form natural cellulose fiber. The crystallinity of the material can range from 40 to 70%, depending on the selection of extraction method. Compared to the crystalline sections, the amorphous regions are less packed and more likely to have interactions with adjacent molecular groups [13–15]. Depending on the molecular orientations, van der Waals, intra- and intermolecular interactions, isolation technique, and treatment procedure, cellulose exists in a variety of polymorphs, including cellulose I, II, III (I), III (II), IV (I), and V. These polymorphs can be interconverted by thermal or chemical treatments [16, 17]. Table 1.1 lists some of the sources of cellulose, including waste paper, wood, grass, herbaceous plants, agricultural crops, animal sources, algae, and bacterial sources. Figure 1.1 illustrates the evolution of cellulose from natural sources to fundamental molecules. Natural resource availability, origin and maturity, pretreatment and processing techniques, and reaction conditions can all affect the qualities of the cellulose that is produced [21–23]. In general,

Table 1.1 Several sources for cellulose fibre synthesis [18–20].

Source	Example
Hardwood	Eucalyptus, Aspen, Balsa, Oak, Elm, Maple, Birch
Softwood	Pine, Juniper, Spruce, Hemlock, Yew, Larch, Cedar
Annual plants/ Agricultural residues	Oil palm, Hemp, Jute, Agave, Sisal, Triticale straw, Soybean straw, Alfa, Kenaf, Coconut husk, Begasse, Corn leaf, Sunflower, Bamboo Canola, Wheat, Rice, Pineapple leaf and coir, Peanut shells, Potato peel, Tomato peel, Garlic straw residues, Mulberry fiber, Mengkuang leaves
Animal	Tunicates, Chordata, Styela clava, Halocynthia roretzi Drasche
Bacteria	Gluconacetobacter, Salmonella, Acetobacter, Azotobacter, Agrobacterium, Rhizobium, Alkaligenes, Aerobacter, Sarcina, Pseudomonas, Rhodobacter
Algae	Cladophora, Cystoseria myrica, Posidonia oceanica

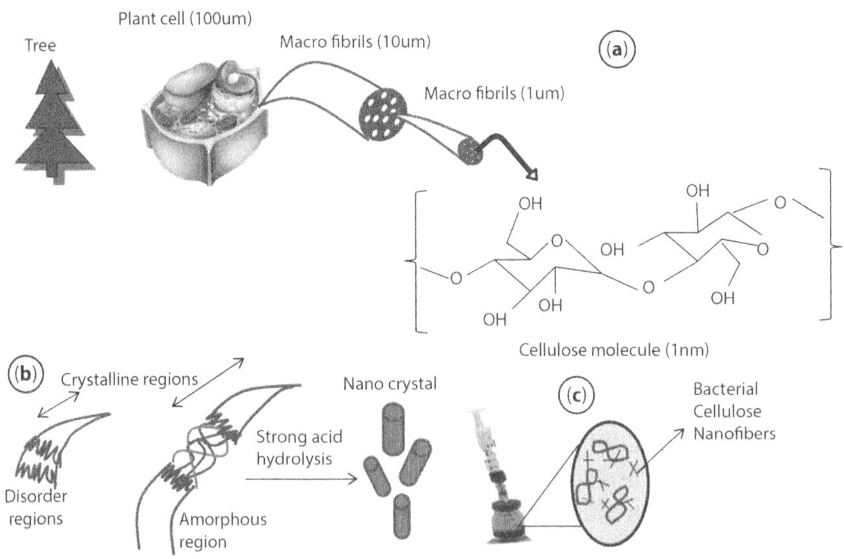

Figure 1.1 The hierarchical structure of cellulose, which extends from the meter to the nanometer scale, is depicted in figure (a). The reaction between cellulose and strong acid to produce nanocellulose is depicted schematically in (b). (c) shows bio nanocellulose grown from cellulose-synthesizing bacteria.

to remove non-cellulosic elements from bleaching, delignification, lignocellulosic sources, and the removal of extractive components (flavonoids, free sugars, fat, resin, tendon, terpenoids, tannins, terpenes, and waxes) are necessary [24, 25]. These pre-treatments can performed by a variety of chemical, physical, biological, and combined approaches [26–29]. They enable lignocellulose's intractability to be overcome as well as its compact structure. More than 40% of the total processing cost goes towards pretreatments [30]. These pretreatments enable access to the cellulose fraction, disrupt the links between cellulose and non-cellulosic substances (lignin and hemicellulose), and permit the separation of pure and microcrystalline cellulose. Some pretreatments may have a negative impact on the process due to the creation of potentially hazardous wastes, ineffective separation, cellulose degradation and loss, as well as the high overall costs of the procedure. Numerous studies are still being done in order to improve the processes' efficacy and effectiveness, lower their costs, and lessen their negative effects on the environment while also better understanding the phenomena that can arise during pretreatments. In order to get pure cellulose from animal cellulose, Trache *et al.* (2017) [18] claim that pretreatments are frequently required. Bacterial cellulose does not include extractives, hemicellulose, or lignin, in contrast to biomass-based cellulose; hence, pretreatments are not necessary. However, the price of producing it on an industrial scale is still quite high [31].

1.2.2 Nomenclature and Types of Nanocellulose

From bionanocomposites to medicine to sensor and biosensing functions, nanotechnology is establishing itself as one of the key drivers of new industrial revolutions in a number of disciplines. Although cellulose has been thoroughly studied for many years, according to three popular databases, PubMed, Web of Science, and ProQuest, nanocellulose has only recently emerged as a significant and distinctive component. Some of the advantageous properties of this nanomaterial are its high surface-to-volume ratio, excellent modulus and tensile properties, low coefficient of thermal expansion, ability to establish hydrogen bonds, biocompatibility, sustainability, homogeneity, and lack of toxicity [32]. According to the literature, a number of terminologies have been used to define nanocellulose or cellulose nanoparticles, which unfortunately leads to misunderstandings and ambiguities [18, 33]. Establishing a standard nomenclature for the family of nanocelluloses is essential because there are still irregularities in the naming of nanocelluloses. Additionally, efforts must be made to rationalise the terminology used in relation to the morphology, size, and synthetic processes of the various compounds. A few years ago, a branch of the Technical Association of the Pulp and Paper Industry (TAPPI) was constituted with the goal of standardising the nomenclature of cellulose nanoparticles [34, 35]. But over the past few years, terms like "cellulose whiskers," "cellulose nanowhiskers," "cellulose nanocrystals," and "nanocrystalline cellulose" have gradually merged into the jargon [36, 37].

There are essentially three different subtypes of nanocellulose: nanofibrillated cellulose (NFCs), nanocrystalline cellulose (NCCs), and bacterial nanocellulose (BNCs). A comparison of the three main varieties of nanocellulose is presented in Table 1.2. This nanocellulose can be created either top-down or bottom-up. Typically, top-down strategies such as enzymatic, chemical, or physical techniques are used to isolate and extract NCCs and NFCs from crops, higher plants, and by-products of forestry and agriculture such as banana stems, rice straws, pineapples, oil palm, rubber, lalang, and sugar palm fibres. Contrarily, bacterial nanocellulose (BNC) is created by a particular bacterial family (*Glueconoacetobacker xylinius*) and is grown as microfibrils in culture media using a bottom-up method [39].

1.3 Extraction of Nanocellulose

Cellulose can be extracted using an abundance of mineral acids. With diameters ranging from 10 to 200 micrometres [40] and an average diameter of roughly 44.28 [41] micrometres, the crystalline phase is nanometer-scale. Cellulose can also be isolated in nanocellulose form. Nanocellulose fibres typically have a diameter of less than 100 nm [42] and a length of several microns. Cellulosic and non-cellulosic components like lignin, hemicellulose, pectin, waxes, and other extracts make up natural plant fibres. Non-cellulosic material must be eliminated in order to extract cellulose as micro- or nano-extracts. In order to extract nanocellulose from cellulosic materials, various methods have been devised. Diverse means of extraction produced divergent nanocellulose with distinct attributes. According to the conditions of the extraction method, the size and aspect ratio of the crystalline area of the cellulose may differ significantly. The types, sizes, and forms of fibrils (micro- or nano-sized) are frequently impacted by this. It is typically non-isometric, though. This section divides the primary extraction methods into two categories: mechanical processes and chemical processes.

1.3.1 Mechanical Extraction

Homogenization under high pressure is an example of a mechanical extraction method. For mass manufacturing of nanocellulose, a piston is

Table 1.2 Comparison between the different types of nanocellulose.

Characteristics	Bacterial nanocellulose (BNCs)	Nano fibrillated cellulose (NFCs)	Nano crystalline cellulose (NCCs)	Ref.
Common sources	Alcohols, low molecular weight sugar, and a number of bacterial species, including gluconacetobacter, agrobacterium, pseudomonas, rhizobium, and sarcin.	Pea hull, hardwood and softwood pulp, recycled pulp, acacia pulp, cotton linter, cotton, cassava bagasse, sugarcane baggase, wheat straw, mengkuang leaves.	Wood, cotton, rice straw, wheat straw, ramie, branch-bark mulberry, black spruce and eucalyptus, tunicate, valonia, kraft wood, and spruce, palm oil, pineapple leaf fibre, rice husk.	[38, 39]
Formation process	Bacterial synthesis and cultivation in medium containing phosphate, glucose, and oxygen in water.	Acid hydrolysis, enzymatic hydrolysis and mechanical shearing process integration by high pressure homogenization, and TEMPO-mediated oxidation.	Prior to delamination of wood pulp by mechanical pressure, acid hydrolysis, it may be treated chemically or enzymatically.	

(Continued)

Table 1.2 Comparison between the different types of nanocellulose. (*Continued*)

Characteristics	Bacterial nanocellulose (BNCs)	Nano fibrillated cellulose (NFCs)	Nano crystalline cellulose (NCCs)	Ref.
Morphology	Ribbon like	Rope like	Needle like	
Crystallite	84-80%	59-64%	54-88%	
Young's modulus	78GPa	180 GPa	150 GPa	
Size	20-100nm in diameter	5-60nm in diameter	5nm in diameter	
Synonyms	Biocellulose,bacterial cellulose, microbial cellulose	Microfibrillated cellulose, nanofibrils, microfibrils, cellulose, nanofibre	Crystallites, whiskers, nanowhiskers, cellulose nanocrystals	

utilised to force the material through an extremely thin channel or orifice at high pressures of 50–2000 MPa [40]. This technique for isolating nanocellulose is eco-friendly [43]. However, this method can mechanically disrupt the crystal structure. Grinding is another mechanical method. Nanocellulose is separated from fibre through grinding, which involves spinning grindstones at a speed of about 1500 rpm to provide shear stress to the fibre [44]. Water evaporates as a result of the heat created by friction during the fibrillation process, which facilitates better extraction [45]. Additionally, cellulose fibre is extracted through crushing. In cold environments, this technique is used to create microcellulose [46]. The manufactured cellulose has a size range of 0.1 to 1 m. Prior to high-pressure homogenization, this procedure can be employed as a pretreatment to produce nanocellulose. The cellulose is extracted using another low-energy technology called steam explosion, which uses steam to explode. It might be regarded as a pretreatment even though lignin is not entirely removed. The resulting fibre needs mechanical adjustment after using this technique.

1.3.2 Chemical Extraction

Alkaline roasting, acid roasting, chemical roasting, chemically assisted natural products (CAN), or degumming are all methods used in chemical extraction operations to remove the lignin component from the fibres. Pectin, hemicellulose, and other non-cellulosic materials are among the various elements of the fibre microstructure that are impacted by these treatments [47–50]. Alkali or acid retting is one application for the chemical extraction technique. Despite being less expensive, mechanical extraction causes less fibre damage [51]. The procedure is finished by heating, cleaning, and soaking the fibre in an alkaline or acidic solution. Using this technique may help improve several fibre properties [52]. Degumming, a chemical extraction technique designed to preserve the ramie fibre's structural integrity, works by eliminating the sticky and pectin components [53]. Another chemical technique is chemical retting. The lignin and water content of fibres are reduced using this technique. Chemical retting is more effective at removing lignin than alkali or acid retting, although it is less effective at removing moisture. A combination of mechanical and chemical extraction methods can be utilised to ensure improved lignin removal efficacy, with mechanical extraction frequently taking place following chemical processing. The mechanical and chemical methods utilised to extract nanocellulose from lignocellulosic biomass are shown in Figure 1.2 [54].

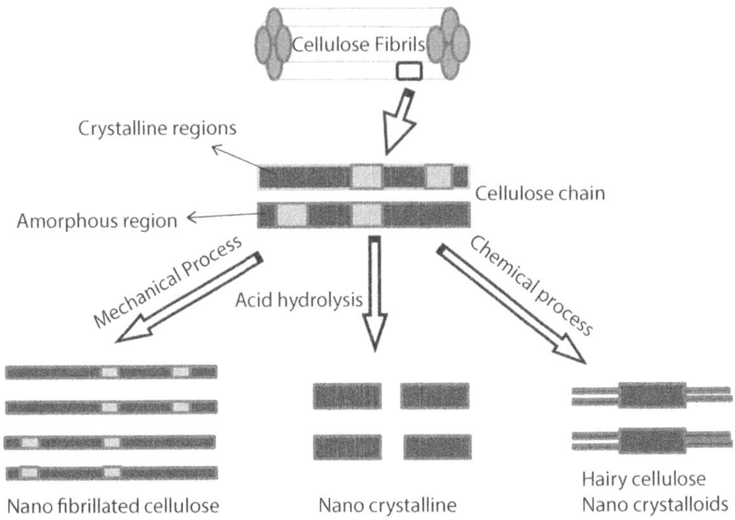

Figure 1.2 Conversion of lignocellulosic biomass into nanocellulose.

1.4 Surface Modification and Functionalization for Nanocelluloses

While there are many benefits to using nanocellulose in polymeric composites, its application is severely limited by a number of serious drawbacks. Low heat stability, moisture absorption, incompatibility with hydrophobic polymers, and poor redispersion are some of these shortcomings. Because of its large surface area, many hydroxyl groups (OH), and strong intra- and intermolecular hydrogen interactions between the hydroxy groups, nanocellulose has a hydrophilic surface and exhibits intense self-aggregation. As a consequence, like the majority of conventional non-polar polymers, it is difficult to distribute nanocellulose uniformly. It should be noted that the samples' poor thermal and dimensional stability may be brought on by the reactive groups on the outward of the nanocellulose. From a different perspective, the many reactive groups on the exterior of nanocellulose provide a variety of potential routes for the chosen alterations, which principally aim to decrease aggregation formation and polarity variances in nanoparticles. Nanocellulose's compatibility, dispersibility, and other qualities can be improved through surface modifications,which in turn opens up the possibilities to a wide range of high-value usage. Nanocellulose endures two main kinds of changes: structural changes that affect how it's organised, and functional changes that affect how its hydroxyl groups operates in. Nanocellulose's surface is often chemically modified to create a covalent link with reactive groups through procedures like silylation, esterification, oxidation, amination, carboxymethylation, sulfonation, polymer grafting, and others [55–59].

1.4.1 Esterification

The esterification reaction is the most prevalent within nanocellulose, and it is the most widely recognised and essential way to reduce the hydroxyl density of nanocellulose. The primary action of acetylation is to replace the hydroxyl groups on nanocellulose with acetyl groups. Typically, the condensation of a carboxylic acid group (COOH) with an alcohol group (OH) results in the introduction of an ester group. When using dilute sulfuric or perchloric acid as a catalyst, typical methods of production require for acid anhydrides or acyl chlorides as acyl donors. It could be done to confirm that the alcohol group of CNC reacts well with the acid anhydride by substituting it with chains of various lengths in the presence of H2SO4. By forming acetyl groups, the acetylation of nanocellulose lowers the hydroxyl density and minimises water absorption [60]. As a result, acetylated nanocellulose has less hydrolytic breakdown and better thermal stability [61]. Esterification, by which the wood-derived CNFs are modified with four different esters (acetyl, benzoyl, myristoyl, and pivaloyl), was discovered by Agustin *et al.* [62] to have the potential to significantly increase the nanofibers' thermal stability. The highly esterified amorphous regions of the nanocellulose are likely to blame for this.

1.4.2 Silylation

Alkoxysilanes or polysiloxanes interact with the hydroxyl groups of nanocellulose in silylation, a technique for altering surfaces. Hexamethyldisilazane (HDMS) [63], 3-aminopropyltriethoxysilane (APTES) [64], trichloromethylsilane (TC) [65, 66], and 3-methacryloxypropyltrimethoxysilane (MPS) [67] are some of the silane derivatives that have been used to make unique silylated nanocellulose. In general, there are several significant steps involved in the silylation of nanocellulose. The unique silylated surface of nanocellulose is produced via covalent bonding, when silane is first hydrolyzed in an aqueous solution to yield silanol, which may then be adsorbed on the surface of nanocellulose. One example of an alkylation modification that can successfully increase the dispersibility of nanocellulose and improve the interfacial contact with the hydrophobic matrix is tetrahydrofuran (THF), an organic solvent with low polarity. The degree of alkylation on the surface of the nanocellulose is directly correlated with the amount of silane present in the reaction. Extensive surface substitution will damage the crystalline regions of nanofibers, decreasing the strength of the nanocellulose. The optimal approach strikes a balance between a certain amount of silylation and the maintenance of nanocellulose structure. The silylation process commonly uses a variety of silylating compounds and reaction conditions to get around this challenge. Goussé *et al.* [68] changed CNCs and CNFs by silylating them using a range of chlorosilanes. They found that a moderate silylation protocol that needed fewer reagents and quicker reaction times resulted in the modified CNCs and CNFs being dispersible in a variety of low-polarity organic solvents and that their morphological integrity had been effectively preserved. The silylated nanocelluloses are employed in a greater variety of biomaterials, coatings, food packaging, optical devices, and polymer reinforcement because of their improved compatibility, hydrophobicity, mechanical strength and thermal stability [58, 69].

1.4.3 Amidation

The most widely used technique for surface modification is amidation, which generates modified nanocellulose with improved surface conductivity and hydrophobicity [70, 71]. An amine and carboxylic moiety reaction can produce the amide bond. Because there are not any carboxy groups on the surface of nanocellulose in general, the carboxy groups are also added through pre-oxidized treatment, such as TEMPO-mediated oxidation [72]. It is possible to combine metal ions or alkylammonium ions to produce nanocellulose with a range of functionalities by employing these carboxylate groups as a powerful platform.

1.4.4 Sulfonation

Sulfonation-esterification can further carry an anionic surface charge Nanocellulose surface. The hydroxyl groups on the surface of the nanocellulose are spontaneously substituted with hydrogens during the hydrolysis of sulfuric acid to produce sulphate half-esters. By preventing the formation of hydrogen bonds by electrostatic repulsion, the sulphate esters on the surface of CNCs or CNFs enable the development of colloidally stable suspensions. Nanocellulose can have carboxymethyl groups added using carboxymethylation to generate an ionic surface, which often leaves the surface negatively charged [73].

1.5 Applications of Nanocellulose in Polymeric Composites

1.5.1 Nanocellulose-Reinforced Polymer Nanocomposites

With synthetic polymer that exhibits distinct chemical and physical properties, nanocellulose can be reinforced [74]. Following that, mixtures of polymer and nanocellulose begin to synthesize materials with totally unique properties yet with relatively approachable qualities from different components. This circumstance is known as a nanocomposite. Polymers have lower modulus and tensile strength as compared to metals and ceramics. PNCs can be determine by

Polymer Matrix - Rg (Polymer radius of gyration),
The Nanoparticle - 2r (Diameter of nanospheres and nanotubes or thickness of nanoplates),
The Composite Morphology - 2ξ (Mean distance between constituents) [75].

This similar (**Rg ~ r ~ ξ**) correlation illustrates a geometrical factor that influences established conceptions of polymer dynamics. In numerous industries, including food packaging, energy and safety, transportation, electromagnetic shielding, military systems, sensors, catalysis, and the information business, polymer nanocomposite generated from both natural and petrochemical resources is used. The biodegradability agents, simple processing, bioabsorbability, minimal cost convincingness, non-abrasive texture, non-toxic, low density, and flammability of nanocellulose-polymer based composites make them advantageous. As a replacement for thermosets and thermoplastics like polypropylene, polystyrene, and high density polyethylene, nanocellulose is a remarkable endeavour. For more reinforcing capability, a higher aspect ratio is necessary. To improve the characteristics of the polymer with a 6% loading volume rate, nanocellulose whiskers or microcrystals derived from tunicate are utilised to enhance styrene and butyl acrylate copolymer latex. When nanocellulose is loaded at such a low rate, it often yields a promising, satisfying outcome that the resulting nanocomposite possessed mechanical properties that were notably better than those of the pure polymer in its elastomeric mode.The bio-based polymers polylactic acid (PLA), polyhydroxy acids (PHA), polyhydroxy butyrate (PHB), polybutylene succinate (PBS), and starch biopolymer are all reinforced by nanocellulose.Direct addition of nanofillers, in situ synthesis, as well as additional methods including electrospinning, self-assembly, phase separation, and template synthesis, can all be used to start creating polymer nanocomposites. Some microscopy techniques to determine polymer nanocomposites are mentioned in Table 1.3.

1.5.1.1 Dispersion of Nanocellulose in the Matrix

(Here polypropylene (PP) and polyethylene (PE) are used as matrix)

Due to their potential to be readily ingested by soil-dwelling microbes and the composite produced by the mixing of matrices and the host component, which results in chains of fragmented polymers with a larger surface area, which can be easily disposed of because they are natural fiber fillers that ultimately decompose into carbon dioxide and water. Polymers made from petroleum, such as polypropylene (PP) and polyethylene (PE), are toxic and non-biodegradable. By increasing the concentration of these natural fillers in the matrix, biodegradability may be expedited, although heavier materials may slow down the process. At the junction of their interactions, nanofiller with a large surface area and small internal load can provide an appropriate dispersion [76]. The load of concentration and rate of dispersion can have an impact on nanocomposites. Insoluble in water polymers may prevent a composite from developing fully during processing. Water soluble polymers can be used to achieve prominent dispersion. The hydrogen bonding between nanocellulose and a water molecule through hydroxyl groups on its surface has been used to detect intercourse activity. Acrylic monomer and ethylene-acrylic oligomer emulsion can encourage the dispersion in the resultant composite, and hydrophobization can reduce the interaction between hydroxyl groups. Solvent casting, melt processing, batch processing, and other processes can be used to develop nanocomposites.However, substantial tensile strength has been subjected to a range of about 70-80%, which is in advantage of future reinforcing potential of nanocellulose composites with matrices.

1.5.1.2 Mechanical Performance of Resulting Nanocomposite [76]

The physiochemical performance can be improved with nanofillers. The stiffness and ductility of a nanocomposite have been mentioned, which ultimately help to enhance the mechanical performance of the composite. As mentioned by their loading capacity, reinforcing elements can act as a stress transfer medium, affecting the crack formation which heading towards inhibitory path by stress absorption. The presence of nanocellulose

Table 1.3 Detection of polymer nanocomposites through microscopy techniques.

SEM (Scanning Electron Microscopy) and TEM (Transmission Electron Microscopy)	**Examine the internal and external features of the material**
WAXED (Wide Angle X-ray Diffraction)	The most ideal and easiest tool for nanocomposite structure determination
SAXS (Small Angle X-ray Diffraction)	Used to determine the structure of polymer composite by matching the contrast or combination of two probes
SPM (Scanning Probe Microscopy)	The small tip of microscope interacts with surface leading material for an image formation
STM (Scanning Tunneling Microscope)	A small sharp tip is settled down of about 0.5nm away from the surface material which can underpass from the tip to the surface material on atomic level
AFM (Atomic Force Microscopy)	Screening of the surface material can happen along with the scanning to ultimately examine its surface features and ruggedness
DSC (Differential Scanning Calorimetry) and DMA (Dynamic Modulus Analyzer)	Both these techniques plays an important role in determination thermal properties of nanocomposites
TGA (Thermo Gravimetric Analyzer)	This tool used to investigate wide variety of polymer properties, thermal stability and polymer improvement has been reported by TGA
TMA (Thermo Mechanical Analyzer)	Measure the co-efficient of thermal expansion (CTE) of nanocomposites
FTIR (Fourier Transform Infrared Spectroscopy)	Analyse the chemical characteristics of Raman spectroscopy using nanocomposites

within the polymer matrix caused it to consume the stresses therein. It may occur at any intersection.

Some Combinations are following which shows a better reinforcing performance;

- Poly(e-caprolactone) or (PCL) and cellulose nanocrystals or (CNCs) resulting nanocomposite give better performance as a reinforcing element which can be achieved through surface modification and better tensile strength.
- Cellulose nanofibrils or (CNFs) and polylactic acid or (PLA) result a better reinforcing nanocomposite through esterification modification can give better tensile strength and higher hydrophobicity.
- Sugar palm crystalline nanocellulose or (SPCNC) reinforced thermoplastic sugar palm starch or (TPS)/polylactic acid or (PLA) blend nanocomposite by melt blending process from solution casting indicate a better dispersed at polylactic acid phase through sonication in advance of starch gelatinization.
- Polylactic acid or (PLA)/polyethylene glycol or (PEG)/lignocellulosic nanofibers or (LCNF) nanocomposites by melt processing resulted in stronger toughness. (LCNFs) used to reinforce (PEG)-plasticized (PLA).

A polymer matrix's quality of the reinforcing element is dictated by the dispersion level. Whiskers and other additives can improve in the creation of composite materials with improved mechanical properties, but excessive bonding can reduce the strength of the nanocomposite at the point of breakdown.

1.5.2 Nanocellulose-Based Hydrogels

On the whole, hydrogels are hydrophilic gels made of crosslinked polymers that have swelled because of the presence of water. The hydrogels' functional groups, such as -NH, -OH, -COOH, and -CONH, among others, enable the retention of a significant quantity of water without becoming dissolved in it. Due to their high-water content, hydrogels are naturally flexible. Because water fills the spaces between macromolecules in hydrogels created by the crosslinking of two or more components, they can swell and hold onto large amounts of water without losing their three-dimensional form. Hydrogels offer a wide range of applications in treatments, drug administration, and particularly in tissue engineering because of their mechanical qualities. Although cellulose-derived materials are environmentally beneficial and hydrogel production is relatively inexpensive, nature may be trusted. The biomedical industry has took note of hydrogels derived from CNCs, CNFs, BNCs due to their greater surface area, improved mechanical strength, variable surface chemistry, amazing biocompatibility, target specific cellular recognition, and biodegradability. In comparison to synthetic superabsorbent hydrogels, cellulose-based superabsorbent hydrogels perform better in terms of absorbance, strength, biodegradability, biocompatibility, and good salt resistance. Nanocellulose-based hydrogels, specific high-quality materials for electronic and sensonic applications, cellulose-based diagnostic chips for biomedical applications, and 3D printing, among others, are all employed in the manufacture of pharmaceutical items and packaging [77]. The performance of printability can be improved by the addition of biopolymers. The swelling ability of hydrogels can be improved by using smart hydrogels that respond to stimuli and are based on a nanocellulose component. In comparison to chemically treated hydrogels, nanocomposite hydrogels exhibit higher viscosity.

Numerous laboratories worldwide are developing hydrogels based on nanocellulose. However, a standardised approach is needed to assess these hydrogels. It becomes challenging to compare the various mechanical properties of hydrogels developed by different labs without a common method to define them.

1.5.3 Nanocellulose-Based Filaments

These primarily fiber-based cellulosic components are created by connecting nanocellulose to its axial direction to make a structure that is continuous and thread-like. Technical techniques can be used to create nanocellulose-based filaments, including electrospinning, wet spinning, melt spinning, dry spinning, microfluidics, and flow focusing, with the solution spinning technique being the most widely used (wet and dry spinning). Interfacial polyelectrolyte complexation, or IPTC, has recently been introduced in the field of spinning techniques. This IPTC spinning approach has been used to fabricate the self-assembled filaments that are specifically derived from nanocellulose at the interface of two oppositely charged polyelectrolytes complex (+ve/-ve). Conductive filaments, antibacterial filaments, and hydrogel threads are some of the major functionally active nanocellulose-based filaments. Nanofilaments can alter their surface electronegativity and charge density properties by altering their surface. Additionally, researchers are also producing nanocellulose fibres using a PVA or poly (vinyl alcohol) mixture. It contributes to a homogenous mixture due to its hydrophilic character. Some boundaries that maintains the resulting filaments quality is -

- The optimum spinning conditions
- Low solids content in dope
- High shear rate in spinneret
- High nanocellulose aspect ratio and
- High draw ratio.

Together, these factors can help the filament become more orientated and aligned with the axis, which improves the filaments' internal properties. Draw ratio, which affects the final filament alignments and is a key component of filament drawing, is the ratio of the spinneret's flow velocity to filament take-up velocity. For this orientation, a draw ratio of approximately >1 is suitable.

1.5.4 Nanocellulose for Biomedical Application

In the biomedical field, nanocellulose has lately been referred to as the "eyes of biomaterial" and has many uses, such as skin substitutes for burns and wounds, drug release mechanisms, blood vessel growth, nerve, gum and dura reconstruction, tissue engineering scaffolds, stent covering and bone reconstruction.

Tissue engineering explores for novel elements and technologies that might interact well with biological tissues, either by rearranging and creating tissue that will soon be transplanted or by acting as an *in vitro* basis for cell proliferation. Additionally, they target new classes of biocompatible, controlled, and particular degradable biopolymers that are more likely to be used as *in vitro* tissue reconstruction or as cell scaffolds.

As previously mentioned, a wide range of biomaterials have lately been produced. The eventual application—tissue regeneration, tissue grafting, or scaffolding, drug holding and release—largely determines their chemical, physical, and mechanical characteristics. The success of the scaffold is heavily dependent on the growth and attachment of cells to the surface, therefore the chemical surface of the biopolymer can control biological response by obstructing cellular adhesion, proliferation, migration, and functioning.

On the other hand, three distinct foundations are used to regenerate tissues: cells, support, and growth factors. Cells create the milieu for new tissues, hold and maintain the ideal environment for growth, while growth factors stimulate and encourage cell regeneration. The material used for implants needs to be biocompatible and cannot cause inflammation or rejection in some people. It should also encourage regeneration and, if necessary, be biodegradable or absorbed over time. For implants to be successful, research on support-cell interactions is essential. The ability of the cell to distinguish between various reactions from various materials and adjust accordingly determines whether the cell will adhere to the surface or not. This is important since it will influence subsequent reactions including cell viability, migration, and proliferation [78].

A biological membrane created by Barud uses bacterial cellulose and a standardized propolis extract. Numerous biological characteristics of propolis include its ability to combat infection and reduce inflammation. The membrane has make it an effective treatment for burns and open wounds.

The use of nanocellulose-based materials for biomedical purposes has produced some intriguingly encouraging results and accomplishments, but more research on the long-term biocompatibility and toxicology should be done in addition to the implementation of the validation of these nanocellulose-based biomaterials using the standards and methodologies used by the competent authorities.

1.6 Summary and Prospects

In various domains, nanocellulose strengthened polymer composites have been researched. These studies all aim to design functional materials with high performance and high added value by utilizing the benefits of nanocellulose, including its lightweight, high strength, low cost, and renewable sources, despite the fact that their study foci, methodologies, and findings differ. Although nanocellulose have significant performance benefits, there are still challenges to their commercialization and industrialization, and some uncharted territory needs to be further investigated.

1.6.1 Advantages of Nanocellulose-Reinforced Polymer Composites

Since Hermann Staudinger developed polymer science, the modification of polymers has been explored, and the field will continue to grow as long as there is a need for high-performance materials. Due to the vast potential for commercialization, scientists have made significant advances in their study of nanocellulose reinforced polymer composites. Nanofiller-reinforced polymers have a higher competitive edge than more conventional methods of modification like blending because they can increase material strength without significantly diminishing or even improving toughness.

Nanocellulose is great nanofiller because of its small particle size, high specific surface area, high modulus, high strength, and ease of modification. Natural biomass is abundant, biodegradable, nontoxic, biocompatible, and environmentally benign; nonetheless, its utilisation depends on the efficiency with which the existing high-pollution, energy-intensive processing procedures may be improved. For these reasons, nanocellulose is a game-changing material in fields as diverse as biodegradable plastics, bioscaffolds, catalysts, and drug carriers.

1.6.2 Current Challenges

Nanocellulose reinforced polymer composites having a number of uses and development prospects. Functionalization and mass production are two issues that must be resolved in order to advance its industrial use as well as utilization.

1.6.2.1 Functionalization

The dispersion of nanocellulose and its fusion with the matrix are significantly impacted by the intrinsic compatibility issue that exists between hydrophilic nanocellulose and the preponderance of hydrophobic polymer matrices. The majority of the research advancements stated above and listed in the preceding article included altering nanocellulose, but one of the goals was really to increase interfacial compatibility. A more varied composite design and improved performance are both possible with modified nanocellulose. This presents both a chance and a difficulty. The surface characteristics of nanocellulose make for a variety of modifications. Not only the modification's intended use must be taken into account, but also how the modification will affect nanocellulose's crystalline structure, surface microstructure, particle morphology, and physical and chemical properties, as well as how these changes will affect the performance of the finished composite. Numerous investigations have additionally demonstrated that the synergistic interaction between nanocellulose and other nanofillers leads to more complex effects. It is necessary to build a multiphase model in order to functionalize techniques that are distinct from the two-phase system of nanocellulose and uniform matrix. This is challenging since the reinforcement processes have not yet been fully understood and require more research.

1.6.2.2 Mass Production

The production cost must be reduced in order to achieve mass production. Nanocellulose takes more to produce and has a lower production capacity than other typical nanofillers like nano montmorillonite and nano silica, which generates higher its cost. Additionally, before combining with the polymer, modification or other pretreatment is frequently required, raising the expense of producing the nanocellulose-reinforced polymer and impeding the commercialization of existing nanocellulose reinforced polymer composites. This is the primary issue impeding mass production. The largest issue facing mass production applications is optimizing the process to lower production costs and improve production capacity to accomplish batch production.

1.6.3 Conclusion and Future Developments

In conclusion, functionalization and mass production are the two development paths for nanocellulose reinforced polymer composites, and they fundamentally address the requirement for better, stronger, more affordable, and more varied materials. The tremendous potential of the sector is really reflected in many of the existing challenges. A small amount of innovation, regardless of theory, technique, or application, can propel an entire sector towards a significant milestone. Engineering plastics, functional films, bioscaffolds, catalysts, drug carriers, adsorption materials, and flexible electronic components are among the many applications being researched for nanocellulose reinforced polymer composites. Nanocellulose reinforced polymer composites are anticipated to surpass all other composite materials in value as a result of current research.

References

1. Pennells, J., Godwin, I.D., Amiralian, N., Martin, D.J., Trends in the production of cellulose nanofibers from non-wood sources. *Cellulose*, 27, 575–593, 2020.
2. Huang, J., Su, J., Weng, M., Xiong, L., Wang, P., Liu, Y., Lin, X., Min, Y., An innovative phase change composite with high thermal conductivity and sensitive light response rate for thermal energy storage. *Sol. Energy Mater. Sol. Cells*, 245, 111872, 2022.
3. Wu, B., Lyu, S., Han, H., Li, T., Sun, H., Wang, J.-K., Li, D., Lei, F., Huang, J., Sun, D., Biomass-based shape-stabilized phase change materials from artificially cultured ship-shaped diatom frustules with high enthalpy for thermal energy storage. *Compos. B. Eng.*, 205, 108500, 2021.
4. McNamara, J.T., Morgan, J.L., Zimmer, J., A molecular description of cellulose biosynthesis. *Annu. Rev. Biochem.*, 84, 895–921, 2015.
5. Ghasemlou, M., Daver, F., Ivanova, E.P., Habibi, Y., Adhikari, B., Surface modifications of nanocellulose: From synthesis to highperformance nanocomposites. *Prog. Polym. Sci.*, 119, 101418, 2021.

6. Heise, K., Kontturi, E., Allahverdiyeva, Y., Tammelin, T., Linder, M.B., Nonappa, Ikkala, O., Nanocellulose: Recent fundamental advances and emerging biological and biomimicking applications. *Adv. Mater.*, 33, 2004349, 2021.
7. Sharma, A., Thakur, M., Bhattacharya, M., Mandal, T., Goswami, S., Commercial application of cellulose nano-composites – a review. *Biotechnol. Rep.*, 21, e00316, 2019.
8. Xing, J., Tao, P., Wu, Z., Xing, C., Liao, X., Nie, S., Nanocellulose-graphene composites: A promising nanomaterial for flexible supercapacitors. *Carbohydr. Polym.*, 207, 447–459, 2019.
9. Noremylia, M.B., Hassan, M.Z., Ismail, Z., Recent advancement in isolation, processing, characterization and applications of emerging nanocellulose: A review. *Int. J. Biol. Macromol.*, 206, 954–976, 2022.
10. Kargarzadeh, H., Ahmad, I., Thomas, S., Dufresne, A., *Handbook of nanocellulose and cellulose nanocomposites*, John Wiley & Sons, Weinheim, Germany, 2017.
11. Huang, Q., Yang, Y., Chen, R., Wang, X., High performance fully paper-based all-solid-state supercapacitor fabricated by a papermaking process with silver nanoparticles and reduced graphene oxide-modified pulp fibers. *EcoMat*, 3, e12076, 2021.
12. Gopi, S., Balakrishnan, P., Chandradhara, D., Poovathankandy, D., Thomas, S., General scenarios of cellulose and its use in the biomedical field. *Mater. Today Chem.*, 13, 59–78, 2019.
13. Wertz, J.L., Mercier, J.P., Bédué, O., *Cellulose science and technology*, CRC Press, Lausanne, 2010.
14. Dufresne, A. and Belgacem, M.N., Cellulose-reinforced composites: From micro-to nanoscale. *Polímeros*, 23, 277–286, 2013.
15. Kargarzadeh, H., Ahmad, I., Thomas, S., Dufresne, A., *Handbook of nanocellulose and cellulose nanocomposites*, John Wiley & Sons, Weinheim, 2017.
16. Thakur, V.K., *Lignocellulosic polymer composites: Processing, characterization, and properties*, John Wiley & Sons, Beverly, MA, 2014.
17. Thakur, V.K., *Nanocellulose polymer nanocomposites: Fundamentals and applications*, John Wiley & Sons, Beverly, MA, 2015.
18. Trache, D., Hussin, M.H., Haafiz, M.M., Thakur, V.K., Recent progress in cellulose nanocrystals: Sources and production. *Nanoscale*, 9, 1763–1786, 2017.
19. Nandi, S. and Guha, P., A review on preparation and properties of cellulose nanocrystal-incorporated natural biopolymer. *J. Package. Technol. Res.*, 2, 149–166, 2018.
20. Kumar, V., Pathak, P., Bhardwaj, N.K., Waste paper: An underutilized but promising source for nanocellulose mining. *Waste Manag. (New York, N.Y.)*, 102, 281–303, 2020.
21. Zhao, Y. and Li, J., Excellent chemical and material cellulose from tunicates: Diversity in cellulose production yield and chemical and morphological structures from different tunicate species. *Cellulose*, 21, 3427–3441, 2014.

22. Ummartyotin, S. and Manuspiya, H., A critical review on cellulose: From fundamental to an approach on sensor technology. *Renew. Sust. Energ. Rev.*, 41, 402–412, 2015.
23. Campano, C., Balea, A., Blanco, A., Negro, C., Enhancement of the fermentation process and properties of bacterial cellulose: A review. *Cellulose*, 23, 57–91, 2016.
24. Pires, J.R., Souza, V.G., Fernando, A.L., Valorization of energy crops as a source for nanocellulose production–current knowledge and future prospects. *Ind. Crops Prod.*, 140, 111642, 2019.
25. Fodil Cherif, M., Trache, D., Brosse, N., Benaliouche, F., Tarchoun, A.F., Comparison of the physicochemical properties and thermal stability of organosolv and kraft lignins from hardwood and softwood biomass for their potential valorization. *Waste Biomass Valori.*, 11, 6541–6553, 2020.
26. Agbor, V.B., Cicek, N., Sparling, R., Berlin, A., Levin, D.B., Biomass pretreatment: Fundamentals toward application. *Biotechnol. Adv.*, *29*, 6, 675–685, 2011.
27. Karimi, K. and Taherzadeh, M.J., A critical review of analytical methods in pretreatment of lignocelluloses: Composition, imaging, and crystallinity. *Bioresour. Technol.*, 200, 1008–1018, 2016.
28. Rabemanolontsoa, H. and Saka, S., Various pretreatments of lignocellulosics. *Bioresour. Technol.*, 199, 83–91, 2016.
29. Hassan, S.S., Williams, G.A., Jaiswal, A.K., Emerging technologies for the pretreatment of lignocellulosic biomass. *Bioresour. Technol.*, 262, 310–318, 2018.
30. Bhutto, A.W., Qureshi, K., Harijan, K., Abro, R., Abbas, T., Bazmi, A.A. *et al.*, Insight into progress in pre-treatment of lignocellulosic biomass. *Energy*, 122, 724–745, 2017.
31. Oun, A.A., Shankar, S., Rhim, J.W., Multifunctional nanocellulose/metal and metal oxide nanoparticle hybrid nanomaterials. *Crit. Rev. Food Sci. Nutr.*, *60*, 3, 435–460, 2020.
32. Salimi, S., Sotudeh-Gharebagh, R., Zarghami, R., Chan, S.Y., Yuen, K.H., Production of nanocellulose and its applications in drug delivery: A critical review. *ACS Sustain. Chem. Eng.*, 7, 9, 15800–15827, 2019.
33. Hussin, M.H., Trache, D., Chuin, C.T.H., Fazita, M.N., Haafiz, M.M., Hossain, M.S., Extraction of cellulose nanofibers and their eco-friendly polymer composites. *Sustainable Polymer Composites and Nanocomposites*, 653–691, 2019.
34. Dufresne, A., Cellulose nanomaterial reinforced polymer nanocomposites. *Curr. Opin. Colloid Interface Sci.*, 29, 1–8, 2017.
35. Kargarzadeh, H., Mariano, M., Gopakumar, D., Ahmad, I., Thomas, S., Dufresne, A., Huang, J., Lin, N., Advances in cellulose nanomaterials. *Cellulose*, 25, 2151–2189, 2018.
36. Brinchi, L., Cotana, F., Fortunati, E., Kenny, J., Production of nanocrystalline cellulose from lignocellulosic biomass: Technology and applications. *Carbohydr. Polym.*, 94, 154–169, 2013.

37. Mariano, M., El Kissi, N., Dufresne, A., Cellulose nanocrystals and related nanocomposites: Review of some properties and challenges. *J. Polym. Sci.*, 52, 791–806, 2014.
38. Ilyas, R.A., Salit, S.M., Sanyang, L.M., Ishak, M.R., Zainudin, E.S., Nanocrystalline cellulose as reinforcement for polymeric matrix nanocomposites and its potential applications: A Review. *Curr. Anal. Chem.*, 14, 203–225, 2018.
39. Klemm, D., Kramer, F., Moritz, S., Lindström, T., Ankerfors, M., Gray, D., Dorris, A., Nanocelluloses: A new family of nature-based materials. *Angewandte Chemie (International Ed. English)*, *50*, 24, 5438–5466, 2011.
40. Kargarzadeh, H., Ioelovich, M., Ahmad, I., Thomas, S., Dufresne, A., Methods for extraction of nanocellulose from various sources, in: *Handbook of Nanocellulose and Cellulose Nanocomposites*, pp. 1–51, 2017.
41. Kian, L.K., Jawaid, M., Ariffin, H., Alothman, O.Y., Isolation and characterization of microcrystalline cellulose from roselle fibers. *Int. J. Biol. Macromol.*, 103, 931–940, 2017.
42. Phanthong, P., Reubroycharoen, P., Hao, X., Xu, G., Abudula, A., Nanocellulose: Extraction and application. *Carbon Resour. Convers.*, 1, 32–43, 2018.
43. Saelee, K., Yingkamhaeng, N., Nimchua, T., Sukyai, P., An environmentally friendly xylanase-assisted pretreatment for cellulose nanofibrils isolation from sugarcane bagasse by high-pressure homogenization. *Ind. Crops Prod.*, 82, 149–160, 2016.
44. Siró, I. and Plackett, D., Microfibrillated cellulose and new nanocomposite materials: A review. *Cellulose*, 17, 459–494, 2010.
45. Jonoobi, M., Harun, J., Tahir, P.M., Zaini, L.H., SaifulAzry, S., Makinejad, M.D., Characteristics of nanofibers extracted from kenaf core. *Bioresources*, 5, 2556–2566, 2010.
46. Alemdar, A. and Sain, M., Isolation and characterization of nanofibers from agricultural residues: Wheat straw and soy hulls. *Bioresour. Technol.*, *99*, 6, 1664–1671, 2008.
47. Deshpande, A.P., Bhaskar Rao, M., Lakshmana Rao, C., Extraction of bamboo fibers and their use as reinforcement in polymeric composites. *J. Appl. Polym. Sci.*, 76, 83–92, 2000.
48. Kushwaha, P.K. and Kumar, R., The studies on performance of epoxy and Polyester-based composites reinforced with bamboo and glass fibers. *J. Reinf. Plast. Compos.*, 29, 13, 1952–1962, 2010.
49. Kushwaha, P.K. and Kumar, R., Studies on performance of acrylonitrile-pretreated bamboo-reinforced thermosetting resin composites. *J. Reinf. Plast. Compos.*, 29, 1347–1352, 2010.
50. Abral, H., Ariksa, J., Mahardika, M., Handayani, D., Aminah, I., Sandrawati, N., Sapuan, S.M., Ilyas, R.A., Highly transparent and antimicrobial PVA based bionanocomposites reinforced by ginger nanofiber. *Polym. Test.*, 81, 106186, 2019.

51. Hyojin, K., Kazuya, O., Toru, F., Kenichi, T., Influence of fiber extraction and surface modification on mechanical properties of green composites with bamboo fiber. *J. Adhes. Sci. Technol.*, 27, 1348–1358, 2013.
52. Zakikhani, P., Zahari, R., Sultan, M.T.H., Majid, D.L., Extraction and preparation of bamboo fibre-reinforced composites. *Mater. Des.*, 63, 820–828, 2014.
53. Rao, K.M.M. and Rao, K.M., Extraction and tensile properties of natural fibers: Vakka, date and bamboo. *Compos. Struct.*, 77, 288–295, 2007.
54. Phong, N.T., Fujii, T., Chuong, B., Okubo, K., Study on how to effectively extract bamboo fibers from raw bamboo and wastewater treatment. *J. Mater. Sci. Res.*, 1, 144–155, 2011.
55. Gousset, C., Chanzy, H., Cerrada, M.L., Fleury, E., Surface silylation of cellulose microfibrils: Preparation and rheological properties. *Polymer*, 45, 5, 1569–1575, 2004.
56. Dadkhah Tehrani, A. and Neysi, E., Surface modification of cellulose nanowhisker throughout graft polymerization of 2-ethyl-2-oxazoline. *Carbohydr. Polym.*, *97*, 1, 98–104, 2013.
57. Chin, K.-M., Sung Ting, S., Ong, H.L., Omar, M., Surface functionalized nanocellulose as a veritable inclusionary material in contemporary bioinspired applications: A review. *J. Appl. Polym. Sci.*, 135, 46065, 2018.
58. Zhang, W., Zhong, L., Wang, T., Jiang, Z., Gao, X., Zhang, L., Surface modification of cellulose nanofibers and their effects on the morphology and properties of polysulfone membranes. *IOP Conference Series Materials Science and Engineering*, vol. 397, p. 012016, 2018.
59. Jamaluddin, N., Kanno, T., Asoh, T.-A., Uyama, H., Surface modification of cellulose nanofiber using acid anhydride for poly(lactic acid) reinforcement. *Mater. Today Commun.*, 21, 100587, 2019.
60. He, Y., Zhu, J., Wang, W., Ni, H., Surface modification of cellulose nanocrystals with different acid anhydrides for improved dispersion in poly (butylene succinate). *RSC Adv.*, 8, 38305–38314, 2018.
61. Sahlin, K., Forsgren, L., Moberg, T. *et al.*, Surface treatment of cellulose nanocrystals (CNC): Effects on dispersion rheology. *Cellulose*, 25, 331–345, 2018.
62. Agustin, M.B., Nakatsubo, F., Yano, H., Improving the thermal stability of wood-based cellulose by esterification. *Carbohydr. Polym.*, 192, 28–36, 2018.
63. Peresin, M.S., Kammiovirta, K., Heikkinen, H., Johansson, L.S., Vartiainen, J., Setälä, H., Österberg, M., Tammelin, T., Understanding the mechanisms of oxygen diffusion through surface functionalized nanocellulose films. *Carbohydr. Polym.*, 174, 309–317, 2017.
64. Tian, X., Wu, M., Wang, Z., Zhang, J., Lu, P., A high-stable soybean-oil-based epoxy acrylate emulsion stabilized by silanized nanocrystalline cellulose as a sustainable paper coating for enhanced water vapor barrier. *J. Colloid Interface Sci.*, 610, 1043–1056, 2022.

65. Cunha, A.G., Lundahl, M., Ansari, M.F., Johansson, L.S., Campbell, J.M., Rojas, O.J., Surface structuring and water interactions of nanocellulose filaments modified with organosilanes toward wearable materials. *ACS Appl. Nano Mater.*, *1*, 9, 5279–5288, 2018.
66. Shen, Y., Li, D., Wang, L., Zhou, Y., Liu, F., Wu, H., Deng, B., Liu, Q., Superelastic polyimide nanofiber-based aerogels modified with silicone nanofilaments for ultrafast oil/water separation. *ACS Appl. Mater. Interfaces*, 13, 20489–20500, 2021.
67. Pacaphol, K. and Aht-Ong, D., The influences of silanes on interfacial adhesion and surface properties of nanocellulose film coating on glass and aluminum substrates. *Surf. Coat. Technol.*, 320, 70–81, 2017.
68. Goussé, C., Chanzy, H., Excoffier, G., Soubeyrand, L., Fleury, E., Stable suspensions of partially silylated cellulose whiskers dispersed in organic solvents. *Polymer*, 43, 2645– 2651, 2002.
69. Ma, Y., Qian, S., Hu, L., Qian, J., Fontanillo, C., Lopez, L., Mechanical, thermal, and morphological properties of PLA biocomposites toughened with silylated bamboo cellulose nanowhiskers. *Polym. Compos.*, 40, 3012–3019, 2019.
70. Ghasemlou, M., Daver, F., Ivanova, E.P., Habibi, Y., Adhikari, B., Surface modifications of nanocellulose: From synthesis to highperformance nanocomposites. *Prog. Polym. Sci.*, 119, 101418, 2021.
71. Chu, Y., Sun, Y., Wu, W., Xiao, H., Dispersion properties of nanocellulose: A review. *Carbohydr. Polym.*, 250, 116892, 2020.
72. Saito, T., Kimura, S., Nishiyama, Y., Isogai, A., Cellulose nanofibers prepared by TEMPO-mediated oxidation of native cellulose. *Biomacromolecules*, 8, 2485–2491, 2007.
73. Wang, J., Wang, C., Wang, W., Li, W., Lou, J., Carboxymethylated nanocellulose-based gel polymer electrolyte with a high lithium-ion transfer number for flexible lithium-ion batteries application. *Chem. Eng. J.*, 428, 132604, 2022.
74. Omran, A.A.B., Mohammed, A.A.B.A., Sapuan, S.M., Ilyas, R.A., Asyraf, M.R.M., Rahimian Koloor, S.S., Petrů, M., Micro- and nanocellulose in polymer composite materials: A review. *Polymers*, *13*, 231, 2021.
75. Krishnamoorti, R. and Vaia, R.A., *Polymer nanocomposites*, Wiley InterScience, United States of America, 2007.
76. Pandey, J.K., Lee, H.T., Takagi, H., Ahn, S.H., Saini, D.R., Misra, M., *Handbook of polymer nanocomposites, processing, performance and application – volume C: Polymer nanocomposites of cellulose nanoparticles*, Springer-Verlag, Berlin Heidelberg, 2015.
77. Athukoralalage, S.S., Balu, R., Dutta, N.K., Roy Choudhury, N., 3D Bioprinted nanocellulose-based hydrogels for tissue engineering applications: A brief review. *Polymers*, *11*, 898, 2019.
78. Guise, C. and Fangueiro, R., *Natural fibres: Advances in science and technology towards industrial applications*, pp. 155–169, Springer, Dordrecht, 2016.

2

Cellulose Nanofibers (CNF) and Nanocrystals (CNC): Pre-Treatment, Preparation, and Characterization

Jiawei Chen[1†], Yu Chen[1†] and Tanushree Ghosh[2*]

[1]*Department of Mechanical Engineering, University of Alberta, Edmonton, Canada*
[2]*Centre for Earth Sciences, Indian Institute of Science, Bangalore, Karnataka, India*

Abstract

The development and use of eco-friendly materials are imposing a huge challenge to researchers and scientists. Studies and scientific exploration of carbohydrate-based organic materials are paving the way towards the replacement of conventional non-renewable materials. Cellulose nanomaterials, derived from widely available plant sources, are sustainable, bio-degradable, bio-compatible and cost-effective materials with multidisciplinary applications in biomedical engineering, food, sensor, packaging, and so on. Crystalline nanocellulose (CNCs) and cellulose nanofibrils (CNFs) are two types of cellulose nanomaterials possess various superior properties, such as large specific surface area, high tensile strength and stiffness, low density, and low thermal expansion coefficient. In this chapter, various methods of preparation of CNCs and CNFs are summarized, including mechanical, chemical, and biological methods of fibre extraction, purification process, sample preparation, and different drying techniques. This chapter also outlines the various physicochemical characterization methods practiced for CNCs and CNFs when used in polymer matrix composites.

Keywords: Crystalline nanocellulose, cellulose nanofibrils, alkali treatment, bleaching, acid hydrolysis, enzymatic hydrolysis

**Corresponding author*: tanushreegho@gmail.com
†*Equal contribution as first author*

Mainak Mukhopadhyay and Debalina Bhattacharya (eds.) Nanocellulose: A Biopolymer for Biomedical Applications, (27–62) © 2024 Scrivener Publishing LLC

2.1 Introduction

Cellulose is the most abundant semicrystalline, biodegradable polymer present in nature. Cellulose nanocrystal (CNC) and cellulose nanofiber (CNF) are viable ecological alternatives to conventional petroleum-based polymers to address the emerging concern of plastic waste management. CNFs are nanosized cellulose fibrils with a diameter lower than 100 nm and lengths ranging from a few nanometers to several micrometers. CNF contains both amorphous and crystalline regions; therefore, it shows lower crystallinity and weaker mechanical properties than CNC in general. Due to the lower crystallinity requirement, the methods to extract CNF from cellulose biomass are predominantly mechanical-based, such as homogenization or steam explosion. Other extraction methods are chemical-based, such as 2,2,6,6-tetra-methylpiperidine-1-oxyl radical (TEMPO) oxidation or biological enzyme hydrolysis. Since mechanical extraction consumes high energy, pre-treatment is generally required to reduce the energy cost, accelerate the process, or achieve better defibrillation. Different from cellulose nanofibers (CNFs), the term cellulose nanocrystals (CNCs) refers to those whisker-shaped nanocelluloses [1]. These nanoparticles possess a high Young's modulus (over 100 GPa), high crystallinity, and good thermal stability and can be used in many applications such as reinforcement for polymer nanocomposites, emulsion stabilizers, self-assembly materials, and other bio-inspired photonic and electronic materials [2, 4]. Due to the promising application potential of CNCs, their extraction process has been widely studied. CNCs can be extracted from renewable, sustainable, and cheap raw materials that contain large amounts of cellulose, such as chestnut shell, cotton cloth waste, edible fungus, and a variety of natural fibers like hemp, sisal, jute, etc. [1, 3–6]. A successful production process usually involves three steps: 1) a pre-treatment process that destroys the molecular or secondary bonds between cellulose and other molecules in the raw materials; 2) a main extraction process such as acid hydrolysis, enzymatic hydrolysis, 2,2,6,6-tetra-methylpiperidine-1-oxyl radical (TEMPO) oxidation, or mechanical processes that apply mechanical energy to disperse the cellulose in the aqueous phase; and 3) a post-treatment process that washes out the solvents and impurities generated during the extraction process and produces clean CNCs for further applications [4, 7]. In this chapter, the general methodologies to produce CNFs and CNCs will be briefly introduced and some examples will be provided.

2.2 Cellulose Nanofibers (CNF)

The methods involved in the preparation of CNF, i.e., pre-treatment, extraction, and post-treatment processes, often vary slightly depending on the raw biomass resources available for treatment. The schematics below (Schemes 2.1 and 2.2) summarize the pre-treatment, extraction, and post-treatment processes in practice for CNF preparation.

2.2.1 Pre-Treatment

CNF is extracted from biomass resources such as wood, seeds, and plant fibers. The pre-treatment methods required for the extraction depend on the raw sources, such as wood, seeds, plant fibers, etc., which undergo

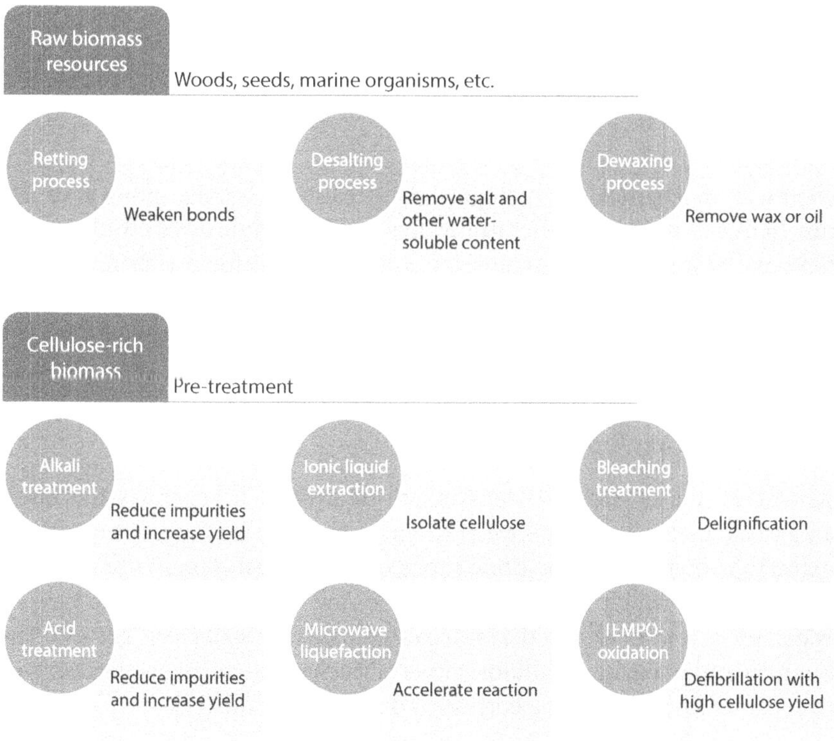

Scheme 2.1 CNF pre-treatment process adoptable for different raw biomass resources (upper panel) and cellulose-rich biomass to enhance CNF yield.

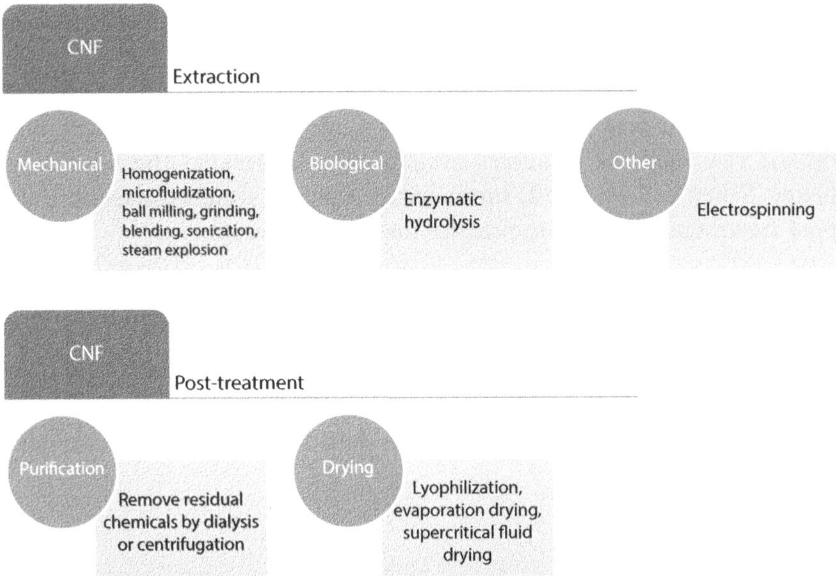

Scheme 2.2 Summary of CNF extraction (upper panel) and post-treatment (lower panel) process.

chemical, mechanical, or biological pre-treatment for the removal of lignin, hemicellulose, and other impurities to yield cellulosic or cellulose-rich biomass. Sometimes, pre-treatment is not required for preparing CNF. It is generally adopted to reduce the cost and energy consumption of the mechanical extraction process to yield CNF with desired properties. The pre-treatment processes are summarized below.

2.2.1.1 *Retting Process*

Retting is one of the first steps in pre-treatment. It separates the fibers and weakens the bonds between them so that the next pre-treatment or extraction method can be done more quickly. Most people use dew and water retting, which involves leaving the raw material out in the field or water for a few weeks so that microorganisms or water can get into the biomass and make lignocellulosic fibers. Mechanical retting is another way to separate fibers using a decorticator or hammer mill [1].

2.2.1.2 *Desalting Process*

In the desalting or desalination processes, the biomass is treated with boiled deionized (DI) water to remove excessive salt. If the raw source contains a large amount of salt, it is necessary to undergo the desalting process during the pre-treatment phase [2].

2.2.1.3 *Dewaxing Process*

The dewaxing process treats the natural fibers in a Soxhlet extractor apparatus at an elevated temperature and is normally followed by extensive water washing and drying to remove water-soluble impurities [3]. The natural fibers are cut into smaller sizes and passed through sieves to ensure uniformity. Then the wax or oil content on the natural fibers is removed by organic solvent mixtures such as ethanol/toluene based on TAPPI T204 [2, 4]. This method is referred to as the dewaxing or degreasing process.

2.2.1.4 *Alkali Treatment*

Alkali treatment involves using an alkaline solution generally consisting of either sodium hydroxide (NaOH) or potassium hydroxide (KOH). The common parameters used for alkali treatment include an alkaline concentration of 2–17.5% at the operating temperature of 25–90 °C for a duration of 0.5–16 h [5–9]. The temperature and duration depend on the alkaline concentration, the raw material of CNF, and other combined pre-treatment steps. Alkali treatment is aimed at reducing lignin, pectin, or hemicellulose content, which facilitates hemicellulose hydrolysis as well as cleaves ester linkages and carboxylic groups between or within lignin and hemicellulose to induce depolymerization of lignin and defibrillation of fiber structure [6, 10]. However, this process cannot entirely remove hemicellulose and lignin content, as a result, some cellulose fibers remain intact even after alkali treatment [10]. The alkaline treatment is crucial for achieving high cellulose yields as it induces swelling of the fibers and opens up more accessible sites for the subsequent extraction processes. But some research claims that alkali treatment prevents cellulose from degradation in mechanical treatment, while other research indicates alkali treatment damages cellulose chains [11, 12]. This process is stopped after rinsing with water until the pH returns to neutral (pH 7).

2.2.1.5 *Ionic Liquid Extraction*

Organic salts with a low melting point (<100°C) and low volatility can interact with hydroxyl groups and dissolve cellulose. Molten organic salts, with 1-methylimidazolium ionic liquid as an example, can isolate cellulose from biomass at an elevated temperature under constant stirring and obtain CNF after mechanical treatment [13].

2.2.1.6 Bleaching Treatment

The bleaching process comes along with other chemical treatments, such as alkali treatment. The bleaching process is necessary to remove lignin and hemicellulose from the cellulosic biomass. Researchers have used three types of bleaching agents: acid chlorite, hydrogen peroxide (H_2O_2), and ozone (O_3). The acid-chlorite treatment, also known as the Wise treatment, uses sodium chlorite solution with acetic acid for delignification. Incubation of biomass with a sodium chlorite/acetic acid mixture is maintained at ~80°C for 1 h [12]. Sodium chlorite and acetic acid solutions can induce lignin oxidation and removal to whiten the cellulose biomass [11]. Bleaching treatment can either be conducted on its own or paired with alkali treatment to attain optimal results. A comparative study reported by Nobuta *et al.* demonstrated CNF pretreated solely with Wise treatment showed better mechanical properties, comparable crystallinity, and higher lignin/ash content than CNF pretreated with both Wise treatment and alkali treatment [12]. Based on the application of CNF, the properties of CNF can be tailored by different combinations of pre-treatment methods. The second bleaching agent is hydrogen peroxide. 30% hydrogen peroxide at 80°C can effectively defibrillate cellulose microfibers and remove lignin [14]. The last bleaching agent is ozone. Ozone can react with lignin to promote lignin fragmentation and extraction, which is particularly effective in a humid environment as the reaction of ozone with water releases hydroxyl radicals. Bahrami *et al.* exposed bagasse fibers to an ozone environment with an ozone concentration of 0.55 g/m^3 at a temperature range of 40–85 °C. The cellulose content increased from 41.5 to 91.39%, and the lignin content decreased from 26.1 to 1.73% [15].

2.2.1.7 Acid Treatment

Acid treatment is another common hydrolysis pre-treatment for CNF. The acids include inorganic acids, e.g., hydrochloric acid, sulfuric acid, nitric acid, and organic acids, e.g., formic acid, acetic acid, or sometimes acid mixtures [4, 9, 16]. Hydrochloric acid can cause CNF agglomeration while sulfuric acid can reduce the thermal stability of CNF, but in combination, pre-treatment with mixed acids yields both thermally stable and uniform CNF. Furthermore, combining alkali treatment with acid treatment sometimes increases both cellulose crystallinity and thermal stability since the removal of amorphous components by alkali treatment makes more hydroxyl groups available for acid hydrolysis on the surface of cellulose. For instance, Corrêa *et al.* found that using 36.5–60% hydrochloric acids or sulfuric acids at 45°C for 1 h yielded CNF with better properties [9].

2.2.1.8 Microwave-Assisted Hydrolysis

Chemical treatment is sometimes accompanied by a microwave treatment to accelerate the reaction and increase cellulose yield. Recently, microwave-assisted hydrolysis has been promoted as an acid-free pre-treatment method for CNF preparation. Microwave energy can reduce processing time and assist with defibrillation. Xie and his research group carried out microwave liquefaction of bamboo biomass at a power range of 0-550 W and a temperature of 120°C for 7 min [17]. Cellulose content increased with microwave duration from 41.72% to 70.74%, while crystallinity increased from 52.3% to 70.6%. Liu also used microwave hydrolysis to assist with alkali treatment at 140°C for 20 min to produce CNF from wheat straw [18]. Cellulose content increased from 44.81% to 94.09%, and the crystallinity index increased from 45.85% to 65.15% after microwave-assisted alkali treatment.

2.2.1.9 TEMPO-Oxidation

The TEMPO (2,2,6,6-tetramethylpiperidine-1-oxyl) oxidation process changes water-insoluble cellulose biomass's C6 primary hydroxyl groups into C6 carboxylate groups or aldehyde groups. The negative charge on the surface of the anionic functional groups makes it easier to separate the cellulose microfibrils. TEMPO oxidation is the most popular method because it is better for the environment than other pre-treatments that use harsh chemicals. The TEMPO-oxidation method is also energy efficient and a high-yielding method. The energy conversion was reported to be lower than 1 kWh/kg and the cellulose yield was over 90% [19, 20] by many researchers. TEMPO oxidation is suitable for cellulose extraction from plant-based biomass, bacterial cellulose, and algae [21]. The general oxidizing agents used in CNF preparation by Isoga and Saito [22] are TEMPO, sodium bromide, and sodium chlorite. In contrast to acid treatment, TEMPO oxidation produces finer and more uniform CNFs after mechanical extraction, which results in more defibrillated CNFs [19]. Kaffashsaie *et al.* reported CNF with a diameter of ~5 nm and a crystallinity index of 67–68% from raw wood treated by bleaching and

TEMPO-oxidation [23]. Huang *et al.* also reported TEMPO-oxidation for producing CNF from bacterial cellulose to obtain CNFs with an average diameter of 32–37 nm and a high crystallinity index of up to 94.3% [24]. In addition, a membrane made from TEMPO-oxidized CNF was found applicable in lithium-ion batteries as a separator with great performance. Therefore, TEMPO-oxidation is a promising green technique to produce CNF at lower energy consumption, but its limitation in the cost of the TEMPO catalyst is the main drawback for achieving a more cost-efficient manufacturing process for CNF [25].

2.2.2 Extraction

The extraction of CNF is predominantly mechanical. After pre-treatment, the bonds between cellulose fibers become weaker, so mechanical extraction techniques such as high-pressure homogenization, microfluidization, ball milling, grinding, blending, cryo-crushing, steam explosion, sonication, or microwave-assisted extraction, can disintegrate the cellulose fibers into CNF.

2.2.2.1 Homogenization

Homogenization was the first technique used to prepare cellulose fibers, introduced by Turbak *et al.* and Herrick *et al.* in 1983 [26]. Homogenizers can either be high-pressure or high-speed. A high-pressure homogenizer introduces high shear forces to cellulose suspension by passing it through narrow openings at high pressure to reduce fiber diameter. Multiple passes through the homogenizer at 50–500 bar were reported to be efficient in increasing repulsion among CNFs to individualize fibers and reduce their lengths [27]. However, high energy cost, clogging problems, and the requirement of repeated passing through the homogenizer to obtain CNFs are the challenges of a high-pressure homogenizer. Pressure and the number of passes can both impact CNF quality positively and negatively. Higher pressure and more passing cycles decrease fiber size, but further increments induce droplet coalescence and damage the cellulose chain, causing an increase in the diameter of CNF and a reduction in crystallinity and thermal stability [28]. High-speed homogenizers achieve defibrillation via high shearing stress as CNF suspension is subjected to a high rotational speed by the homogenizer rotor. For instance, upon undergoing an alkali pre-treatment, bleaching, and subsequent acetic acid/sodium chlorite treatment, Arecanut husk fibers produced CNFs exhibiting a diameter of 3-5 nm, and aspect ratio of 120–150 using a high-speed homogenizer at 12000 rpm for a duration of 2 h [5].

2.2.2.2 *Microfluidization*

A microfluidizer is an alternative to a high-pressure homogenizer, consisting of an intensifier pump for increasing pressure to around 1200-2000 bar and single or multiple interaction chambers to pass cellulose suspension through channels with a fixed orifice size of 100–200 μm so that shear and impact forces induced by a collision can result in defibrillation. Multiple passes are normally required to achieve nanoscale fibers. Jiménez-López *et al.* produced CNF with either mechanical pre-treatment or TEMPO oxidation using microfluidization and yielded CNF with a crystallinity index of 77–86% and a uniform average diameter of 2.6–6.1 nm, depending on the pre-treatment methods and sources of elm [29]. Similar to the high-pressure homogenizer, large-sized cellulose fibers can induce clogging in the channels, so downsizing cellulose fibers with pre-treatment is generally necessary for microfluidization.

2.2.2.3 *Ball Milling*

Ball milling is a mechanical grinding process that applies shear and impact on cellulose suspension by collision with solid balls within an enclosed space. Reduction in the size of cellulose is dependent on the size of the balls, the number of balls, the mass of the sample, and the centrifugal force in rotation. Sofla *et al.* compared CNC treated with acid hydrolysis and CNF treated with ball milling from bleached bagasse fibers [3]. The ball milling was conducted for 1 hour using cerium-doped zirconia balls with a diameter of 0.5 mm and a mass of 20 g. CNF produced by ball milling showed an average diameter of 50 nm, length of ~2 μm, and crystallinity index of 68%, compared to CNC with an average diameter of 20–30 nm, length of 160–400 nm, and crystallinity index of 73%.

2.2.2.4 *Grinding*

In the grinding process, cellulose microfibers were fed in between two stones, and the grinding stone exerted shear force on the fibers to achieve defibrillation. A disc grinder comprises a stationary stone and a rotating stone. Even though the process is simple and achieves a high yield, it is found to have a negative influence on the crystallinity and thermal stability of the CNF produced.

2.2.2.5 *Blending*

Blending is another mechanical treatment similar to the high-speed homogenizer. It uses blades to apply shearing force on the cellulose fibers at a high rotational speed. The apparatus is easily and commercially available as a household blender and is sufficient for defibrillating pretreated cellulose fibers [21]. The typical speed and duration for blending reported are 48000 rpm and 5–10 min [14, 30].

2.2.2.6 *Sonication*

Ultrasound energy waves can penetrate the cellulose fibers to reduce the size of fibrils. The cellulose suspension can either be submerged in an ultrasonic bath or exposed to an ultrasonic probe. The diameter of the fiber is dependent on the intensity and duration of sonication. Sonication can be considered a post-treatment after homogenization or microfluidization to reduce agglomeration.

2.2.2.7 *Steam Explosion*

At high pressure and temperature, cellulose can be defibrillated by an explosion of steam, which is also called hydrothermal treatment. Kalita *et al.* steam exploded rice husks in an autoclave with a pressure of 20 lb and a temperature of 210°C for 8 h, combining with alkali treatment and bleaching to yield CNF with a diameter of 30–40 nm and cellulose content increasing from 37.1% to 94.2% [31]. 'lhe main drawback of steam explosions is their high energy consumption and cost.

2.2.2.8 *Enzymatic Hydrolysis*

Enzymatic hydrolysis is the process of isolating cellulose by using enzymes or microorganisms that can secrete enzymes that target the right parts of cellulose biomass. Hemicellulase can break down hemicellulose, laccase can break down lignin; pectinase can break down pectin, and cellulase can break down cellulose. Cellobiohydrolase and endoglucanase are two types of cellulases that can break down crystalline and amorphous parts of cellulose, respectively. Enzymatic hydrolysis is a bioprocess that is thought to be one of the most environmentally friendly ways to get cellulose out of plants because it doesn't change the structure of the cellulose. Enzymatic hydrolysis has a mild effect on defibrillating cellulose fibers and can cause fibers to tangle.

To improve nanofiber production, multiple enzymes can be used at the same time or other pre-treatments can be added. Since reducing the size of fibers to the nanoscale with enzymatic hydrolysis alone is hard, it is usually followed by sonication or other mild shearing treatments. De Campos *et al.* used hemicellulase and pectinase endoglucanase to treat bleached bagasse and curauá fibers at 50°C for 72 h followed by sonication to produce CNFs with an average diameter of 55–109 nm and a crystallinity index of 75% [32]. Enzymatic hydrolysis can also be used as a pre-treatment to weaken the bonds in the cellulose fibers. Janardhnan and Sain treated kraft softwood pulp with fungus extracted from Dutch elm trees for 4 days and defibrillated it with a refiner to obtain CNFs with a diameter below 250 nm, with an average of around 50–75 nm [33]. The enzymes were found to interact with hydroxyl groups and reduce intermolecular hydrogen bonding. Enzymatic hydrolysis shows advantages in cost and sustainability but presents drawbacks in lengthy processes.

2.2.3 Post-Treatment

Most extracted CNF is a suspension in water, so it needs to be dried in order to be characterized, used in certain ways, or stored.

2.2.3.1 Dialysis

After acid treatment, dialysis post-treatment can be used to remove residual acids from CNF after resuspending CNF in water. In this process, a CNF suspension is put through dialysis tubing (which has a cellulose membrane) until the pH is at a neutral level [34].

2.2.3.2 Centrifugation

High rotational speeds (10000–12000 rpm for 15–20 mins) can remove excessive chemicals from CNF [3, 9, 34]. After centrifugation, the precipitate is resuspended in water to achieve uniform suspension.

2.2.3.3 Drying

The drying process can be either evaporation drying, lyophilization, or supercritical drying, depending on how the conventional oven, frozen water sublimation, or supercritical fluid works.

2.3 Cellulose Nanocrystals (CNC)

Different from cellulose nanofibers (CNFs), the term cellulose nanocrystals (CNCs) refers to those whisker-shaped nanocelluloses [35]. These nanoparticles possess a high Young's modulus (over 100 GPa), high crystallinity, and good thermal stability and can be used in many applications

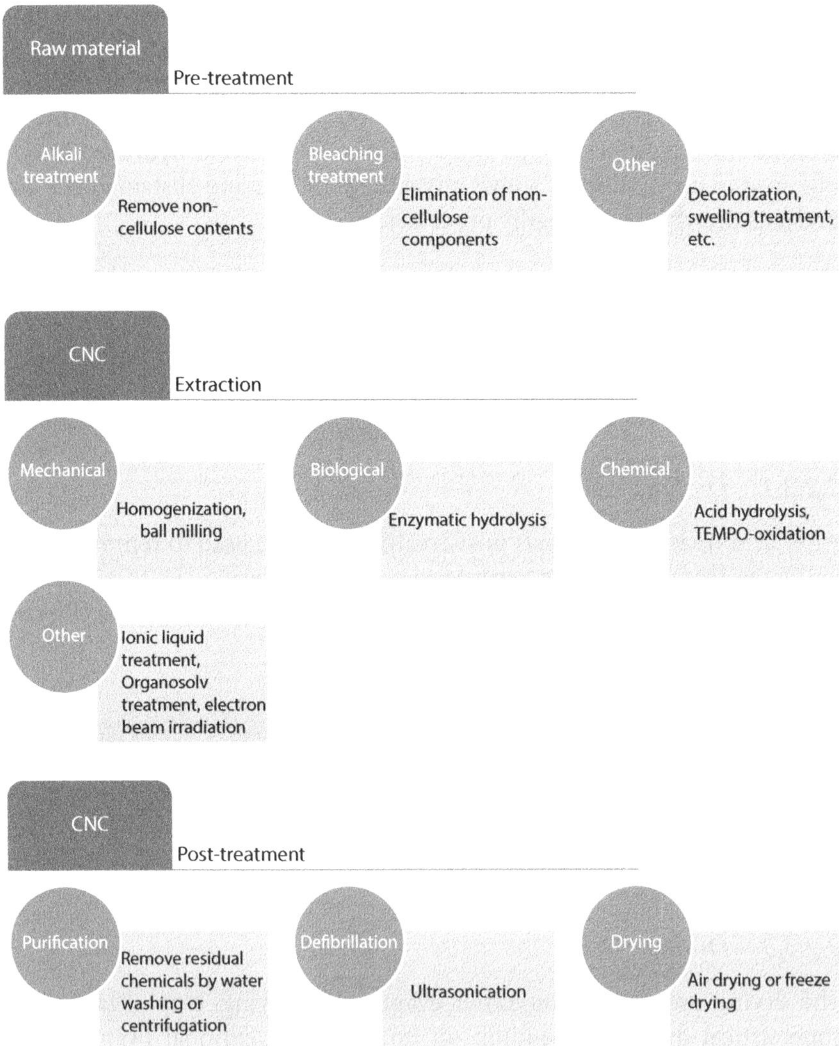

Scheme 2.3 Summary of CNC pre-treatment (upper panel) extraction (middle panel) and post-treatment (lower panel) process.

such as reinforcement for polymer nanocomposites, emulsion stabilizers, self-assembly materials, and other bio-inspired photonic and electronic materials [36, 37]. Due to the promising application potential of CNCs, the extraction process has been widely studied. CNCs can be extracted from renewable, sustainable, and cheap raw materials that contain large amounts of cellulose, such as chestnut shell, cotton cloth waste, edible fungus, and a variety of natural fibers like hemp, sisal, jute, etc. [35, 38–40]. A successful production process usually involves three steps as shown in Scheme 2.3: 1) a pre-treatment process that destroys the molecular or secondary bonds between cellulose and other molecules in the raw materials; 2) a main extraction process such as acid hydrolysis, enzymatic hydrolysis, TEMPO oxidation, or mechanical processes that apply mechanical energy to disperse the cellulose in the aqueous phase; and 3) a post-treatment process that washes out the solvents and impurities generated during the extraction process and produces clean CNCs for further applications [37, 41]. In this section, the general methodologies used in these steps will be briefly introduced, and some examples will be provided.

2.3.1 Pre-Treatment

During the pre-treatment process, some or all of the materials that hold the cellulose in the raw materials can be removed. This leaves a product that is mostly made of cellulose [41]. For better pre-treatment results, it usually involves either alkali treatment or bleaching treatment, or a combination of the two.

2.3.1.1 Alkali Treatment

Alkali treatment is a process in which the cellulose-rich raw materials are immersed in a certain concentration of aqueous alkali solutions for a pre-defined temperature and length of time [42]. For example, Ozkan *et al.* immersed horse chestnut seed shells and chestnut seed shells in a 10% (w/v) NaOH/distilled water solution at 60°C for 6 hours [38]. Alothman *et al.* immersed date palm fibers in an 8% (w/v) NaOH/distilled water solution at 25°C for 30 min at a ratio of 2 g fiber to 100 mL solutions [43]. The purpose of this process is to remove the non-cellulose contents such as lignin, hemicellulose, wax, oils, and pectin that are generally covered on the surface of the materials. Therefore, the temperature and time of this treatment are highly dependent on the composition of the raw material and need to be optimized every time a new raw material is involved.

2.3.1.2 Bleaching

Bleaching is another important pre-treatment process involved in the CNC extraction, and it is usually applied before or after the alkali treatment [40, 43]. Bleaching is applied to ensure the complete elimination of all other components such as lignin and hemicellulose that contain chromophore groups and phenolic compounds [42]. It is usually conducted by repeatedly filtering and washing the alkali-treated solutions, then adding aqueous sodium chloride solution ($NaClO_2$) at a certain temperature for a period of time [39]. For example, before the alkali treatment process, Alothman *et al.* bleached the raw materials in a 10% (w/v) $NaClO_2$ solution at 80°C for 1 h [43]. On the contrary, the bleaching process for Ozkan *et al.* was performed after the alkali treatment [38].

2.3.1.3 Other Pre-Treatments

Depending on the nature of the raw materials, some other pre-treatment such as decolorization, swelling treatment, and hydroperoxide treatment may also be needed to remove some substances and residuals from the raw materials to further expose cellulose [39, 44, 45].

2.3.2 Extraction

2.3.2.1 Acid Hydrolysis

Acid hydrolysis is the most common and convenient way to extract CNCs out of all the other ways [46]. The procedure starts with the diffusion of acid molecules into the cellulose microfibrils after the pre-treatment [47]. The acid molecules can remove the amorphous part of the pretreated cellulose materials and break the glycosidic bonds between the crystalline parts of the cellulose, which breaks the cellulose microfibrils into nanofibrils or CNCs [35]. The selective cleavage of cellulose polymer chains originates from the difference in the kinetics of hydrolysis between paracrystalline and crystalline regions [36]. In general, the properties of CNCs are mainly dependent on the acid type, temperature, time, and concentration of the acid agent [38, 39, 45, 48]. The preferred acid used for acid hydrolysis is usually sulfuric acid. For instance, Maiti *et al.* used 47% sulfuric acid solutions to extract CNCs from China cotton, South African cotton, and waste tissue paper [49]. The raw materials were mixed with sulfuric acid and vigorously stirred at 60°C for 2 hours until an acidic suspension of cellulose was achieved. After the reaction, CNC suspensions are usually washed with deionized water, centrifuged, and neutralized by 0.5 N NaOH solutions [41]. A mixed acid (organic and inorganic acids) solution may also be applied to control the final properties and boost the yield [44].

2.3.2.2 Enzymatic Hydrolysis

Although widely used by a variety of works, the major drawback of the acid hydrolysis technique is that it involves hazardous, toxic, and corrosive acids that require corrosion-resistant reactors and extreme precautions during handling [37]. In addition, due to the elevated environmental concerns, the acid hydrolysis requires a large amount of energy and yields a great amount of chemical waste, which prevents it from being used in large-scale industrial production [35]. To resolve these issues, enzymatic hydrolysis has been proposed since the use of enzymes usually does not require harsh chemical reagents, solvents, or large amounts of energy [50]. As an example, Anderson *et al.* used a type of commercial enzyme (cellulases, which is a mixture of endoglucanases, exoglucanases, and cellobiodrikases) and prepared another type of enzyme from *Spodoptera frugiperda* insect cells to produce CNCs using an enzymatic hydrolysis process [51]. The pre-treated raw materials were mixed with a concentration of 1 mg/mL enzymes of different types and incubated for up to 72 h at 35°C. Compared with the temperature used for acid hydrolysis (such as 60°C), enzyme hydrolysis only requires a little more than half of the temperature for reactions. It has also been discovered that enzymes can selectively degrade the amorphous regions of cellulose while preserving the crystalline regions. As a result, CNCs produced by enzymatic hydrolysis possess hydroxyl group surface chemistry, which has the potential for further chemical manipulation [41]. However, the major drawback of enzymatic hydrolysis is that it is very time-consuming. For example, the enzymatic hydrolysis reported by Anderson *et al.* [51] takes 72 h, while the acid hydrolysis reported by Maiti *et al.* [49] took only 2 h for complete hydrolysis of cellulose biomass. A few investigations by Zhang *et al.* also revealed that the rate of enzymatic hydrolysis may be accelerated by the rough surfaces of the raw materials after pre-treatment and the slightly elevated temperature during the reaction [52]. In addition, this technique is still hindered by the higher cost of cellulose enzymes compared with the available acids [41]. It has also been reported that the CNCs extracted by enzymatic hydrolysis usually possess a lower strength and have a higher tendency to agglomerate [50].

2.3.2.3 *TEMPO Oxidation*

Compared with acid hydrolysis and enzymatic hydrolysis, the TEMPO oxidation method was only introduced recently by Saito and Isogai in 2006 and is very popular for use in the lab-scale synthesis of CNCs [53]. In the oxidation reaction, the 2,2,6,6-tetra-methylpiperidine-1-oxyl radical (TEMPO) is used as a catalyst with NaClO or $NaClO_2$ as the oxidant to react with the hydroxyl groups on the cellulose surfaces in the presence of NaBr at alkaline conditions. The reaction was found to only occur in the amorphous domains, and as the carboxyl content is increased to a certain level, the crystalline cellulose starts to disperse in the aqueous solution [35, 41]. As an example, Cao *et al.* produced CNCs from jute fibers using the TEMPO/NaBr/NaClO system followed by mechanical homogenization [54]. The TEMPO oxidation can be performed at room temperature, and it only takes about 1-2 hours to complete. They reported a very high yield of about 80% of CNCs with sizes in the range of 3–10 nm [54]. However, this methodology still exhibits serious drawbacks, such as the toxicity and high costs of the TEMPO chemicals, as well as the limited oxidation at C_6 primary hydroxyls [41].

2.3.2.4 *Mechanical Process*

Mechanical processes such as ultrasonication, high-pressure homogenization, and ball milling can also be used to produce CNCs. CNC extraction by dissolving the raw material in an ionic liquid has also caught the attention of many researchers. Li *et al.* reported CNC extraction from sugarcane bagasse using ionic liquid for dissolution and then passed it through a high-pressure homogenizer at 80 MPa for 30 cycles under optimized conditions [55]. However, low crystallinity and low thermal stability of CNCs were reported due to the breakdown of the intermolecular and intramolecular hydrogen bonding of cellulose at high pressure. Zhang *et al.* used the ball milling method on alkali-treated softwood pulp and CNCs with an average diameter of 100 nm [56]. However, similar to what is reported by Li *et al.*, under heavy mechanical deformation, the crystallinity cannot be preserved, and the high energy consumption involved in these mechanical processes also impedes the wide application of these mechanical processes [55].

2.3.3 Post-Treatment

CNCs are suspended in water when they go through processes like acid hydrolysis or enzymatic hydrolysis.Therefore, some post-treatment is needed to extract the CNCs from the suspensions. Typical processes are water washing, centrifugation, ultrasonication, and air drying or freeze drying [35, 38–40, 44, 48]. It is worth noting that, for the acid hydrolysis process, free acid molecules may still exist in the suspensions and need to be eliminated by adjusting the pH to 9 using sodium hydroxide and washing with distilled water until reaching neutrality [57]. Although it is simple, the post-treatment process can take up to three days until clean CNC powders have been achieved.

2.4 Characterization

The characterization of CNC and CNF mostly includes morphological, physical, chemical, and mechanical characterizations.

2.4.1 Morphology

Morphologies of CNC and CNF can be observed using scanning electron microscopy (SEM) and transmission electron microscopy (TEM).

2.4.1.1 Microscopy

The purpose of performing microscopy for CNC and CNF is mainly to look at the fiber surface, structure, and determination of fiber diameter. The samples need to be gold coated for scanning electron microscopy to avoid electrostatic charge buildup and burning of the samples during imaging. Another important measure is sample dispersion for microscopic analysis. For both SEM and TEM, sample dispersion in ethanol and sonication for about 30 minutes would lead to a better spatial resolution of single cellulose fibers. Morphological characterization also helps to compare the fiber structure before and after several treatment processes, as well as between different treatment methods. Figure 2.1 depicts SEM and TEM images of CNC and CNF extracted from sugarcane bagasse [4]. The smoothness of the original fiber surface was due to the presence of lignin, wax, pectin, and oil. During the pre-treatment and extraction processes, those impurities were separated and resulted in needle-shaped CNC, which are evident in TEM. The diameters of CNC and CNF were also analyzed from SEM and TEM images using image processing software.

2.4.2 Physical Characterization

Physical characterization of CNC and CNF mainly includes determination of crystallinity, zeta potential, and thermal properties.

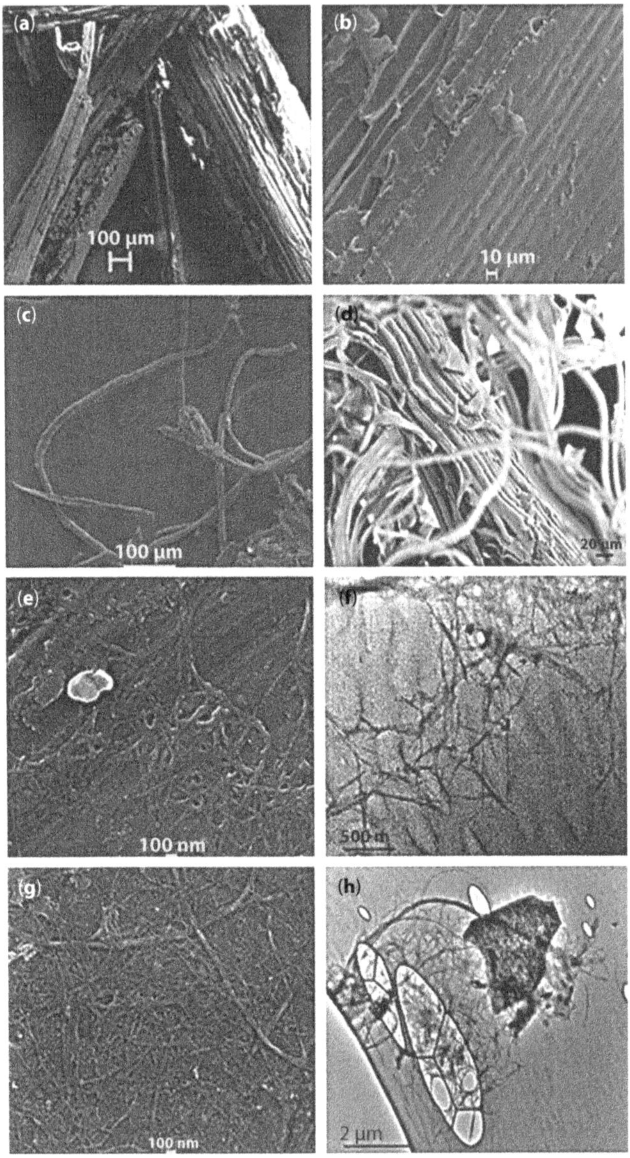

Figure 2.1 FE-SEM images of sugarcane bagasse as a raw material at (a) 100 µm scale and (b) 10 µm scale; extracted cellulose at (c) 100 µm scale and (d) 20 µm scale; CNF extracted by mill balling in (e) SEM and (f) TEM images; acid hydrolysed cellulose (CNC) in (g) SEM and (h) TEM images. CNC and CNF samples were dispersed at 0.5% concentration. (Reprinted with permission from ref [4]).

2.4.2.1 XRD

X-ray diffraction (XRD) analysis can be used to find out how CNC and CNF are made of crystals. XRD spectra can be measured with an X-ray diffractometer equipped with Cu-Kα radiation at a wavelength of 0.154 nm, a voltage of 40 kV, and a current of 40 mA.

The crystallinity index of the samples can be calculated using the Segal method [79]:

$$\text{Crystallinity Index} = (I_{200} - I_{am})/I_{200} \times 100\% \tag{2.1}$$

where I_{200} is the maximum intensity (in arbitrary units) of the diffraction from the 200 planes, and I_{am} is the intensity of the background scatter. Simply, I_{200} is the intensity of the peak around 22°, and I_{am} is the intensity around 18°, representing the crystallinity and amorphous regions in the cellulose samples, respectively. The crystallinity of CNC and CNF can also be determined with a Raman spectroscope or ^{13}C solid-state nuclear magnetic resonance spectroscopy (NMR). XRD provides the total crystallinity of CNC/CNF, while NMR can provide further detailed information on crystalline polymorphs such as cellulose I_{α}, I_{β}, and II and their concentration [80].

As a very simple example, XRD patterns of CNCs isolated from various sources are depicted in Figure 2.2. The cellulose sources, such as TS-*teff* straw, EF-*enset* fiber, SB-sugarcane bagasse, CH-coffee hull, and CC-CNCs

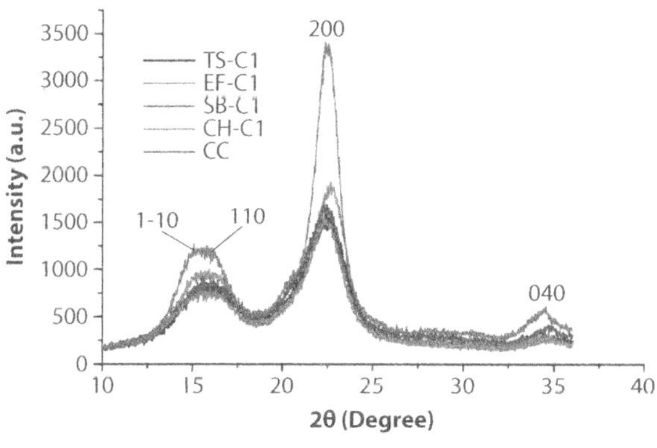

Figure 2.2 XRD patterns of CNCs isolated from cellulose (C1) precursors. (Key: TS-*teff* straw; EF-*enset* fiber; SB-sugarcane bagasse; CH-coffee hull; CC-CNCs isolated from commercial cellulose (Reprinted with permission from ref [81]).

isolated from commercial cellulose, were characterized and compared using XRD. A typical crystal lattice structure was predicted by deconvoluting 2θ signals using a Gaussian profile. The main diffraction signals around 2θ values of 15°, 16°, 22°, and 34° were assigned to the crystallographic planes of 1–10, 110, 200, and 040, respectively [81].

2.4.2.2 Zeta Potential

The zeta potential of CNC and CNF can be characterized with a zetasizer. It indicates the aqueous dispersibility of CNC or CNF on account of surface charges, especially when compared before and after surface modification. The electrostatic repulsion caused by negative functional groups on the surface facilitates stable colloidal dispersion of CNC/CNF in water. If zeta potential is low, van der Waals attraction force can cause agglomeration of CNC/CNF. Depending on the pre-treatment used, the zeta potential of CNF was reported to range from -89.9 mV to -14.0 mV [27].

2.4.2.3 TGA

Thermal properties of CNC and CNF can be characterized using thermogravimetry analysis (TGA), which shows change in mass with temperature as shown in Figure 2.3. TGA is normally conducted in an inert environment with temperatures changing from room temperature (22°C–25°C) to 500°C at a heating rate of 10°C/min [82]. The weight loss below 100°C is attributed to the evaporation of moisture. Depending on the crystallinity of CNC, the initial degradation temperature can occur around 150–280°C,

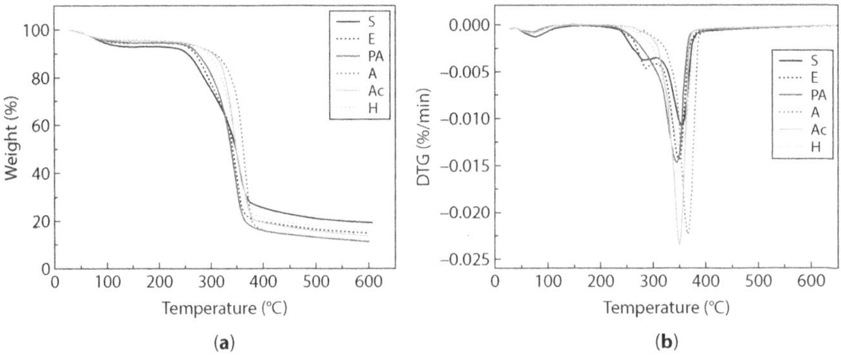

Figure 2.3 TGA thermograms (a), and DTG (b) curves for raw sisal fibers and samples obtained after a succession of chemical treatments in the nitrogen atmosphere (Reprinted with permission from ref [82] and ref [31]).

caused by the larger surface area susceptible to elevated temperature and weakening the crystal structure. Pyrolysis of CNC occurs around 200–360 °C with substantial weight loss. The weight of char residue after the heating process can be affected by different pre-treatment methods, inducing variation in surface functional groups and influencing the thermal stability of CNC [82].

2.4.3 Chemical Characterization

The general methods to characterize chemical composition and functional groups of CNC and CNF are summarized below.

2.4.3.1 FTIR

Fourier-transformed infrared spectroscopy (FTIR) is a technique using infrared light radiating through CNC/CNF to record the absorbance of light at different wavelengths and correspond the absorption peaks to vibrations of specific molecular structures. For CNC/CNF, the FTIR spectra are normally obtained in the wavelength range of 400–4000 cm^{-1} at a scanning resolution of 2 cm^{-1}, as shown in Figure 2.4. Based on the extraction methods and composition of the raw cellulose biomass, the typical peaks observed in FTIR for CNC/CNF are 3200-3500 cm^{-1} relating to O–H stretching, 2700–2900 cm^{-1} relating to C–H stretching, 1430 cm^{-1}

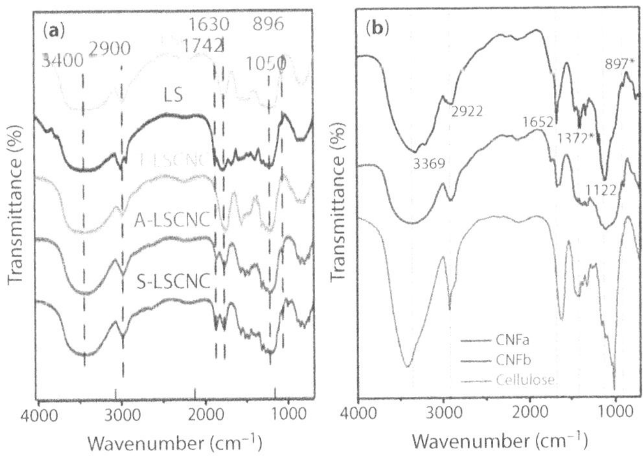

Figure 2.4 FTIR spectra of CNCs (a) from various sources and CNFs (b) extracted from rice husk autumn (CNFa) and summer (CNFb) varieties and compared with commercial cellulose (Reprinted with permission from ref [82] and ref [31]).

relating to CH_2 bending, 1050–1170 cm^{-1} relating to C–O stretching, and 890 cm^{-1} relating to C–H deformation vibration [3]. The presence of lignin, hemicellulose, and other content may cause the FTIR spectrum to vary for CNC and CNF.

2.4.3.2 XPS

The surface chemistry of CNC and CNF can be characterized with X-ray photoelectron spectroscopy (XPS), which is generally used for modified CNC to determine the change in surface elemental composition after treatment. Surface oxygen and carbon functional groups such as C-OH, C=O, and O-C-O can be obtained from XPS at binding energies between 286 and 533 eV. The atomic ratio between C and O can be used to demonstrate changes in chemical composition on the surface of CNC/CNF [82].

2.4.3.3 Elemental Analysis

Elemental analysis of CNC and CNF can be measured with either energy-dispersive X-ray spectroscopy (EDS), electron energy loss spectroscopy (EELS), or CHNSO elemental analysis. EDS is not as sensitive as EELS for organic material composed of mainly low-atomic-number elements such as C and O. For CNC/CNF doping with metals to form composite materials, EDS is sufficient to quantify the elemental fraction. CHNSO analysis is a combustion test to determine the compositional percentages of C, H, N, S, and O.

2.4.4 Mechanical Characterization

Since CNC and CNF are nanoscale materials, their mechanical properties are calculated theoretically or with atomic force microscopy (AFM). With AFM, the cantilever with a known spring constant can make contact with the surface of the CNC/CNF in an inert gas environment. The adhesion force and stiffness can be measured from the contact force and the displacement, which is the sum of the indentation CNC/CNF depth and cantilever bending displacement. The characterization process is humidity-sensitive. Lahiji *et al.* have reported using Jump-mode AFM to measure the mechanical properties of CNC and also measured increasing mechanical properties from adhesion from 13.0 nN to 14.7 nN and effective stiffness from 1.67 N/m to 1.77 N/m as relative humidity increases from 0.1%

to 30% [83]. The authors attributed the increment to water adsorption on surfaces and further calculated the transverse elastic modulus of CNC to be 18.1-39.5 GPa.

When CNF or CNC is made into nanopaper, the mechanical properties of the nanopaper can be evaluated with tensile tests. Sehaqui *et al.* obtained CNF-based nanofibers treated with TEMPO-oxidation and cold drawing to achieve partial alignment, which yielded an elastic modulus of 10.3-33.3 GPa, a tensile strength of 185-428 MPa, and 1.79–5.26% strain at break [84].

2.5 Conclusion

There are three parts to the process of making CNF: pre-treatment, extraction, and post-treatment. Pre-treatment, extraction, and post-treatment, based on the treatment agent used, can be divided into chemical, biological, and mechanical. Table 2.1 summarizes the extraction process of CNF/CNC w.r.t. energy or chemical consumption, advantages, and disadvantages. The purpose of pre-treatment is to reduce energy for the extraction stage by modifying the intermolecular or intramolecular bonds to achieve a higher cellulose yield, assist with defibrillation, or speed up the reaction process. The quality of the extracted CNF depends on the raw resource, the pre-treatment, and the extraction methods adopted. Post-treatment is mainly the removal of chemicals used in the preparation procedures or the drying of CNF from suspension. To tailor CNF with desired properties, suitable combinations of pre-treatment, extraction, and post-treatment methods need to be selected based on the composition of the cellulose biomass, the functional groups on the cellulose content, and the manufacturing cost. Currently, the most commonly used processes, both at the lab scale and in industry, to produce CNCs are still acid hydrolysis and enzymatic hydrolysis. Depending on the raw materials and requirements for the properties and geometries of the final production, pre-treatments such as alkali treatment and bleaching are needed before the actual extraction. The properties of the produced CNCs are closely related to the raw materials, pre-treatment processes, and extraction processes. A combination of extraction processes, such as the mechanical process plus acid hydrolysis, may also be used, depending on the end target. Finally, a post-treatment process is usually needed to obtain dry CNC powders from aqueous suspensions.

Table 2.1 Comparative extraction process of CNF/CNC w.r.t. energy or chemical consumption, advantages and disadvantages.

Method	Energy or chemical consumption	Advantage	Disadvantage	Ref
Retting process	0.59 GJ/ton	Low cost and normally environmental-friendly	Limited to environment; long duration	[58]
Desalting process	-	Low cost and typically no chemical used	Can only remove most water-soluble contents	[2]
Dewaxing process	400 ml ethanol/toluene for 40 g raw biomass	Effective in removing wax or oil	Normally use harsh solvent; relatively long duration	[59]
Alkali treatment	$1.21 USD per 30g sodium hydroxide used	High cellulose yield; can reduce non-cellulose content	Strong chemical	[60]
Bleach (sodium hypochlorite)	$1.51 USD per 128.57 mL sodium hypochlorite used	Effective in removing lignin and other non-cellulose content	Strong chemical	[60]
Bleach (H_2O_2)	120 ml bleach agent containing H2O$_2$ for 2in by 4in raw material	Effective in removing lignin and other non-cellulose content	May cause cellulose depolymerization	[61]

(Continued)

Table 2.1 Comparative extraction process of CNF/CNC w.r.t. energy or chemical consumption, advantages and disadvantages. (*Continued*)

Method	Energy or chemical consumption	Advantage	Disadvantage	Ref
Bleach (Organosolv)	6000 kg/hr ethanol/water used per 402.7 g cellulose produced	Environmental-friendly; effective in removing lignin and other non-cellulose content	Normally require large amount of solvent	[62]
Bleach (ozone)	24.9 kWh/kg O_3 used	Environmental-friendly; can reduce lignin and ash	Relatively challenging to control	[63]
Ionic liquid extraction	1/20 biomass/ionic liquid ratio	Can reduce lignin and other non-cellulose content	Limited extraction; normally need to be combined with other pre-treatment or extraction	[64]
Microwave liquefaction	0.142-0.624 kWh for temperature of 120-200°C and duration of 0-30min	Can speed up pre-treatment reaction	Normally can not be used as extraction methods on its own	[65]
Acid hydrolysis	$9.06 USD per 168.4 mL sulfuric acid used; 0.7 kWh/kg of nanocellulose produced	High cellulose crystallinity	Strong chemical	[60, 66]

(*Continued*)

Table 2.1 Comparative extraction process of CNF/CNC w.r.t. energy or chemical consumption, advantages and disadvantages. (*Continued*)

Method	Energy or chemical consumption	Advantage	Disadvantage	Ref
TEMPO-oxidation	44-140.8 €/kg CNF produced; 0.5 kWh/kg of nanocellulose produced	Low cost; environmental-friendly; fine and uniform CNC/CNF	High-cost chemicals and limited oxidation	[66, 67]
Homogenization	6.5 kWh/kg of nanocellulose produced	Efficient in defibrillation	High energy consumption; multiple pass required; clogging problem	[68]
Microfluidization	200-630 kJ/kg cellulose produced per pass	Efficient in defibrillation	High energy consumption; multiple pass required; clogging problem	[69]
Ball milling	1.7 kWh/kg CNF produced	Efficient in defibrillation	High energy consumption; dimension of CNF/CNC dependent on ball size	[70]
Grinding	3.1-7.1 kWh/kg of nanocellulose produced	Efficient in defibrillation	High energy consumption; negative effect on crystallinity and thermal stability	[66, 68]

(*Continued*)

Table 2.1 Comparative extraction process of CNF/CNC w.r.t. energy or chemical consumption, advantages and disadvantages. (*Continued*)

Method	Energy or chemical consumption	Advantage	Disadvantage	Ref
Blending	1.3-3.1 kWh/kg CNF produced at speed of 1440-3200 rpm	Good thermal stability and efficient in defibrillation	High energy consumption	[71]
Sonication	6 kWh/kg CNF produced	Efficient in defibrillation when used after pre-treatment or extraction	High energy consumption	[72]
Steam explosion	0.2-0.6 MJ/kg lignocellulose produced	Typically no strong chemical used	High energy consumption	[73]
Enzymatic hydrolysis	0.6432 €/kg CNF produced; 1.91 kWh/kg of nanocellulose produced	Low cost and environmental-friendly	Can induce fiber entanglement; long duration	[67, 68]
Electron beam irradiation	500-3000 kGy	Short duration and typically no chemical used	Normally need to be combined with other pre-treatment or extraction	[74]

(*Continued*)

Table 2.1 Comparative extraction process of CNF/CNC w.r.t. energy or chemical consumption, advantages and disadvantages. (*Continued*)

Method	Energy or chemical consumption	Advantage	Disadvantage	Ref
Electrospinning	0.4 Wh	Low cost; low energy consumption	High voltage required; hard to tailor length of CNF	[75]
Dialysis	$5.38 USD per dialysis bag used	Can remove excessive chemical	Long duration	[60]
Centrifugation	65.34 kWh/kg sample	Can remove excessive chemical in shorter time	High energy consumption	[76]
Lyophilization	0.66-3.06 kWh/kg	Less likely cause agglomeration	Higher energy consumption than evaporation drying	[77]
Evaporation drying (oven)	0.245 kWh/hr at 70°C	Low cost	Can cause agglomeration; long duration	[60]
Supercritical fluid drying	47.89-926.01 kWh/ton raw biomass used	Typically cause no agglomeration	High energy consumption	[78]

References

1. Nagarajan, K.J., Balaji, A.N., Ramanujam, N.R., Extraction of cellulose nanofibers from cocos nucifera var aurantiaca peduncle by ball milling combined with chemical treatment. *Carbohydr. Polym.*, 212, 312–22, 2019. https://doi.org/10.1016/j.carbpol.2019.02.063.
2. Paniz, O.G., Pereira, C.M.P., Pacheco, B.S., Wolke, S.I., Maron, G.K., Mansilla, A. *et al.*, Cellulosic material obtained from Antarctic algae biomass. *Cellulose*, 27, 113–26, 2020. https://doi.org/10.1007/s10570-019-02794-2.
3. Rahimi Kord Sofla, M., Brown, R.J., Tsuzuki, T., Rainey, T.J., A comparison of cellulose nanocrystals and cellulose nanofibres extracted from bagasse using acid and ball milling methods. *Adv. Nat. Sci. Nanosci. Nanotechnol.*, 7, 035004, 2016. https://doi.org/10.1088/2043-6262/7/3/035004.
4. Mondragon, G., Fernandes, S., Retegi, A., Peña, C., Algar, I., Eceiza, A. *et al.*, A common strategy to extracting cellulose nanoentities from different plants. *Ind. Crops Prod.*, 55, 140–8, 2014. https://doi.org/10.1016/j.indcrop.2014.02.014.
5. Julie Chandra, C.S., George, N., Narayanankutty, S.K., Isolation and characterization of cellulose nanofibrils from arecanut husk fibre. *Carbohydr. Polym.*, 142, 158–66, 2016. https://doi.org/10.1016/j.carbpol.2016.01.015.
6. Abraham, E., Deepa, B., Pothen, L.A., Cintil, J., Thomas, S., John, M.J. *et al.*, Environmental friendly method for the extraction of coir fibre and isolation of nanofibre. *Carbohydr. Polym.*, 92, 1477–83, 2013. https://doi.org/10.1016/j.carbpol.2012.10.056.
7. Chen, M., Liu, T., Zhang, X., Zhang, R., Tang, S., Yuan, Y. *et al.*, Photoinduced Enhancement of uranium extraction from seawater by MOF/Black phosphorus quantum dots heterojunction anchored on cellulose nanofiber aerogel. *Adv. Funct. Mater.*, 31, 2100106, 2021. https://doi.org/10.1002/adfm.202100106.
8. Dilamian, M. and Noroozi, B., A combined homogenization-high intensity ultrasonication process for individualizaion of cellulose micro-nano fibers from rice straw. *Cellulose*, 26, 5831–49, 2019. https://doi.org/10.1007/s10570-019-02469-y.
9. Corrêa, A.C., de Teixeira, E.M., Pessan, L.A., Mattoso, L.H.C., Cellulose nanofibers from curaua fibers. *Cellulose*, 17, 1183–92, 2010. https://doi.org/10.1007/s10570-010-9453-3.
10. Li, M., Wang, L.J., Li, D., Cheng, Y.L., Adhikari, B., Preparation and characterization of cellulose nanofibers from de-pectinated sugar beet pulp. *Carbohydr. Polym.*, 102, 136–43, 2014. https://doi.org/10.1016/j.carbpol.2013.11.021.
11. Jonoobi, M., Khazaeian, A., Tahir, P.M., Azry, S.S., Oksman, K., Characteristics of cellulose nanofibers isolated from rubberwood and empty fruit bunches of oil palm using chemo-mechanical process. *Cellulose*, 18, 1085–95, 2011. https://doi.org/10.1007/s10570-011-9546-7.
12. Nobuta, K., Teramura, H., Ito, H., Hongo, C., Kawaguchi, H., Ogino, C. *et al.*, Characterization of cellulose nanofiber sheets from different refining processes. *Cellulose*, 23, 403–14, 2016. https://doi.org/10.1007/s10570-015-0792-y.

13. Menezes, D.B., Diz, F.M., Romanholo, L.F., Luiz, R.V.B., Maria, P.C., Jose, Y.C. *et al.*, Starch-based biocomposite membrane reinforced by orange bagasse cellulose nanofibers extracted from ionic liquid treatment. *Cellulose*, 0123456789, 4137–49, 2021. https://doi.org/10.1007/s10570-021-03814-w.
14. Feng, Y.H., Cheng, T.Y., Yang, W.G., Ma, P.T., He, H.Z., Yin, X.C. *et al.*, Characteristics and environmentally friendly extraction of cellulose nanofibrils from sugarcane bagasse. *Ind. Crops Prod.*, 111, 285–91, 2018. https://doi.org/10.1016/j.indcrop.2017.10.041.
15. Bahrami, B., Behzad, T., Zamani, A., Heidarian, P., Nasri-Nasrabadi, B., Optimal design of ozone bleaching parameters to approach cellulose nanofibers extraction from sugarcane bagasse fibers. *J. Polym. Environ.*, 26, 4085–94, 2018. https://doi.org/10.1007/s10924-018-1277-5.
16. Nuruddin, M., Hosur, M., Uddin, M.J., Baah, D., Jeelani, S., A novel approach for extracting cellulose nanofibers from lignocellulosic biomass by ball milling combined with chemical treatment. *J. Appl. Polym. Sci.*, 133, 1–10, 2016. https://doi.org/10.1002/app.42990.
17. Xie, J., Hse, C., De Hoop, C.F., Hu, T., Qi, J., Shupe, T.F., Isolation and characterization of cellulose nanofibers from bamboo using microwave liquefaction combined with chemical treatment and ultrasonication. *Carbohydr. Polym.*, 151, 725–34, 2016. https://doi.org/10.1016/j.carbpol.2016.06.011.
18. Liu, Q., Lu, Y., Aguedo, M., Jacquet, N., Ouyang, C., He, W. *et al.*, Isolation of high-purity cellulose nanofibers from wheat straw through the combined environmentally friendly methods of steam explosion, microwave-assisted hydrolysis, and microfluidization. *ACS Sustain. Chem. Eng.*, 5, 6183–6191, 2017. https://doi.org/10.1021/acssuschemeng.7b01108.
19. Soni, B., Barbary, E., Mahmoud, B., Chemical isolation and characterization of different cellulose nanofibers from cotton stalks. *Carbohydr. Polym.*, 134, 581–9, 2015. https://doi.org/10.1016/j.carbpol.2015.08.031.
20. Isogai, A. and Bergström, L., Preparation of cellulose nanofibers using green and sustainable chemistry. *Curr. Opin. Green Sustain. Chem.*, 12, 15–21, 2018. https://doi.org/10.1016/j.cogsc.2018.04.008.
21. Tanpichai, S. and Wimolmala, E., Facile single-step preparation of cellulose nanofibers by TEMPO-mediated oxidation and their nanocomposites facile single-step preparation of cellulose nanofibers by TEMPO-mediated oxidation and their nanocomposites. *J. Nat. Fibers*, 19, 10094–110, 2022. https://doi.org/10.1080/15440478.2021.1993483.
22. Saito, T. and Isogai, A., TEMPO-mediated oxidation of native cellulose. The effect of oxidation conditions on chemical and crystal structures of the water-insoluble fractions. *Biomacromolecules*, 5, 1983–9, 2004.
23. Kaffashsaie, E., Yousefi, H., Nishino, T., Matsumoto, T., Mashkour, M., Madhoushi, M. *et al.*, Direct conversion of raw wood to TEMPO-oxidized cellulose nanofibers. *Carbohydr. Polym.*, 262, 117938, 2021. https://doi.org/10.1016/j.carbpol.2021.117938.

24. Huang, C., Ji, H., Yang, Y., Guo, B., Luo, L., Meng, Z. *et al.*, TEMPO-oxidized bacterial cellulose nano fiber membranes as high- performance separators for lithium-ion batteries. *Carbohydr. Polym.*, 230, 115570, 2020. https://doi.org/10.1016/j.carbpol.2019.115570.
25. Serra, A., Gonz, I., Oliver-ortega, H., Reducing the amount of catalyst in TEMPO-oxidized cellulose nanofibers: Effect on properties and cost. *Polymers (Basel)*, 9, 11, 557, 2017. https://doi.org/10.3390/polym9110557.
26. Raj, P., Mayahi, A., Lahtinen, P., Varanasi, S., Garnier, G., Martin, D. *et al.*, Gel point as a measure of cellulose nanofibre quality and feedstock development with mechanical energy. *Cellulose*, 23, 3051–64, 2016. https://doi.org/10.1007/s10570-016-1039-2.
27. Andrade-Mahecha, M.M., Pelissari, F.M., Tapia-Blácido, D.R., Menegalli, F.C., Achira as a source of biodegradable materials: Isolation and characterization of nanofibers. *Carbohydr. Polym.*, 123, 406–15, 2015. https://doi.org/10.1016/j.carbpol.2015.01.027.
28. Wang, Y., Wei, X., Li, J., Wang, F., Wang, Q., Zhang, Y. *et al.*, Homogeneous isolation of nanocellulose from eucalyptus pulp by high pressure homogenization. *Ind. Crops Prod.*, 104, 237–41, 2017. https://doi.org/10.1016/j.indcrop.2017.04.032.
29. Clone, D.U., Jim, L., Eugenio, E., Ibarra, D., Darder, M., Mart, J.A. *et al.*, Cellulose nanofibers from a Dutch Elm. *Polymers (Basel)*, 12, 11, 2450, 2020.
30. Yang, W., Feng, Y., He, H., Yang, Z., Environmentally-friendly extraction of cellulose nanofibers from steam-explosion pretreated sugar beet pulp. *Materials (Basel)*, 11, 1160, 2018. https://doi.org/10.3390/ma11071160.
31. Kalita, E., Nath, B.K., Deb, P., Agan, F., Islam, R., Saikia, K., High quality fluorescent cellulose nanofibers from endemic rice husk: Isolation and characterization. *Carbohydr. Polym.*, 122, 308–13, 2015. https://doi.org/10.1016/j.carbpol.2014.12.075.
32. De Campos, A., Carolina, A., David, C., Eliangela, C., Obtaining nanofibers from curaua fibers using enzymatic hydrolysis followed by sonication. *Cellulose*, 20, 1491–500, 2013. https://doi.org/10.1007/s10570-013-9909-3.
33. Janardhnan, S. and Sain, M., Isolation of cellulose nanofibers: Effect of biotreatment on hydrogen bonding network in wood fibers. *Int. J. Polym. Sci.*, 2011, 279610, 2011. https://doi.org/10.1155/2011/279610.
34. Teixeira, E.D.M., Corre, A.C., Manzoli, A., Leite, F.D.L., Henrique, L., Mattoso, C., Cellulose nanofibers from white and naturally colored cotton fibers. *Cellulose*, 17, 595–606, 2010. https://doi.org/10.1007/s10570-010-9403-0.
35. Kumar, R., Rai, B., Gahlyan, S., Kumar, G., A comprehensive review on production, surface modification and characterization of nanocellulose derived from biomass and its commercial applications. *Express Polym. Lett.*, 15, 104–120, 2021. https://doi.org/10.3144/expresspolymlett.2021.11.
36. Dufresne, A., *Nanocellulose: From nature to high performance tailored materials*, Walter de Gruyter GmbH, Berlin, Germany, 2012.

37. Hamad, W.Y., *Cellulose nanocrystals properties, production, and applications*, John Wiley & Sons, Chichester, England, 2017.
38. Çiçek Özkan, B. and Güner, M., Isolation, characterization, and comparison of nanocrystalline cellulose from solid wastes of horse chestnut and chestnut seed shell. *Cellulose*, 29, 6629–6644, 2022. https://doi.org/10.1007/s10570-022-04682-8.
39. Doan, T.K.Q. and Chiang, K.Y., Characteristics and kinetics study of spherical cellulose nanocrystal extracted from cotton cloth waste by acid hydrolysis. *Sustain. Environ. Res.*, 32, 26, 2022. https://doi.org/10.1186/s42834-022-00136-9.
40. He, Q., Yang, Y., Liu, Z., Shao, D., Jiang, D., Xing, L., Pan, Q., Shan, H., Preparation and characterization of cellulose nanocrystals from spent edible fungus substrate. *J. Sci. Food Agric.*, 102, 2761–2772, 2022. https://doi.org/10.1002/jsfa.11617.
41. Trache, D., Hussin, M.H., Haafiz, M.K.M., Thakur, V.K., Recent progress in cellulose nanocrystals: Sources and production. *Nanoscale*, 9, 1763–1786, 2017. https://doi.org/10.1039/c6nr09494e.
42. Jawaid, M., Thariq, M., Saba, N., *Mechanical and physical testing of biocomposites, fibre-reinforced composites and hybrid composites*, Woodhead Publishing, Duxford, England, 2019.
43. Alothman, O.Y., Kian, L.K., Saba, N., Jawaid, M., Khiari, R., Cellulose nanocrystal extracted from date palm fibre: Morphological, structural and thermal properties. *Ind. Crops Prod.*, 159, 113075, 2021. https://doi.org/10.1016/j.indcrop.2020.113075.
44. Zhu, P., Feng, L., Ding, Z., Bai, X., Preparation of spherical cellulose nanocrystals from microcrystalline cellulose by mixed acid hydrolysis with different pretreatment routes. *Int. J. Mol. Sci.*, 23, 10764, 2022. https://doi.org/10.3390/ijms231810764.
45. Wang, S., Yi, Z., Iqbal, Y., Chen, Z., Xue, S., Fu, T., Li, M., Polyploid miscanthus lutarioriparius: A sustainable and scalable biomass feedstock for cellulose nanocrystal preparation in biorefinery. *Agronomy*, 12, 1057, 2022. https://doi.org/10.3390/agronomy12051057.
46. Kumar, A., Singh Negi, Y., Choudhary, V., Kant Bhardwaj, N., Characterization of cellulose nanocrystals produced by acid-hydrolysis from sugarcane bagasse as agro-waste. *J. Mater. Phys. Chem.*, 2, 1–8, 2020. https://doi.org/10.12691/jmpc-2-1-1.
47. Mariano, M., El Kissi, N., Dufresne, A., Cellulose nanocrystals and related nanocomposites: Review of some properties and challenges. *J. Polym. Sci. B Polym. Phys.*, 52, 791–806, 2014. https://doi.org/10.1002/polb.23490.
48. Kassa, H.S., Jabasingh, S.A., Mohammed, S.A., Baek, S.Y., Park, S.Y., Extraction and characterization of cellulose nanocrystals from anchote (Coccinia abyssinica) bagasse. *Macromol. Res.*, 31, 11, 776–782, 2022. https://doi.org/10.1007/s13233-022-0088-7.

49. Maiti, S., Jayaramudu, J., Das, K., Reddy, S.M., Sadiku, R., Ray, S.S., Liu, D., Preparation and characterization of nano-cellulose with new shape from different precursor. *Carbohydr. Polym.*, 98, 562–567, 2013. https://doi.org/10.1016/j.carbpol.2013.06.029.
50. Rana, A.K., Frollini, E., Thakur, V.K., Cellulose nanocrystals: Pretreatments, preparation strategies, and surface functionalization. *Int. J. Biol. Macromol.*, 182, 1554–1581, 2021. https://doi.org/10.1016/j.ijbiomac.2021.05.119.
51. Anderson, S., Esposito, D., Gillette, W., Zhu, J., Baxa, U., Mcneil, S., Enzymatic preparation of nanocrystalline and microcrystalline cellulose. *Cellulose Nanocrystals*, 13, 5, 35–42, 2014. http://www.sigmaaldrich.com.
52. Zhang, J., Zhang, B., Zhang, J., Lin, L., Liu, S., Ouyang, P., Effect of phosphoric acid pretreatment on enzymatic hydrolysis of microcrystalline cellulose. *Biotechnol. Adv.*, 28, 613–619, 2010. https://doi.org/10.1016/j.biotechadv.2010.05.010.
53. Saito, T. and Isogai, A., Introduction of aldehyde groups on surfaces of native cellulose fibers by TEMPO-mediated oxidation. *Colloids Surf A Physicochem. Eng. Asp.*, 289, 219–225, 2006. https://doi.org/10.1016/j.colsurfa.2006.04.038.
54. Cao, X., Ding, B., Yu, J., Al-Deyab, S.S., Cellulose nanowhiskers extracted from TEMPO-oxidized jute fibers. *Carbohydr. Polym.*, 90, 1075–1080, 2012. https://doi.org/10.1016/j.carbpol.2012.06.046.
55. Li, J., Wei, X., Wang, Q., Chen, J., Chang, G., Kong, L., Su, J., Liu, Y., Homogeneous isolation of nanocellulose from sugarcane bagasse by high pressure homogenization. *Carbohydr. Polym.*, 90, 1609–1613, 2012. https://doi.org/10.1016/j.carbpol.2012.07.038.
56. Zhang, L., Tsuzuki, T., Wang, X., Preparation of cellulose nanofiber from softwood pulp by ball milling. *Cellulose*, 22, 1729–1741, 2015. https://doi.org/10.1007/s10570-015-0582-6.
57. Wang, N., Ding, E., Cheng, R., Thermal degradation behaviors of spherical cellulose nanocrystals with sulfate groups. *Polymers (Guildf)*, 48, 3486–3493, 2007. https://doi.org/10.1016/j.polymer.2007.03.062.
58. Dissanayake, N.P.J., Summerscales, J., Grove, S.M., Singh, M.M., Energy use in the production of flax fiber for the reinforcement of composites. *J. Nat. Fibers*, 6, 331–46, 2009. https://doi.org/10.1080/15440470903345784.
59. Kunusa, W.R., Abdullah, R., Bilondatu, K., Tulie, W.Z., Analysis of cellulose isolated from sugar bagasse: Optimization and treatment process scheme. *J. Phys. Conf. Ser.*, 1422, 012040, 2020. https://doi.org/10.1088/1742-6596/1422/1/012040.
60. Wahib, S.A., Da'na, D.A., Al-Ghouti, M.A., Insight into the extraction and characterization of cellulose nanocrystals from date pits. *Arab. J. Chem.*, 15, 103650, 2022. https://doi.org/10.1016/j.arabjc.2021.103650.
61. Brooks, R.E. and Moore, S.B., Alkaline hydrogen peroxide bleaching of cellulose. *Cellulose*, 7, 263–86, 2000. https://doi.org/10.1023/A:1009273701191.

62. Alriols, M.G., García, A., Llano-ponte, R., Labidi, J., Combined organosolv and ultrafiltration lignocellulosic biorefinery process. *Chem. Eng. J.*, 157, 113–20, 2010. https://doi.org/10.1016/j.cej.2009.10.058.
63. Lee, J.M., Jameel, H., Venditti, R.A., Effect of ozone and autohydrolysis pretreatments on enzymatic digestibility of coastal bermuda grass. *Bioresources*, 5, 1084–101, 2010.
64. Fu, D., Mazza, G., Tamaki, Y., Lignin extraction from straw by ionic liquids and enzymatic hydrolysis of the cellulosic residues. *J. Agric. Food Chem.*, 58, 2915–22, 2010. https://doi.org/10.1021/jf903616y.
65. Matharu, A.S., de Melo, E.M., Remón, J., Wang, S., Abdulina, A., Kontturi, E., Processing of citrus nanostructured cellulose: A rigorous design-of-experiment study of the hydrothermal microwave-assisted selective scissoring process. *ChemSusChem*, 11, 1344–53, 2018.
66. Malucelli, L.C., Matos, M., Jordão, C., Lomonaco, D., Lacerda, L.G., Carvalho Filho, M.A.S. *et al.*, Influence of cellulose chemical pretreatment on energy consumption and viscosity of produced cellulose nanofibers (CNF) and mechanical properties of nanopaper. *Cellulose*, 26, 1667–81, 2019. https://doi.org/10.1007/s10570-018-2161-0.
67. Delgado-Aguilar, M., González, I., Tarrés, Q., Alcalà, M., Pèlach, MÀ, Mutjé, P., Approaching a low-cost production of cellulose nanofibers for papermaking applications. *BioResources*, 10, 5330–44, 2015. https://doi.org/10.15376/biores.10.3.5330-5344.
68. Bharimalla, A.K., Deshmukh, S.P., Patil, S., Nadanathangam, V., Saxena, S., Development of energy efficient nanocellulose production process by enzymatic pretreatment and controlled temperature refining of cotton linters. *Cellulose*, 30, 2, 833–847, 2022. https://doi.org/10.1007/s10570-022-04959-y.
69. Spence, K.L., Venditti, R.A., Rojas, O.J., Habibi, Y., Pawlak, J.J., A comparative study of energy consumption and physical properties of microfibrillated cellulose produced by different processing methods. *Cellulose*, 18, 1097–111, 2011. https://doi.org/10.1007/s10570-011-9533-z.
70. Pennells, J., Chaléat, C., Martin, D.J., Benchmarking the production of cellulose nanofibres: Biomass feedstock, mechanical processing, and nanopaper performance. *J. Polym. Environ.*, 125, 1–27, 2022. https://doi.org/10.1007/s10924-022-02672-2.
71. Jonoobi, M., Mathew, A.P., Oksman, K., Producing low-cost cellulose nanofiber from sludge as new source of raw materials. *Ind. Crops Prod.*, 40, 232–8, 2012. https://doi.org/10.1016/j.indcrop.2012.03.018.
72. Gallo Stampino, P., Riva, L., Punta, C., Elegir, G., Bussini, D., Dotelli, G., Comparative life cycle assessment of cellulose nanofibres production routes from virgin and recycled raw materials. *Molecules*, 26, 1–20, 2021.
73. Liu, L.Y., Chandra, R.P., Tang, Y., Huang, X.Y., Bai, F.W., Liu, C.G., Instant catapult steam explosion: An efficient preprocessing step for the robust and cost-effective chemical pretreatment of lignocellulosic biomass. *Ind. Crops Prod.*, 188, 115664, 2022. https://doi.org/10.1016/j.indcrop.2022.115664.

74. Lee, M., Heo, M.H., Lee, H., Lee, H.H., Jeong, H., Kim, Y.W. *et al.*, Facile and eco-friendly extraction of cellulose nanocrystals: Via electron beam irradiation followed by high-pressure homogenization. *Green Chem.*, 20, 2596–610, 2018. https://doi.org/10.1039/c8gc00577j.
75. Mollahosseini, A., Rastegari, M., Hatefi, N., Electrospun polyacrylonitrile as a new coating for mechanical stir bar sorptive extraction of polycyclic aromatic hydrocarbons from water samples. *Chromatographia*, 83, 549–58, 2020. https://doi.org/10.1007/s10337-020-03874-3.
76. Guldhe, A., Misra, R., Singh, P., Rawat, I., Bux, F., An innovative electrochemical process to alleviate the challenges for harvesting of small size microalgae by using non-sacrificial carbon electrodes. *Algal Res.*, 19, 292–8, 2016. https://doi.org/10.1016/j.algal.2015.08.014.
77. Karzar Jeddi, M., Laitinen, O., Liimatainen, H., Magnetic superabsorbents based on nanocellulose aerobeads for selective removal of oils and organic solvents. *Mater. Des.*, 183, 108115, 2019. https://doi.org/10.1016/j.matdes.2019.108115.
78. Albarelli, J.Q., Santos, D.T., Ensinas, A.V., Marechal, F., Cocero, M.J., Meireles, M.A.A., Product diversification in the sugarcane biorefinery through algae growth and supercritical CO2 extraction: Thermal and economic analysis. *Renew. Energy*, 129, 776–85, 2018. https://doi.org/10.1016/j.renene.2017.05.022.
79. Segal, L., Creely, J.J., Martin Jr., A.E., Conrad, C.M., An empirical method for estimating the degree of crystallinity of native cellulose using the X-ray diffractometer. *Text. Res. J.*, 29, 786–94, 1959.
80. Sacui, I.A., Nieuwendaal, R.C., Burnett, D.J., Stranick, S.J., Jorfi, M., Weder, C. *et al.*, Comparison of the properties of cellulose nanocrystals and cellulose nanofibrils isolated from bacteria, tunicate, and wood processed using acid, enzymatic, mechanical, and oxidative methods. *ACS Appl. Mater. Interfaces*, 6, 6127–38, 2014.
81. Gabriel, T., Belete, A., Hause, G., Neubert, R.H.H., Gebre-Mariam, T., Isolation and characterization of cellulose nanocrystals from different lignocellulosic residues: A comparative study. *J. Polym. Environ.*, 29, 2964–77, 2021.
82. Zhang, H., Chen, Y., Wang, S., Ma, L., Yu, Y., Dai, H. *et al.*, Extraction and comparison of cellulose nanocrystals from lemon (Citrus limon) seeds using sulfuric acid hydrolysis and oxidation methods. *Carbohydr. Polym.*, 238, 116180, 2020.
83. Lahiji, R.R., Xu, X., Reifenberger, R., Raman, A., Rudie, A., Moon, R.J. *et al.*, Atomic force microscopy characterization of cellulose nanocrystals. *Langmuir*, 26, 4480–8, 2010. https://doi.org/10.1021/la903111j.
84. Sehaqui, H., Ezekiel Mushi, N., Morimune, S., Salajkova, M., Nishino, T., Berglund, L.A., Cellulose nanofiber orientation in nanopaper and nanocomposites by cold drawing. *ACS Appl. Mater. Interfaces*, 4, 1043–9, 2012. https://doi.org/10.1021/am2016766.

3

Synthesis and Characterization of Bacterial Nanocellulose

Moushali Sil, Kamolika Mukhopadhyay, Durba Roy, Sampriti Kundu and Debalina Bhattacharya*

Department of Microbiology, Maulana Azad College, Kolkata, West Bengal, India

Abstract

Bacterial nanocellulose is a well-known biopolymer produced by certain bacterial species. It is a linear, unbranched exopolysaccharide that has β-D-glucopyranose units linked by β-1,4 glycosidic bonds. Its structure is like ultrafine nanofibrils, forming a 3D web-shaped construction. Nowadays, bacterial nanocellulose has a multifaceted importance in the textile, biomedical and agro-waste industry. Production of nanocellulose by bacteria is driving more attention as compared to normal plant-based nanocellulose despite sharing the same molecular formula ($(C_6H_{10}O_5)_n$) due to the higher tensile strength, better fibrillary texture, thermal stability etc. Stabilization of such a structure is done by intermolecular and intramolecular hydrogen bonds. Through the coupling of the two processes polymerization and crystallisation, bacterial nanocellulose production is carried out in bacteria. Cellulose synthase is the prime enzyme for bacterial nanocellulose production. The enzyme has four subunits - BcsA, BcsB, BcsC and BcsD. They are encoded by three (bcsAB, bcsC and bcsD) or four (bcsA, bcsB, bcsC and bcsD) genes. Moreover, there are several factors upon which nanocellulose production depends, some of which are - temperature, carbon source, moisture content, pH of the medium, presence of organic acid, etc. Therefore, in this chapter we have compared and analysed the diversified nature of bacterial nanocellulose production,

**Corresponding author*: debalina.bhattacharya13@gmail.com

Mainak Mukhopadhyay and Debalina Bhattacharya (eds.) Nanocellulose: A Biopolymer for Biomedical Applications, (63–88) © 2024 Scrivener Publishing LLC

mainly concentrating upon the following bacterial genera - *Escherichia coli*, *Komagataeibacter* sp., *Agrobacterium* sp., and *Lactobacillus sp.* In addition, characterization of bacterial nanocellulose by several techniques such as AFM, FTIR, XRD, SEM has also been discussed.

Keywords: Bacterial nanocellulose, cellulose synthase, Bcs operons, bacterial cellulose biosynthesis, nanocellulose characterization

3.1 Introduction

The cell wall of plants is primarily constituted of cellulose, the most common natural polymer on the planet. Other organisms, particularly Gram negative bacteria, may produce cellulose as well. Bacterial nanocellulose (BNC) is an exopolysaccharide that is linear, unbranched, heterogeneous, and comprises β-D-glucopyranose units, linked by β-1,4 glycosidic linkages. Its construction resembles ultrafine nanofibrils and takes the form of a three-dimensional (3D) web [1]. Disaccharide cellobiose makes up its monomeric components. In ecologically diverse bacteria, cellulose, a characteristic exopolysaccharide component of the extracellular polymeric substances, encourages cell-cell communication, cell adhesion, and biofilm formation. Because cellulose slows down the motility brought on by flagella, an increase in cellulose synthesis primes the bacteria for surface colonisation. Also, cellulose biosynthesis provides protection to bacteria against chlorine treatment [1].

BNC is found to be chemically identical to plant cellulose, although there lies a significant difference in terms of their properties such as purity, macromolecular structure, and physical characteristics. The percentage of purity of plant cellulose is found to be less as compared to BNC since it consists of molecules like lignin, hemicellulose, and pectin [2]. The subsequent internal and external hydrogen bonding between the anhydrous glucose units in BNC helps it in forming a better crystalline structure. When compared to plant cellulose, BNC has a greater thermal stability, more water vapour transmission rate, malleability (due to large elastic modulus), porosity and pore size etc. Compared to plants, BNC has a higher hydrophilicity since it has more hydroxyl (OH) groups on its surface. Due to its numerous special properties, including stability, rigidity, high water retention capacity, stretchability, and non-toxic nature, bacterial nanocellulose has been found to be particularly promising in the environmental biome, biological field, medical field, food industry, cosmetic industry, and in the biomedical industry to create implants such as vascular stents, artificial heart valve stents, and artificial bones, etc. In order to regenerate and repair tissue, it is also employed to create a precise 3D scaffold [2].

There have been reports of several bacterial genera producing nanocellulose such as *Komagataeibacter sp.* (formerly called *Gluconacetobacter sp.), Lactobacillus sp., Escherichia sp., Agrobacterium sp., Enterobacter sp., Rhizobium sp.,* and *Erwinia sp.*, although the quantity and quality of their production may vary from organism to organism [2–5]. Recent studies indicate that species of *Komagataeibacter sp.* confer a higher BNC production as compared to the other genera and are more cost-effective in terms of genetic engineering [2, 3].

Therefore, in this chapter we have compared and analysed the diversified nature of bacterial nanocellulose production, mainly concentrating upon the following bacterial genera - *Escherichia coli., Komagataeibacter* sp., *Agrobacterium* sp. and *Lactobacillus sp.* In addition, characterization of bacterial nanocellulose by several techniques such as AFM, FTIR, XRD, and SEM has also been discussed.

3.1.1 Production of Bacterial Nanocellulose

Bacterial nanocellulose is produced by two steps: (1) β-1,4-glucan chain synthesis and (2) cellulose crystallisation [5, 6].

The process of conversion from glucose into cellulose involves four enzymatic steps:

1. In order to make glucose-6-phosphate (G6P), glucokinase phosphorylates glucose.
2. Glucose-6-phosphate (G6P) is isomerised to glucose-1-phosphate (G1P) by phosphoglucomutase (PGM).
3. UTP-glucose-1-phosphate uridylyltransferase (EC 2.7.7.9) catalyses the reaction between glucose-1-phosphate and UTP to produce uridine 5′-diphosphate-D-glucose (UDP-glucose).
4. As a member of the family of glycosyltransferases (GT), cellulose synthase converts UDP-glucose into cellulose by transporting glucosyl units from the precursor β-D-1,4-glucan chain [6].

The reaction pathway of bacterial nanocellulose synthesis has been described through Figure 3.1.

UDP glucose phosphorylase (UGPase) is discovered to be present in all cellulose producers as opposed to non-cellulose generating bacteria,

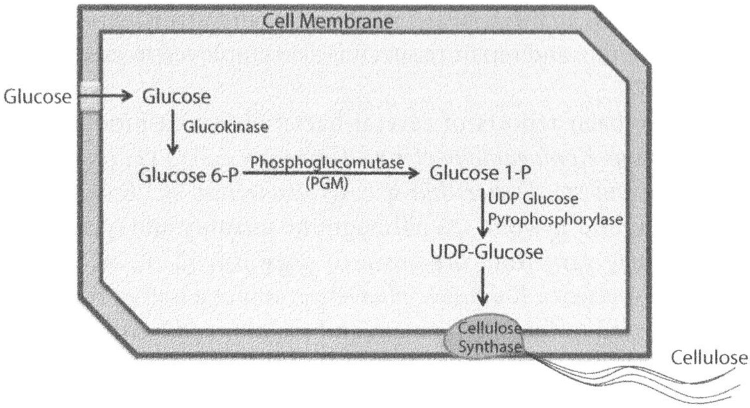

Figure 3.1 A simple representation of cellulose production pathway from glucose to cellulose in cellulose synthesizing bacteria.

where it plays a minor role in cellulose synthesis [7]. When sucrose and maltose are employed as carbon sources during the manufacture of bacterial nanocellulose, they are hydrolysed to produce monosaccharides like glucose and fructose. Although the processes for producing uridine diphosphoglucose (UDP-glucose) are well recognised, more research is required to comprehend the mechanism of the polymerization of glucose in long, unbranched chains.

3.1.2 Classification of Bcs Operon

BcsA, BcsB, BcsC, and BcsD are the four subunits of the bacterial cellulose synthase, and they are each coded by three separate (bcsAB, bcsC, and bcsD) or four distinct (bcsA, bcsB, bcsC, and bcsD) genes. Based on the gene order, gene content, and subtypes, the bacterial cellulose synthase (bcs) operon has been classified into three primary classes (Class I, II, and III) [8]. According to Table 3.1 these operons are present in the bacteria *Komagataeibacter xylinus*, *Escherichia coli*, and *Agrobacterium tumefaciens*, respectively.

Figure 3.2 shows the gene content and gene order of the three major classes (Class I, II, and III), illustrates the diversified character of the bcs operon.

Class I – This class of bcs operon is evidenced by the existence of bcsD gene whose expression and product aids in the extrusion of glucan chains to the external medium by directing them towards the pore in the outer membrane [9]. This class is common among the members of alpha, beta and gamma *Proteobacteria*.

Table 3.1 Characteristic features of Bcs operon.

Name of organism	Class of Bcs operon	Proteins involved	Functional annotation
Komagataeibacter xylinus	Class I	BcsA	• Exhibits β-1,4 glycosyltransferase activity. • Catalyses the polymerization reaction of UDP-Glucose monomers to β-1,4-glucan chains. • Forms the central core of the bcs complex
		BcsB	• Forms the periplasmic portion of the bcs complex. • Involved in the transportation of nascent β-1,4-glucan chains to the external environment.
		BcsC	• Aids in the extrusion of the nascent glucan out of the cell by forming a channel.
		BcsD	• Involved in arrangement and formation of crystalline portion of cellulose chain by aiding in hydrogen bond formation between the nascent glucan chains.
		CcpAx	• Likely known to interact with BcsD subunit. • Functions as a mediator of protein-protein interactions. • Plays critical role in locating the bcs complex [13, 15].
		CmcAx	• Exhibits β-endo-1,4-glucanase activity. • Involved regulation of cellulose fibril packaging.

(Continued)

Table 3.1 Characteristic features of Bcs operon. (*Continued*)

Name of organism	Class of Bcs operon	Proteins involved	Functional annotation
		BglAx	• Exhibits β-glucosidase activity.
Escherechia coli	Class II	BcsA	• Exhibits β-1,4 glycosyltransferase activity. • Catalyses the polymerization reaction of UDP-Glucose monomers to β-1,4-glucan chains. • Forms the central core of the bcs complex.
		BcsB	• Forms the periplasmic portion of the bcs complex. • Involved in the transportation of nascent β-1,4-glucan chains to the external environment.
		BcsC	• Aids in the extrusion of the nascent glucan out of the cell by forming a channel.
		BcsE	• Binds to c-di-GMP and is involved in the regulation of cellulose production.
		BcsF	• Function is yet to be known.
		BscG	• Function is yet to be known.
		BcsQ	• Likely to be involved in cellular localization of respective Bcs complexes.

(*Continued*)

Table 3.1 Characteristic features of Bcs operon. (*Continued*)

Name of organism	Class of Bcs operon	Proteins involved	Functional annotation
		BcsR	• Involved in cellulose biosynthesis (the proper mechanism is yet to known). • Positive regulator of adhesion/invasion into the host cell and biofilm formation [34].
		BcsZ	• Exhibits β-glucosidase activity.
Agrobacterium tumefaciens	Class III	BcsA	• Exhibits β-1,4 glycosyltransferase activity. • Catalyses the polymerization reaction of UDP-Glucose monomers to β-1,4-glucan chains. • Forms the central core of the bcs complex.
		BcsB	• Forms the periplasmic portion of the bcs complex. • Involved in the transportation of nascent β-1,4-glucan chains to the external environment.
		CelG	• Interacts with peptidoglycan layer and organizes cellulose secretion complex.
		CmcAx	• Exhibits β-endo-1,4-glucanase activity. • Involved regulation of cellulose fibril packaging.

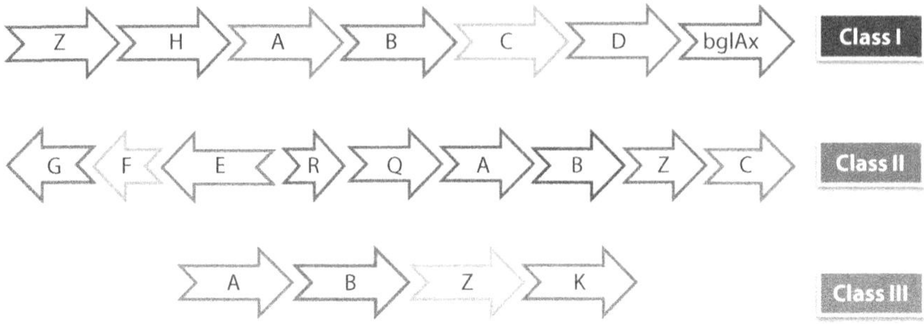

Figure 3.2 Diversity of bcs operon; Class I bcs operon of *Komagataeibacter xylinus*; Class II bcs operon of *Escherechia coli*; Class III bcs operon of *Agrobacterium tumefaciens*. **Figure annotation:** A-bcsA, B-bcsB, C-bcsC, D-bcsD, E- bcsE, F-bcsF, G- bcsG, H-ccpAx, K-celG, Q-yhjQ, R-bcsR and Z- cmcAx.

Class II – This class is indicated by the absence of the bcsD gene in contrast to the Class I operon and the inclusion of the bcsE and bcsG genes along with the two short genes bcsF and bcsQ [8]. *Escherichia coli* and members of the β and γ *Proteobacteria* are the main hosts of this operon.

Class III – The bcsA and bcsB genes make up the third class of bcs operons; bcsD, bcsE, or bcsG genes are absent [8]. Such operons typically contain an extra gene, such as bcsZ, commonly referred to as bcsK, which codes for a protein with a TPR (Tetratricopeptide repeat) domain similar to bcsC [8].

The absence of bcsB and other bcs genes is characteristic of Class IV, a subclass of bcs-like operons. These operons were discovered to be present in two cellulose-producing cyanobacteria, *Thermosynechococcus vulcanus* and *Anabaena sp.* PCC7120 [10]. The membrane-fusion family protein HlyD and the bcsW gene product, which are fixed in the same operons as bcsA, are responsible for binding the nascent glucan polymer produced by bcsA to the external surface of the membrane and responsible for its outward translocation [10]. This phenomenon is prevalent because the bcsB gene product is absent.

3.2 Structure and Functions of Proteins Involved in Bcs Operon

BcsA and BcsB

BcsA forms a 1:1 stoichiometric complex with BcsB and traverse approximately 150Å perpendicular and 55Å parallel to the membrane.

A catalytically active inner membrane protein called BcsA is a glycosyltransferase. It has two large cytoplasmic domains, two transmembrane (TM) domains, a glycosyltransferase domain in the centre, and a cyclic-di-GMP-binding PilZ domain at the C-terminus [2, 6]. The C-terminal end of a single transmembrane helix links the periplasmic protein BcsB to the cytoplasmic membrane. In the periplasm, the flagroxin (FD) domain and the repetitive carbohydrate-binding domain (CBD) are linked twice [11]. The transmembrane helices 3–8 of the BcsA subunit and the CBD domains of the BcsB subunit work cooperatively to construct the export pathway of the resultant polysaccharide from the cytoplasm [11]. The connection between BcsA-BcsB was documented by the crystal structure of the BcsA-BcsB complex found in *Rhodobacter sphaeroides* [6].

BcsC

A periplasmic protein known as BcsC contains an N-terminal α-helical region made up of multiple tetratricopeptide repeat (TPR) domains and a C-terminal section with the same structure as the β-barrels found in outer membrane proteins [6, 11]. The C-terminal domain of BcsC creates a channel that aids in the extrusion of the developing glucan out of the cell while the N-terminal portion of BcsC interacts with peptidoglycan and other bcs components [12].

BcsD

A 90 Å diameter octamer is created by the oligomerization of the periplasmic protein BcsD. The four dimers that make up the octamer are AB, CD, EF, and GF. With the help of these four dimers, a cylinder-shaped structure with an internal channel and a big central pore that can accommodate up to four distinct glucan molecules is created. The arrangement of the bcs complex is greatly influenced by BcsD and the longitudinal cell axis [13]. Due to their interactions with one another, these glucan chains rearrange, forming a crystal with a well-defined structure [9].

CcpAx

The ccpAx gene, which also goes by the name cellulose complementing factor (Ccp) in *Acetobacter xylinum* (CcpAx), is located in the upstream region of the bcs operon in *Komagataeibacter xylinus* and *Komagataeibacter hansenii* [14]. Since the ccpAx gene's mutation prevented the production of cellulose *in vivo*, Standal *et al.* were the first to notice its role in cellulose biosynthesis [15]. The exceptionally high activity of the bacterial cellulose synthase in *K. xylinus* has been linked to the colocalization of CcpAx and BcsD in the periplasm, where they interact with one another [15].

CmcAx

The endo-β-1,4-glucanase gene, also referred to as the bcsZ gene, encodes a globular protein with about 370 residues. It is a member of the eight-glycoside hydrolase family. The hydrolytic activity of CmcAx controls the production of cellulose. It was found that the BcsZ protein can effectively break down individual glucan chains in an *in vitro* cellulose biosynthesis experiment, but it was unable to break down the crystalline polymers that are present in the majority of commercial celluloses [11]. Additionally, it has been shown that *K. xylinus* produces more BNC when endogenous CmcAx is overexpressed or when exogenous CmcAx is present in high levels [16]. On the other hand, when bcsZ gene in *K. xylinus* was disrupted a reduced production of cellulose was observed with irregular packing of its fibril [16].

BglAx

The bglAx gene, which can be found in the downstream area of the bcs operon, encodes a β-glucosidase enzyme. It is one of the three glycoside hydrolases of the GH3 family. When secreted, BglAx has the ability to hydrolyze oligosaccharides with more than three residues into a single β-D-glucose unit [17]. Although BglAx is not required to make BNC, *K. hansenii* ATCC 23769 bglAx mutant produces less cellulose than the wild-type strain [17].

3.3 Diverse Nature of Bcs Operon

For bacterial cellulose production in *Komagataeibacter sp.*, two forms of bcs operons (Type I and Type II) have so far been identified [8, 17]. Only one of these, Type I, is structurally complete and consists of cellulose synthase, flanking genes for endo-1,4-glucanase (CmcAx), along with the complementary cellulose factor (CcpAx), and a β-glucosidase (BglAx).

The cellulose synthase operon of *Escherichia coli* and *Salmonella enterica* consists of bcsA, bcsB, and bcsC genes along with two additional genes – yhjQ and bcsZ. yhjQ helps in cellulose biosynthesis and has been renamed as bcsQ [18]. A divergent operon called bcsEFG, necessary for the cellulose synthesis, is also present at this locus [17]. Further analysis also revealed that an additional short gene, yhjR is present prior to the yhjQ-bcsA-BZC operon [19]. *Agrobacterium tumefaciens* and *Rhizobium leguminosarum* are two examples of bacteria with cellulose synthase operons that have bcsA and bcsB genes but no bcsC gene [19, 20]. The bcs locus of *A. tumefaciens* contains two convergent operons, celABCG and celDE, unlike *E. coli* and *S. enterica* [19]. The first three genes, called celA, celB, and celC, are homologs of bcsA, bcsB, and bcsZ, however the next three genes, called celG, celD, and celE, are specific to this kind of operon [19]. Among them, celE is essential for the synthesis of cellulose. Members of the phyla *Actinobacteria* and *Firmicutes*, which lack the outer membrane and appear to use cellulose secretion machinery different from that of *Proteobacteria*, can also be found to have similar operons [8].

3.4 Regulation of Bacterial Nanocellulose Biosynthesis

Both transcriptional and post-translational levels are equally controlled in the biosynthesis of bacterial nanocellulose (BNC). In this section cyclic-di-GMP regulated BNC synthesis has been discussed. Different regulators in different bacteria regulate the bcs gene expression. During biofilm formation bcs gene expression is stimulated as compared to the log phase. Initially, it was believed that GTP was the activator needed for bacterial cellulose biosynthesis, but in an experiment conducted by Ross *et al.* revealed that cyclic-di-GMP, which is enzymatically created from two GTP molecules, is an effective activator of cellulose biosynthesis [5]. Cyclic-di-GMP detection allowed for the initial purification of BcsA and the discovery of the UDP-glucose-binding subunit by product entrapment [5]. Another *in vivo* investigation revealed that cyclic-di-GMP, the allosteric regulator of the bcs, as well as both of its subunits are required for bcs activity [21].

The catalytic activity of the BcsA subunit of bacterial cellulose synthase contains a C-terminal cytosolic domain, PilZ, which is in charge of binding cyclic-di-GMP, as was mentioned earlier in this chapter [6]. A "RXXXR" motif with conserved arginine residues encircling one guanine base and a "DXSXXG" motif encircling the second guanine base were the two motifs discovered in the C-terminal PilZ domain of cyclic-di-GMP [21, 22]. The first motif, "RXXXR," is located within a bendable loop, and the second motif, "DXSXXG," is located on the surface of a β-sheet or β-barrel. While the second molecule (cyclic-di-GMP B) is stabilised by ð-ð stacking interactions with cyclic-di-GMP A, the first molecule (cyclic-di-GMP A) interacts with the 'DXSXXG' motif [6]. The gating-loop moves, exposing the UDP-glucose binding site in the GT domain and activating cellulose synthase when two molecules of cyclic-di-GMP connect. However, BcsA is catalytically inactive and unable to carry out the polymerization step without the help of an allosteric agent. The gating-loop interacts with the first arginine residue of the PilZ domain's RXXXR motif and the backbone of the gating-loop to limit the substrate's entry to the glucosyltransferase domain (GT) in the absence of cyclic-di-GMP [22]. Multiple conformational alterations brought on by cyclic-di-GMP binding through the PilZ regulatory domain ultimately reveal the active site [22].

By controlling the cellular levels of cyclic-di-GMP, the antagonistic actions of two enzymes, diguanylate cyclase (DGC) and cyclic-di-GMP diesterase (PDE), affect the cellulose production in *Komagataeibacter sp* [21]. Cyclic-di-GMP is hydrolyzed into 5′ phosphoguanylyl (pGpG) or GMP by the conserved (GGDEF) domain-specific phosphodiesterases (PDEs) that contain the EAL or HD-GYP domains [23, 24]. Therefore, both the pde and dgc genes must be expressed in a coordinated manner in order to obtain the ideal cyclic-di-GMP concentration, which is necessary for the rate of cellulose formation with environmental conditions.

3.5 Genetic Manipulation of BNC Producing Strains to Increase Yield

The significant quality of bacterial nanocellulose, its thickness, texture, can be emphasized by various genetic manipulations. Genera like *Komagataeibacter* (formerly called *Gluconacetobacter*), *Lactobacillus, Escherichia, Agrobacterium, Enterobacter, Rhizobium, Erwinia* can produce BNC but their texture and yield varies from genera to genera. Till date most studies have been done on different strains of *Komagataeibacter* like *K. hansenii, K. xylinum,* and *K. pasteurianus* for BNC production [2, 3, 13, 25]. Thus, selecting a proper producer strain is highly important to bring out the significant quality and overall yield.

For nanocellulose production strains of *Komagataeibacter* are much efficient for genetic modification in terms of its cost effectiveness as well as high productivity but still it poses some demerits of genomic instability [25]. Research is getting more centred on the selection and extraction of novel strains, optimal utilization of media, proper culture conditions (pH, temperature, aerobic or anaerobic environment etc.), and design of bioreactors to yield more BNC. Species which gets spontaneously converted to Cel$^-$ mutants doesn't aid in producing BNC to much extent. A major challenge to improvise high throughput BNC production from genetically unstable strain has been focused [25].

Many observations showed that formation of Cel⁻ mutants for genetically unstable strain during shaking (Eg: *Komagataeibacter xylinus* E25) utilizes all the carbon source (lactose from whey) of the culture to divide and produce progeny, but they fail to produce bacterial nanocellulose [3, 15]. But there can be a significant conversion of the phenotypic trait of Cel^- mutant to Cel^+ without shaking of the culture medium, leading to the increased production of BNC [25]. Therefore, this method can be termed as reversible shift only on phenotypic scale. However, if spotlight is bought over genotypic mutation of Cel^-, it cannot be converted to Cel^+ just by some reversible reaction, because of the involvement of genes, and mutation has taken place within the genes. Therefore Cel^- genotypic strains are now incapable of BNC production unless & until any genetic modification is performed. Recent investigation by Hur *et al.* has shown that an IS (insertion sequence) element inserted in the bcsA gene of the bcs operon is responsible for giving a defective Cel^- genotype incapable of BNC production [26]. Genetic engineering was thereby performed in bcsA gene by introducing an engineered strain (SAIT-IS) in bcsA, which showed a 1.7-fold rise in BNC yield. This effective engineering led to a success in getting a greater yield to manufacture cellulose by just engineering an unstable strain of *Komagataeibacter* to a stable one [26].

In recent studies, Liu *et al.* reported that during static culture condition regulation of oxygen tension affects BNC production and cell growth in *Komagataeibacter xylinus* CGMCC 2955 [27]. But on depletion of the concentration of oxygen there was a significant drop in quality of the cellulose pellicles. Moreover, it appeared more fragile & softer due to this change. It was also noticed that addition of Vitreoscilla hemoglobin (VHb) in correlation with high oxygen tension increased the yield of BNC production from *Komagataeibacter xylinus* by 25.5% in comparison to that of the parental strain. Liu *et al.* also noted that under aerobic conditions cells having VHb expression allowed BNC production in the stationary phase [27].

Diguanylate cyclase (positive regulator) facilitates the catalysis of cyclic-di-GMP and stimulates cellulose synthase for BNC production in *Komagataeibacter* strains. In this context an experiment was conducted to observe the difference in yield in BNC production between the dgc1 disrupted mutant and the wild-type strain of *Komagataeibacter xylinus* BPR2001 by using different culture methods [27]. At first it was observed that, dgc1 disruption didn't bring any significant change in BNC production compared to wild type strain, in both static & shaking culture conditions as wild type strains gave superior yield than genetically modified strain [27]. But on stirred tank bioreactors a significant drop of about 36% was observed in case of wild type strains in comparison to dgc1- disrupted mutants. However morphologically wild type strains gave larger BNC pellets than dgc1-disrupted mutants. Therefore, based on culture conditions, yield of BNC production varies [27]. The average size of BNC pellets in both the wild-type strain and the dgc1-disrupted mutants became nearly identical in another research when 0.4% agar was added to stirred culture medium [28]. Despite the fact that cyclic-di-GMP is required for cellulose synthase activation, it was inferred from this investigation that mutation of the dgc1 gene did not likely prevent BNC production in stirred tank bioreactors [28].

Some *Komagataeibacter* sp. strains trigger the production of BNC via the endo-β-1,4-glucanase enzyme (CmcAx) [10]. According to Kawano *et al.* introducing a small amount of His-tagged CmcAx to the culture medium causes a 1.2-fold increase in BNC production [10]. The yield of BNC production increased in *Komagataeibacter hansenii* ATCC 23769 when cmcAx was overexpressed. Therefore, it can be assumed that BNC hydrolyzing activity of CmcAx may have a regulating impact on the synthesis of cellulose [10].

The cytoplasmic membrane of *K. xylinus* contains glucose dehydrogenase (GDH). By lowering the pH of the medium, it oxidises glucose to gluconic acid and prevents the conversion of glucose to bacterial cellulose (BC). Now Kuo *et al.* created two mutant genes with glucose dehydrogenase removal activity to lessen the generation of the byproduct gluconic acid from glucose [29]. The mutant genes were GDH-KO in *K. xylinus* BCRC 12334 and GD-1 in *K. xylinus* BPR2001, respectively [29]. Interestingly, it was shown that GD-1 mutants produced twice as much BNC as wild-type strains did, although GDH (glucose dehydrogenase)-deficient mutants did not [29].

motA and motB genes in *K. hansenii* play different roles in BNC synthesis, cell division, and motility. Studies obtained three overexpression mutants (motA+, motB+, and motAB+) and two disrupted mutants (motA:kan and motAB:: kan) from *K. hansenii* [16]. In comparison to the wild-type strain, the motA:kan and motAB::kan mutants were smaller, less motile, had a compact membrane structure, and had better strength. In contrast to the wild type strain, cells of the motA+, motB+, and motAB+ strains were elongated, shown increased motility, produced more cellulose, and formed membranes with a relaxed appearance [16]. As a result, depending on the mutation, the products of this genetically modified organism offered a wide variety of benefits in the medical area for the construction of artificial bones, vascular stents, and artificial heart valves. Additionally, they work effectively in creating three-dimensional scaffolds with precise boundaries for tissue regeneration and repair.

In a different work, Fang *et al.* expressed the crdS gene in a plasmid to introduce the curdlan synthesis gene from *Agrobacterium* spp. ATCC31749 into *K. xylinus* AY201 [30]. This caused internal UDP-glucose to polymerize under its own direction, releasing curdlan and cellulose fibres [30]. The characteristics of the composite surface of the resulting product showed a clear difference. Curdlan was used to coat the cellulose pellicles to lessen water permeability. Curdlan secretion was thought to have low priority with regard to BNC synthesis, though the curdlan protein only altered the properties of cellulose and had very little impact on the crystalline structure of cellulose pellicles [30].

Three genes (NAG5, AGM1, and UAP1) from the yeast *Candida albicans* were inserted into *K. xylinus* by Yadav *et al.* [31]. The genes encoded a metabolic process that caused the cell to transform N-acetylglucosamine (GlcNAc), the monomer of chitin, into UDP GlcNAc. By inducing cellulose synthase, these activated cytoplasmic UDP-GlcNAc monomers contribute to the formation of a chimaera polymer containing both glucose and GlcNAc. The produced copolymer degrades naturally, in contrast to BNC [31].

Several studies have also focused on the good productivity of BNC from the strains of *Lactobacillus* [4, 32]. *Lactobacillus* has a crucial role in maintaining a healthy gastrointestinal tract (GI) in humans [32]. *Lactobacillus mali* JCM1116 grown on Y-agar enrichment medium (Y comes from S. Yamanaka) showed a distinguished amount of cellulose production [33]. There was also an increased productivity to some extent when they were co-cultured in corn steep liquor (CSL)/sucrose solid medium, whereas the liquid medium showed a decreased production of BNC. Strains of *Lactobacillus mali* possessing a mutation on its exocellular polysaccharide (EPS) showed a defective production of BNC [33]. However, co-culturing them with a strain of *Gluconoacetobacter xylinus* st-6-12 resulting in a cell-cell interaction allowed formation of EPS promoting BNC production [33]. Increased productivity of upto 3 fold was seen on co-culturing these two strains in a CSL/sucrose medium [33].

3.6 Factors for BNC Production

Methods of BNC production can be of two types: 1. static culture 2. stirred culture [2]. By static culture method white coloured bulk amount of bacterial cellulose was obtained at the junction of air and liquid interface and by stirred culture method bacterial nanocellulose was obtained in the medium in a scattered manner forming pellets having no shape or fibres [34]. Which method will be used for desired production of bacterial nanocellulose solely depends upon the application of the product as synthesized BNC changes its physical, morphological, and mechanical characteristics according to the method of preparation. Table 3.2 is provided for difference of BNC production between static culture and stirred culture.

We should consider some important conditions during production of BNC because a component's physical properties and chemical properties can affect the production of BNC in many ways such as – 1. cost should be low, 2. yield should be high, 3. components and amount of the media must be considered, 4. medium should contain a carbon source (C), a nitrogen source (N) and other macro and micronutrients sources like phosphorus (P), potassium (K), magnesium (Mg) etc.

Hestrin and Schramm (HS) first described how a typical media for production of BNC should be like what is known as HS media [35]. This media contains glucose, peptone, yeast extract, anhydrous Na_2HPO_4 and citric acid monohydrate but this media is not appropriate for industrial production scale as it is not cost-effective enough [35]. That's why several studies have been performed to find out what changes should be done to get an abundant amount of BNC in a short time in a cost effective way [36, 37].

In a study, it was observed that in HSL media (HS media with 1% lignosulfonate) more than five strains of *Gluconacetobacter xylinus* gave higher yield than those which were grown in HS media [38]. This is

Table 3.2 Differences between static culture and stirred culture.

Static culture	Stirred culture
1. Produced BNC is of high mechanical strength.	1. Produced BNC is of low mechanical strength.
2. Higher yields and lower chance of mutations.	2. Lower yields and higher chance of mutations.
3. Needs large area for cultivation.	3. Requires small area for cultivation.
4. Takes longer period of time for production.	4. Does not take that much time for production.

because lignosulfonate inhibits formation of gluconic acid as it has anti-oxidant compounds in it and helps in getting more yield. Another experiment was performed with *Komagataeibacter saccharivorans* which can utilize different waste materials as a source of nutrients [39]. Using waste materials for industrial production of BNC can be helpful for us in two ways 1) it helps in waste reduction and 2) it reduces the cost of the production. *Komagataeibacter saccharivorans* was grown in a HS medium as a control and in three different media containing wastes of dates, figs, and sugarcane molasses [39]. Highest yield was obtained from sugarcane molasses as BNC produced from the sugarcane molasses has the highest water holding capacity. Some other experiments were also done where different fruit juices or rotten fruits in the media were used to increase the yield [40, 41]. According to some studies by using alternate sources of carbon like mannitol, glycerol, fructose etc. or in the presence of different organic acids like acetic acid, citric acid etc. yields of BNC production get higher [7, 42]. In another experiment HS medium supplemented with thin stillage (rich in organic acids) can promote the growth of *G. xylinus* and can achieve surprisingly higher yield under certain conditions [33]. In another experiment more than five types of alcohols were used to test whether it helps in getting more yield or not with *Acetobacter xylinum* [36]. Results showed that where n-butanol was used, highest yield was obtained in respect to other alcohols used in the experiment [36]. Other study also reported the use of olive oil residues, waste beer yeast, molasses etc. for better morphological qualities of BNC [42]. Kuo *et al.* studied the effects of acetate buffered medium for BNC production and concluded that it can balance satisfactory pH to produce BNC [29]. Wastewater of candied jujube and lipid fermentation was also used as a source of nutrients [39].

Most popular energy source for BNC production is glucose as it is an important precursor of the biosynthetic process of cellulose. But in the presence of glucose dehydrogenase pH becomes low as glucose is converted into gluconic acid and as a result the production yield will be low.

Several types of agricultural wastes like rice husk, dried algal biomass, sugarcane bagasse, etc. were used by many groups of scientists as raw materials for BNC production [39–41]. This helps in getting higher yields of BNC. Along with this, pH, temperature, dissolved O_2 concentration and type of cultivation method are also very important factors. Therefore, it is clearly observed that if the volume and surface area of a medium will be increased, it will increase the yield of BNC along with less production time and cost.

There are two methods through which we can create BNC matrix composites – *in situ* and *ex situ*. Secondary components are added in the medium at the starting of BNC production in *in situ* method and secondary components are added into the matrix of BNC production in *ex situ* method [43]. Many more experiments should be conducted to identify what method will be the best for BNC production and what parameter changes must be done to increase the yield at industrial level. Hopefully future studies will surely help us to know more about this.

3.7 Characterization of Bacterial Nanocellulose

There are several techniques to characterize the bacterial nanocellulose – 1) spectroscopic techniques such as fourier transform infrared (FTIR), nuclear magnetic resonance (NMR), and x-ray diffraction spectroscopy (XRD), 2) microscopic methods such as atomic force microscopy (AFM), scanning electron microscopy (SEM), and transmission electron microscopy (TEM), 3) thermo analytical methods such as thermogravimetric analysis (TGA) and differential scanning calorimetry (DSC) etc. In this section, we have briefly discussed different characterization techniques of bacterial nanocellulose synthesis. Figure 3.3 represents an overview of different techniques for BNC characterization.

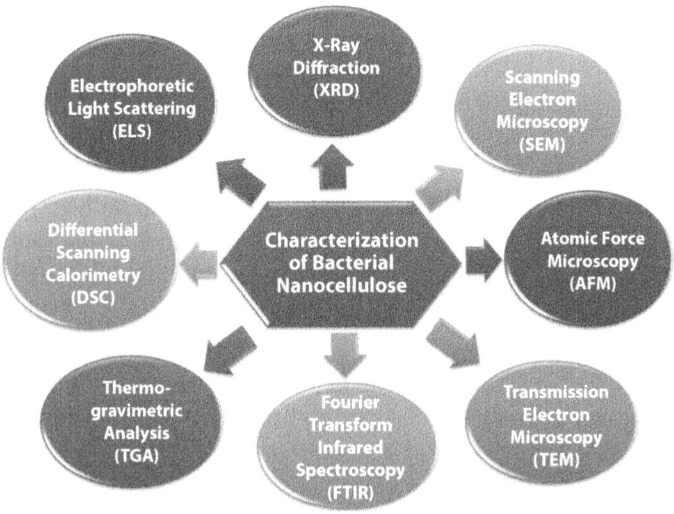

Figure 3.3 An overview of characterization of bacterial nanocellulose.

3.7.1 X-Ray Diffraction Spectroscopy (XRD)

XRD is a common method to measure the crystallinity index (CrI) or degree of crystallinity. The crystallinity index (CrI) is a ratio between crystallinity and amorphous region by using Cu K_α radiation at 40 kV and 30 mA in X-ray diffractometer. It also determines the arrangement of crystal patterns and crystal structure. Many researchers use the Segal method to measure CrI using the formula [43]:

$$CrI\% = \{(I_{200} - I_{am}) \times 100\}/\ I_{200}$$

where I_{200} = the intensity at (200) peak ($2\theta = 22.3°$) and I_{am} = the minimum intensity between (100) and (200). It is only providing a rough estimate of the CrI.

3.7.2 Nuclear Magnetic Resonance (NMR)

NMR helps to determine the lateral size of BNC particle and various kind of crystallinity. The regions of the crystalline and amorphous C-4 signals in the NMR spectra were used to calculate the percentage of crystallinity [44].

3.7.3 Fourier Transform Infrared (FTIR)

FTIR are used to analyse the bonding pattern of the functional group of the BNC films [45]. Normally BNC samples are pressed with pure potassium bromide (KBr) and then squeezed to get small tablets. After that 4000 to 500 cm^{-1} spectral range is used for signal [46, 47]. To see the effect of alkaline treatment, presence of impurities, and formation of hydrogen bonds between cellulose fibres and other materials in BNC, FTIR is used [48].

3.7.4 Scanning Electron Microscopy (SEM)

This is a very common method to study the surface characterization of BNC. Usually, the surface of the samples which are analysed by SEM are electrically conductive, but BNC is non-conductive. Therefore, to make BNC more conductive, their films are coated with a thin layer of heavy atoms (gold, platinum, or osmium). Low voltage mode of SEM is used to analyse uncoated non-conductive BNC [49]. BNC exhibited a diverse interlinked network arrangement made up of nanocellulose that was randomly orientated. SEM often helps to identify 3D-microstructure of BNC, their porosity, dimension, uniformity and possible alignment of crystals of BNCs [49, 50].

3.7.5 Transmission Electron Microscopy (TEM)

TEM provides 2D shape, size, internal structure, morphology, crystallization, and magnetic domains of BNC [51]. TEM can be used to study the biosynthesis mechanism and degradation of nanocellulose [52].

For sample preparation, BNC is treated with sulphuric acid, washing it with molecular biology grade water followed by 15 mins of sonication and placed in a clean carbon-coated copper grid. After that the overnight dried sample is stained with 2.5% gadolinium triacetate and observed under TEM [52].

3.7.6 Atomic Force Microscopy (AFM)

Freeze dried samples of BNC are used to classify and invasion the variation of 3D surface microstructures (surface roughness), and pore morphology. The operation of scanning is done in contactless mode, with a very high resolution [51]. In AFM, the ubiquitous character of repulsive force between the tip and the sample are utilized to study the topography of BNC without disrupting their durability [53].

3.7.7 Electrophoretic Light Scattering (ELS)

To obtain the value of zeta potential of BNC colloidal solution, ELS is carried out [54]. The negative zeta absolute potential value infers the stability of the colloidal solution of BNC and the rate of aggregation of rapid particles [48]. When the zeta potential value is near 0 mV, it suggests that aggregations occur very often. In general, zeta potential values less than +25 mV or larger than -25 mV will aggregate due to interparticle interactions like Van der Waals forces or hydrophobic interactions between nanoparticles [54].

3.7.8 Thermogravimetric Analysis (TGA)

TGA can determine thermal, physical, and chemical stability of BNC [45]. For analysis, 5-15 mg samples can be taken and heated in an aluminium pan at 20 °C per minute in 26-600 °C range of temperature [45]. The amount of deduction of weight depends on the primary moisture content of the sample [55]. The following equation can be used to determine the water content [45]:

$$[(W_t - W_0) / W_t] \times 100\ \%$$

where W_0 = weight of dried nanocellulose, W_t = weight of wet nanocellulose

3.7.9 Differential Scanning Calorimetry (DSC)

Differential scanning calorimetry (DSC) gives information about the thermal degradation of a BNC component. Normally the DSC curve of BNC contains several peaks due to the loss of water because of thermal effect or the glass transition, crystallization or degradation of BNC or melting of crystalline phase etc [56]. For performing DSC of BNC sample, it should be sealed in aluminium pan followed by heating within 25 °C to 400 °C at a rate of 10 °C/min and cooled at a rate of 5 °C/min under a constant stream of nitrogen (20mL/min) [57]. Vazquez *et al.* observed that around up to 200 °C BNC was stable but at 338.13 °C it started to decompose [58].

3.7.10 Particle Size Analysis

BNC's particle size is examined using a particle size analyser (PSA). It cannot be compared to the sizes measured by SEM because it only provides a relative value of sizes. At a water concentration of 0.05 percent, the dispersion of particle size is assessed [59].

3.8 Conclusion

Nowadays bacterial nanocellulose is being favoured more by industrialists and biologists, yet there is another source of cellulose production taking plant origin into consideration. However, the cellulose obtained from plants is of low percentage purity and sub-standard durability in comparison to bacterial nanocellulose. As discussed earlier in this chapter, for different bacteria, different genes are involved to produce bacterial nanocellulose such as *Esherichia Coli, A. tumefaciens, K. xylinus,* and many more share the common genes bcsA and bcsB, but they may also contain additional genes for example bcsE and bcsG pre-dominant to *E. coli*. The utilisation of genetically modified strains in conjunction with chemical and physical alterations of BNC significantly raises the importance of BNC.

Genes that synthesise cellulose can be precisely controlled by genetic engineering, and BNC fibres with particular qualities can be produced. Incorporation of new sources for culture medium in addition to various techniques like maintaining a static culture or stirring culture, use of air reactors can also help in increasing the yield of BNC. Bacterial nanocellulose production involves the usage of several agro-industrial wastes, and organic wastes as substrates thus calling for a zero-waste technology. From bacterial nanocellulose characterization, it is also ensured that nano size cellulose is obtained which can be used in medical purpose as coating material for rapid diffusion.

References

1. Trache, D., Tarchoun, A.F., Derradji, M., Hamidon, T.S., Masruchin, N., Brosse, N., Hussin, M.H., Nanocellulose: From fundamentals to advanced applications. *Front. Chem.*, 8, 392, 2020 May 6.
2. Cannazza, P., Rissanen, A.J., Sarlin, E., Guizelini, D., Minardi, C., Losoi, P., Molinari, F., Romano, D., Mangayil, R., Characterization, genome analysis and genetic tractability studies of a new nanocellulose producing *Komagataeibacter intermedius* isolate. *Sci. Rep.*, 12, 1, 20520, 2022.
3. Florea, M., Reeve, B., Abbott, J., Freemont, P.S., Ellis, T., Genome sequence and plasmid transformation of the model high-yield bacterial cellulose producer *Gluconacetobacter hansenii* ATCC 53582. *Sci. Rep.*, 6, 23635, 2016.
4. Kousheh, S.A., Moradi, M., Tajik, H., Molaei, R., Preparation of antimicrobial/ultraviolet protective bacterial nanocellulose film with carbon dots synthesized from lactic acid bacteria. *Int. J. Biol. Macromol.*, 155, 216–225, 2020.
5. Ross, P., Mayer, R., Benziman, M., Cellulose biosynthesis and function in bacteria. *Microbiol. Rev.*, 55, 1, 35–58, 1991.
6. Morgan, J.L., Strumillo, J., Zimmer, J., Crystallographic snapshot of cellulose synthesis and membrane translocation. *Nature*, 493, 7431, 181–6, 2013.
7. Samyn, P., Meftahi, A., Geravand, S.A., Heravi, M.E.M., Najarzadeh, H., Sabery, M.S.K., Barhoum, A., Opportunities for bacterial nanocellulose in biomedical applications: Review on biosynthesis, modification and challenges. *Int. J. Biol. Macromol.*, 231, 123316, 2023.
8. Römling, U. and Galperin, M.Y., Bacterial cellulose biosynthesis: Diversity of operons, subunits, products, and functions. *Trends Microbiol.*, 23, 9, 545–57, 2015.
9. Hu, S.Q., Gao, Y.G., Tajima, K. *et al.*, Structure of bacterial cellulose synthase subunit D octamer with four inner passageways. *Proc. Natl. Acad. Sci. U.S.A.*, 107, 42, 17957–61, 2010.

10. Kawano, Y., Saotome, T., Ochiai, Y., Katayama, M., Narikawa, R., Ikeuchi, M., Cellulose accumulation and a cellulose synthase gene are responsible for cell aggregation in the cyanobacterium *Thermosynechococcus vulcanus* RKN. *Plant Cell Physiol.*, 52, 6, 957–966, 2011.
11. McNamara, J.T., Morgan, J.L., Zimmer, J., A molecular description of cellulose biosynthesis. *Annu. Rev. Biochem.*, 84, 895–921, 2015.
12. Keiski, C.L., Harwich, M., Jain, S., Neculai, A.M., Yip, P., Robinson, H., Whitney, J.C., Riley, L., Burrows, L.L., Ohman, D.E., Howell, P.L., AlgK is a TPR-containing protein and the periplasmic component of a novel exopolysaccharide secretin. *Structure*, 18, 2, 265–273, 2010.
13. Mehta, K., Pfeffer, S., Brown, R.M., Characterization of an acsD disruption mutant provides additional evidence for the hierarchical cell-directed self-assembly of cellulose in Gluconacetobacter xylinus. *Cellulose*, 22, 119–137, 2014.
14. Sunagawa, N., Fujiwara, T., Yoda, T., Kawano, S., Satoh, Y., Yao, M., Tajima, K., Dairi, T., Cellulose complementing factor (Ccp) is a new member of the cellulose synthase complex (terminal complex) in *Acetobacter xylinum. J. Biosci. Bioeng.*, 115, 607–612, 2013.
15. Standal, R., Iversen, T.G., Coucheron, D.H., Fjaervik, E., Blatny, J.M., Valla, S., A new gene required for cellulose production and a gene encoding cellulolytic activity in Acetobacter xylinum are colocalized with the bcs operon. *J. Bacteriol.*, 176, 3, 665–72, 1994.
16. Jacek, P., Dourado, F., Gama, M., Bielecki, S., Molecular aspects of bacterial nanocellulose biosynthesis. *Microb. Biotechnol.*, 12, 4, 633–649, 2019.
17. Abidi, W., Torres-Sánchez, L., Siroy, A., Krasteva, P.V., Weaving of bacterial cellulose by the Bcs secretion systems. *FEMS Microbiol. Rev.*, 46, 2, fuab051, 2022.
18. Serra, D.O., Richter, A.M., Hengge, R., Cellulose as an architectural element in spatially structured *Escherichia coli* biofilms. *J. Bacteriol.*, 195, 24, 5540–5554, 2013.
19. Thompson, M.A., Onyeziri, M.C., Fuqua, C., Function and regulation of *Agrobacterium tumefaciens* cell surface structures that promote attachment. *Curr. Top. Microbiol. Immunol.*, 418, 143–184, 2018.
20. Ausmees, N., Jonsson, H., Höglund, S., Ljunggren, H., Lindberg, M., Structural and putative regulatory genes involved in cellulose synthesis in *Rhizobium leguminosarum bv. trifolii. Microbiol. (Reading)*, 145, Pt 5, 1253–1262, 1999.
21. Chou, S.H. and Galperin, M.Y., Diversity of Cyclic Di-GMP-Binding Proteins and Mechanisms. *J. Bacteriol.*, 198, 1, 32–46, 2016.
22. Junsang, K., Kyoung-Seok, R., Henna, K., Jae-Sun, S., Jie-Oh, L., Chaejoon, C., Byong-Seok, C., Structure of PP4397 reveals the molecular basis for different c-di-GMP binding modes by Pilz domain proteins. *J. Mol. Biol.*, 398, 1, 97–110, 2010.

23. Cohen, D., Mechold, U., Nevenzal, H., Yarmiyhu, Y., Randall, T.E., Bay, D.C., Rich, J.D., Parsek, M.R., Kaever, V., Harrison, J.J., Banin, E., Oligoribonuclease is a central feature of cyclic diguanylate signaling in *Pseudomonas aeruginosa*. *Proc. Natl. Acad. Sci. U.S.A.*, 112, 36, 11359–64, 2015.
24. Orr, M.W., Donaldson, G.P., Severin, G.B., Wang, J., Sintim, H.O., Waters, C.M., Lee, V.T., Oligoribonuclease is the primary degradative enzyme for pGpG in *Pseudomonas aeruginosa* that is required for cyclic-di-GMP turnover. *Proc. Natl. Acad. Sci. U.S.A.*, 112, 36, E5048–57, 2015.
25. Moradi, M., Jacek, P., Farhangfar, A., Guimarães, J.T., Forough, M., The role of genetic manipulation and *in situ* modifications on production of bacterial nanocellulose: A review. *Int. J. Biol. Macromol.*, 183, 635–650, 2021.
26. Hur, D.H., Choi, W.S., Kim, T.Y., Lee, S.Y., Park, J.H., Jeong, K.J., Enhanced production of bacterial cellulose in Komagataeibacter xylinus via tuning of biosynthesis genes with synthetic RBS. *J. Microbiol. Biotechnol.*, 30, 9, 1430–1435, 2020.
27. Liu, M., Li, S., Xie, Y., Jia, S., Hou, Y., Zou, Y., Zhong, C., Enhanced bacterial cellulose production by Gluconacetobacter xylinus via expression of Vitreoscilla hemoglobin and oxygen tension regulation. *Appl. Microbiol. Biotechnol.*, 102, 3, 1155–1165, 2018.
28. Bae, S.O., Sugano, Y., Ohi, K., Shoda, M., Features of bacterial cellulose synthesis in a mutant generated by disruption of the diguanylate cyclase 1 gene of Acetobacter xylinum BPR 2001. *Appl. Microbiol. Biotechnol.*, 65, 3, 315–22, 2004.
29. Kuo, C.H., Teng, H.Y., Lee, C.K., Knock-out of glucose dehydrogenase gene in Gluconacetobacter xylinus for bacterial cellulose production enhancement. *Biotechnol. Bioprocess Eng.*, 20, 18–25, 2015.
30. Fang, J., Kawano, S., Tajima, K., Kondo, T., *In vivo* curdlan/cellulose bionanocomposite synthesis by genetically modified Gluconacetobacter xylinus. *Biomacromolecules*, 16, 3154–3160, 2015.
31. Yadav, V., Paniliatis, B.J., Shi, H., Lee, K., Cebe, P., Kaplan, D.L., Novel *in vivo*-degradable cellulose-chitin copolymer from metabolically engineered Gluconacetobacter xylinus. *Appl. Environ. Microbiol.*, 76, 18, 6257–65, 2010.
32. Bao, W., He, Y., Liu, W., Diversity analysis of bacterial and function prediction in Hurunge from Mongolia. *Front. Nutr.*, 9, 835123, 2022.
33. Seto, A., Saito, Y., Matsushige, M., Kobayashi, H., Sasaki, Y., Tonouchi, N., Tsuchida, T., Yoshinaga, F., Ueda, K., Beppu, T., Effective cellulose production by a coculture of Gluconacetobacter xylinus and Lactobacillus mali. *Appl. Microbiol. Biotechnol.*, 73, 4, 915–21, 2006.
34. Gao, J.X., Li, P., Du, X.J., Han, Z., Xue, R., Liang, B., Wang, S., A negative regulator of cellulose biosynthesis, *bcsR*, affects biofilm formation, and adhesion/invasion ability of *Cronobacter sakazakii*. *Front. Microbiol.*, 8, 1839, 2017.

35. Hestrin, S. and Schramm, M., Synthesis of cellulose by Acetobacter xylinum. II. Preparation of freeze-dried cells capable of polymerizing glucose to cellulose. *Biochem. J.*, 58, 2, 345–52, 1954.
36. Aswini, K., Gopal, N.O., Uthandi, S., Optimized culture conditions for bacterial cellulose production by Acetobacter senegalensis MA1. *BMC Biotechnol.*, 20, 1, 46, 2020.
37. Gökhan, G., Esra, E.K., Alper, K., Ayhan, G., Deniz, A., Nejla, A., Production of bacterial cellulose fibers in the presence of effective microorganism. *J. Nat. Fibers*, 16, 4, 567–575, 2019.
38. Keshk, S. and Sameshima, K., Influence of lignosulfonate on crystal structure and productivity of bacterial cellulose in a static culture. *Enzyme Microb. Technol.*, 40, 1, 4–8, 2006.
39. Güzel, M. and Akpınar, Ö., Preparation and characterization of bacterial cellulose produced from fruit and vegetable peels by Komagataeibacter hansenii GA2016. *Int. J. Biol. Macromol.*, 162, 1597–1604, 2020.
40. Revin, V., Liyaskina, E., Nazarkina, M., Bogatyreva, A., Shchankin, M., Cost-effective production of bacterial cellulose using acidic food industry by-products. *Braz. J. Microbiol.*, 49 Suppl 1, Suppl 1, 151–159, 2018.
41. Yu, S., Sun, J., Shi, Y., Wang, Q., Wu, J., Liu, J., Nanocellulose from various biomass wastes: Its preparation and potential usages towards the high value-added products. *Environ. Sci. Ecotechnol.*, 5, 100077, 2021.
42. Abol-Fotouh, D., Hassan, M.A., Shokry, H., Bacterial nanocellulose from agro-industrial wastes: Low-cost and enhanced production by *Komagataeibacter saccharivorans* MD1. *Sci. Rep.*, 10, 3491, 2021.
43. Yang, Y., Jia, J., Xing, J., Chen, J., Lu, S., Isolation and characteristics analysis of a novel high bacterial cellulose producing strain Gluconacetobacter intermedius CIs26. *Carbohydr. Polym.*, 92, 2, 2012–7, 2013.
44. Larsson, P.T., Westermark, U., Iversen, T., Determination of the cellulose Iα allomorph content in a tunicate cellulose by CP/MAS 13C-NMR spectroscopy. *Carbohydr. Res.*, 278, 2, 339–343, 1995.
45. Akhter, S., Khan, M.A., Mahmud, S., Biki, S.P., Shamsuzzoha, Md., Hasan, S.M.K., Ahmed, M., Biosynthesis and characterization of bacterial nanocel lulose and polyhydroxyalkanoate films using bacterial strains isolated from fermented coconut water. *Process Biochem.*, 122, 1, 214–223, 2022.
46. Lin, S.P., Liu, C.T., Hsu, K.D., Hung, Y.T., Shih, T.Y., Cheng, K.C., Production of bacterial cellulose with various additives in a PCS rotating disk bioreactor and its material property analysis. *Cellulose*, 23, 1, 367–377, 2016.
47. Fan, M., Dai, D., Huang, B., Fourier transform infrared spectroscopy for natural fibers, in: *Fourier Transform*, S. Mohammed (ed.), Intech Open, London, UK, 2012.
48. Sijabat, E.K., Nuruddin, A., Aditiawati, P., Purwasasmita, B.S., Synthesis and characterization of bacterial nanocellulose from banana peel for water filtration membrane application. *J. Phys.: Conf. Ser.*, 1230, 012085, 2019.

49. Janaina, A.A., Nathália, C., Rebelo, M., Marco, C., Denise, G., Marli, G., Mahendra, R., Priscila, M., Angela, J., Bacterial nanocellulose loaded with bromelain: Assessment of antimicrobial, antioxidant and physical-chemical properties. *Sci. Rep.*, 7, 18031, 2017.
50. Sabu, T., Visakh, P.M., Mathew, A.P., Advances in natural polymers, in: *Composites and Nanocomposites*, Springer, Berlin, Springer USA, 2012.
51. Ashrafi, Z., Lucia, L., Krause, W., Bioengineering tunable porosity in bacterial nanocellulose matrices. *Soft Matter*, 15, 45, 9359–9367, 2019.
52. Gupte, Y., Kulkarni, A., Raut, B., Sarkar, Choudhury, R., Chawande, A., Kumar, G.R.K., Bhadra, B., Satapathy, A., Das, G., Vishnupriya, B., Dasgupta, S., Characterization of nanocellulose production by strains of Komagataeibacter sp. isolated from organic waste and Kombucha. *Carbohydr. Polym.*, 266, 118176, 2021.
53. De, R.R.L., Albuquerque, D.A.C., Cruz, T.G.S., Yamaji, F.M., Leite, F.L., Measurement of the nanoscale roughness by atomic force microscopy: Basic principles and applications, in: *Atomic Force Microscopy -Imaging, Measuring and Manipulating Surfaces at the Atomic Scale*, InTech Open, London, UK, 2012.
54. Kumar, A. and Dixit, C.K., Methods for characterization of nanoparticles, in: *Advances in Nanomedicine for the Delivery of Tera-peutic Nucleic Acids*, pp. 43–58, Woodhead Publishing, Cambridge, 2017.
55. Abouelkheir, S.S., Kamara, M.S., Atia, S.M., Amer, S.A., Youssef, M.I., Abdelkawy, R.S., Khattab, S.N., Sabry, S.A., Novel research on nanocellulose production by a marine Bacillus velezensis strain SMR: A comparative study. *Sci. Rep.*, 10, 1, 14202, 2020.
56. El-Naggar, N.E.A., El-Malkey, S.E., Abu-Saied, M.A. *et al.*, Exploration of a novel and efficient source for production of bacterial nanocellulose, bioprocess optimization and characterization. *Sci. Rep.*, 12, 18533, 2022.
57. Phruksaphithak, N., Kaewnun, C., Thong, S.O., Characterization of bacterial cellulose from oil palm shoot juices and coconut juice/poly (ethylene glycol) biocomposite. *J. Renew. Mater.*, 7, 5, 493–504, 2019.
58. Vazquez, A., Foresti, M.L., Cerrutti, P., Galvagno, M., Bacterial cellulose from simple and low-cost production media by Gluconacetobacter xylinus. *J. Polym. Environ.*, 21, 2, 45–554, 2013.
59. Phanthong, P., Reubroycharoen, P., Hao, X., Xu, G., Abudula, A., Guan, G., Nanocellulose: Extraction and application. *Carbon Resour. Convers.*, 1, 1, 32–43, 2018.

4

Process and Applications of Electrospinning

Anushka Lather* **and Shubhangi Paliwal**

Dept. of Bioscience & Biotechnology, Banasthali Vidyapith, Rajasthan, India

Abstract

Electrospinning is an efficient and simple technique for producing uninterrupted nanoscale fibres using a high-voltage power source. Utilising an electric field, a charged polymer solution discharge is produced. As the jet travels through the air, the solvent dissipates, leaving behind a charged fibre that can be collected on a metal screen or electrically deflected. Electrospun fibres and nonwoven webs created via this method have attracted a great deal of attention from scientists and other researchers because of its exceptional properties, which include a high degree of porosity, smaller diameters, proficient porous connections, and significant surface-to-volume ratios. Numerous natural and artificial polymeric materials, such as single and mixed polymers, are electrospun to fibres which are suitable for filtrating and insulation purposes, and additionally for the making of protective apparel, sensors, conducting devices, wound dressings, and scaffolds that are employed in tissue engineering. The content of this chapter provides a concise description of electrospinning, including its fundamentals, methods, materials, and applications.

Keywords: Electrospinning, electrospun fiber, applications of electrospinning

4.1 Introduction

The natural world is filled with fibres, which can be either continuous strands or elongated formations. Spiders have been entangling their victims in webs of fiber for more than 1.4 million years. Silk filaments with widths ranging from 2 to 5 micrometers make up these webs. Silkworms

**Corresponding author*: anushkalather@gmail.com

Mainak Mukhopadhyay and Debalina Bhattacharya (eds.) Nanocellulose: A Biopolymer for Biomedical Applications, (89–112) © 2024 Scrivener Publishing LLC

are renowned for their extraordinary ability to produce silk threads for making cocoons. The production of synthetic fibers has been significantly influenced by all of these natural systems [1–3]. It is no secret that since the beginning of civilization, fibers have been an essential part of human life.

The curiosity of research fraternity was quickly sparked when Dupont announced nylon as the first synthetic material that was commercially viable in 1938 [4, 5]. The development of further polyester types and synthetic polymers for the manufacture of synthetic fibres followed that [6]. These synthetic fibres significantly decrease the general public's need for natural fibres while dramatically increasing the number of applications.

The first study on electrospinning dates back to 1600, when William Gilbert observed that a cone-shaped water droplet formed when subjected to an electric field [7]. Numerous businesses nowadays have created and implemented industrial production lines that can produce a significant amount of electrospun nanofibers. From commercial applications like automobile filters to basic household items like face masks and water filters, these nanofibers are frequently utilized in water and air filtration. Additionally, these nanofibers are being created in labs with distinct compositions and cutting-edge architectures to give them extra capabilities for more complex applications [8].

4.2 A Brief History of Electrospinning

William Gilbert first proposed the use of electrospinning in 1600 after noticing that water droplets take on a conical shape when an electric charge is applied [8]. According to Abbe Nollet's electrospraying research from 1747, water can spray into an aerosol when subjected to an electrostatically charged vessel.

High voltage is used in both electrospraying and electrospinning to emit viscous and viscoelastic liquid jets. Electrospinning, as opposed to electrospraying, constantly ejects polymer solution to produce fibers as opposed to dissolving it into droplets.

Charles V. Boys discovered in 1887 that when a viscoelastic substance is exposed to an electric field, fibers may be formed. He used an insulated dish that was wired to an electrical power source for his experiment [9]. A well-known technique for creating ultra-thin filaments with widths ranging from 10 nanometers to 100 micrometers is electrospinning.

John Zeleny was working to develop a mathematical model for liquid drops subject to electrostatic forces ten years after John Cooley and William Morton patented an electrospinning apparatus [8, 10]. Formhals attempted to electrospin fibers in the 1930s but ran into challenges like drying because the nozzle and collection were too far apart [4, 11]. Sir Geoffrey Ingram Taylor created the theoretical framework for electrospinning during this time, contributing to its development by simulating the hopper shape from which liquid droplets were created by electric forces. His work with JR Melcher resulted in the expansion of the "leaky dielectric model" for conducting liquids. The electrospinning procedure was reintroduced in the early 1990s by researchers, in particular those under the direction of Darrel Reneker and Gregory Rutledge. They showed that a variety of organic polymers could be electrospun at the nanoscale. Darrel Reneker created thin fibers with a diameter lower than 5 m for the first time by energizing the polymer mixture with a high voltage [8, 12–14].

This electrospinning method has shown to be quite successful since it can produce fibres with a diameter ranging from micrometres to nanometers (Figure 4.1). The development of techniques for modifying the order and structure of the electrospun nanofibers opened up possibilities

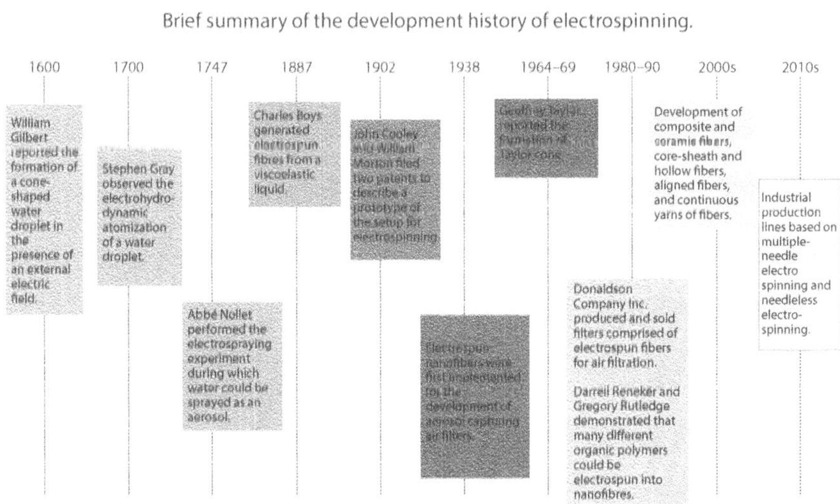

Figure 4.1 Timeline of the history of electrospinning.

for energy-related and medical uses. The nanofibers were also aligned using additional techniques to improve their size, formation, shape, content, porosity, and electrical conductivity. Coaxial electrospinning is an illustration of one such procedure; it produces continuous core-sheath and hollow nanofibers [15–20].

4.3 Setup for the Experiment

Electricity is used to power the process of electrospinning, which combines the principles of hydrodynamics with a high-voltage current. Using this technique, a polymer solution is manipulated to create a droplet that is stretched, extended, and electrified to create fibres. These fibres have a width that varies between a few nanometers and many micrometres. This technique has the benefit of being flexible in its use, enabling the creation of a wide variety of structures and configurations [8, 21].

Figure 4.2 displays a schematic of a typical electrospinning configuration. A high-voltage power source, a spinneret (a metallic needle), and a grounded collector make up the setup's fundamental elements. Figure 4.3 also depicts the three different types of metallic collectors that are widely used: a fixed flat plate, a spinning drum, and a revolving disc.

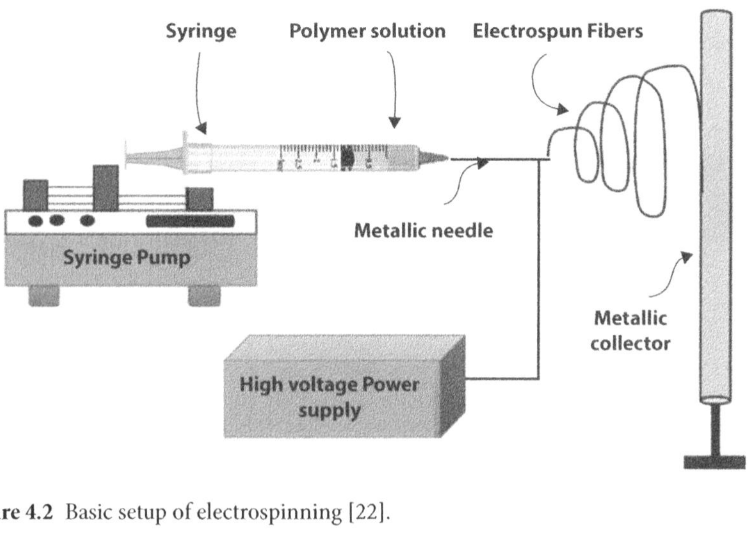

Figure 4.2 Basic setup of electrospinning [22].

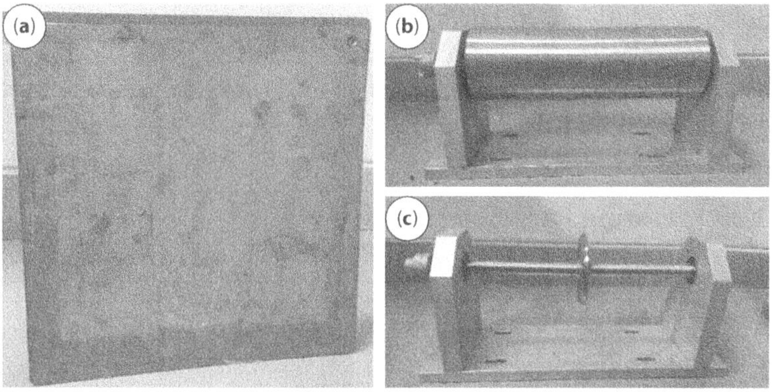

Figure 4.3 Metallic collectors. (a) Stationary flat plate, (b) drum collector, and (c) rotating disc. [22].

4.4 Principle of the Process

An electroconductive, low-molecular-weight liquid droplet with a positive charge is placed in a vacuum to better understand the basic principle underlying the electrospinning procedure. The droplet is being acted upon by two forces: a surface tension that helps to keep the droplet's spherical shape, and a disintegrative repulsion.

Powerful voltage is supplied from the polymer melt or solution onto the liquid droplet at the spinneret's end during the electrospinning process. The liquid droplet will begin to change into a Taylor cone-shaped structure as the voltage is steadily increased. Electrostatic repulsion overrides surface tension to cause elongation. Once the Taylor cone has formed, the charged liquid jet is directed towards a metallic receiver. This material may be a melted polymer, polymer solution, or emulsion. Depending on the liquid viscosity, solid fibres either emerge as the melt cools or the solvent evaporates during the whipping action that occurs during the melt's flight from the Taylor cone to the collector after the nonwoven fibre mat is put to the Taylor cone [23–26].

4.5 Factors Affecting the Process of Electrospinning

Operating conditions, material qualities, and environmental elements all have an impact on the shape of fiber. The operational parameters include the polymer solution feed rate, the distance between the metallic needle's tip and the collector, the applied voltage or electric field, and the needle size. The operating settings can be changed in a way that significantly alters fiber shape [16, 19, 27–30].

4.5.1 Operating Parameters

4.5.1.1 Applied Voltage

The voltage employed directly affects how many charges are carried by the jet, how strongly electrostatic attraction exists between the charges, and how much the jet interacts with the electric field around it. Although increasing the voltage encourages the growth of thinner fibres, it can also result in a higher fluid release, which results in fibres with thicker diameters [27, 31].

4.5.1.2 Flow Rate

In order to keep the Taylor cone constant when using the electrospinning technique, it is essential to adjust the speed of the spinning liquid in line with the applied voltage. When the Taylor cone is stable and homogeneous, it generates fibers with a modest spread. More material passes through the tip at a higher flow rate, resulting in fibers with a larger diameter. The gravitational pull causes the polymeric jet to lose stability at an excessively high flow rate, and it tends to become an electrospray [31].

4.5.1.3 The Separation between the Metallic Needle's Tip and the Collector

Although not as significantly as other elements, the distance between the nozzle and the collector can influence the width and structure of nanofibers. In order for the solvent to evaporate before the fibre reaches the collector, the electrospinning process needs a certain amount of space. Shorter distances have led to the production of thinner fibres. However, if the separation was either too great or too little, beads would form [8, 18].

4.5.2 Material Parameters

Both the electrospinning process and the structure of the fibre are significantly influenced by the properties of the material being used.

The solvent's and polymer's particular properties, as well as the concentration, viscosity, and surface tension of the melt or solution of the polymer, are among these [8, 18].

4.5.2.1 *Polymer Concentration*

The amount of polymer in a material is the most important factor in preserving its fibrous structure as it influences other properties including surface tension, electrical conductivity, and solution viscosity [8, 18].

4.5.2.2 *Viscosity*

The viscosity of the polymer mixture has a big impact on the size and form of the electrospun fibre. Along with other polymer characteristics, the atomic weight and concentration of the polymer mixture affect it. The viscosity of a mixture increases when the amount of polymer is increased. If the mixture's viscosity is too high, it could be challenging to syphon through the syringe pump or it might dry at the needle tip before the electrospinning process can begin. A smaller deposition area and larger measured filaments are caused by increased viscosity [8, 18].

4.5.2.3 *Surface Tension*

Surface tension, the force that pulls molecules in a liquid towards one another, is influenced by intermolecular interactions. The charges in the polymer mixture must be high enough during electrospinning to overcome the surface tension of the solution. A polymer jet is produced at the tip of a needle during electrospraying, stretched in the direction of a collector, and then fragmented into smaller droplets as a result of a reduction in surface tension. When the surface tension is high and the polymer concentration is low, beads form. However, the jet will form at a lower voltage if the surface tension is low. By adding surfactants or changing the solvent used, the surface tension of the polymer solution may be changed. [19, 32].

4.5.2.4 *Conductivity*

In the electrospinning process, a liquid that has received an electrical charge is stretched and forms fibers as a result of the charge's repulsive properties. The jet carries more charges as the electroconductivity of the solution rises, making the fiber thinner. A tiny amount of polyelectrolyte (like salt) can be added to help with the jet's elongation and fiber formation. High voltages make it difficult to electrospin polymer solutions, and the lack of conductivity in the solution makes fiber production impossible [8, 18].

4.5.3 Properties of Solvents

Since the vapor pressure and permittivity of the solvent can have an impact on the process' efficiency, choosing the right solvent is crucial while electrospinning [8, 18].

4.5.3.1 Vapor Pressure

The solvent's level of volatility or vapor pressure affects the rate of evaporation, which in turn affects the rate at which the jet solidifies. It becomes challenging to spin fibers when the volatility is too high because the jet rapidly solidifies as it leaves the spinneret. The fibers will still be damp when they reach the collector if the vapor pressure is too low. The solvent's volatility directly affects the shape of the fibers and their nano-membrane structures [8, 18].

4.5.3.2 Allowability

The electrospinning procedure and the structure of the resultant fibres are significantly influenced by a solvent's permittivity, or capacity to store electrical energy. When electrospinning is carried out in a solution with a greater permittivity, the number of beads produced and the diameter of the fibres are both reduced. A longer jet path and more bending instability of the electrospinning jet, as well as thinner fibres and a bigger deposition area, are all effects of increased permittivity. N, N-dimethylformamide (DMF) is a solvent that may be used to raise the permittivity of a polymer solution [8, 18].

4.5.4 Environmental Aspects

The local climate, including the temperature and humidity, can affect the process of electrospinning as well as the final shape of the fibers. The parameters of the electrospinning solution change as a result of a decrease in temperature because a decrease in temperature is associated with a decrease in the viscosity of the solution. Furthermore, because humidity affects the evaporation of the solvent, it can have an effect on the size of fiber pores [8, 18].

4.6 Electrospinning Variations

Based on the electrospinning configuration, the two forms of electrospinning—needle-based and needleless—are distinguished. For needle-based electrospinning, there are four further categories: single nozzle, coaxial, tri-axial, and multichannel spinnerets.

Melt electrospinning, emulsion electrospinning, and solution electrospinning are the three subcategories of electrospinning based on the condition of the spinning material.

4.6.1 Needle-Based Electrospinning

In this form of electrospinning, a spinneret resembling a needle is employed, and the needle's tip has a sharp "cone shape" when the electric field strikes it.

4.6.1.1 Single Nozzle

A single needle is used as the spinneret in this method of electrospinning, which is the simplest and is categorically shown in Figure 4.2.

Side-by-side electrospinning is a modification of the traditional single-nozzle electrospinning method. In this process, Janus-shaped fibers are produced using two parallel syringes, as shown in Figure 4.4. Due to the repellency of the two solutions, the same voltage is supplied to both solutions, and the fibers are often kept apart. Due to how difficult it is to create

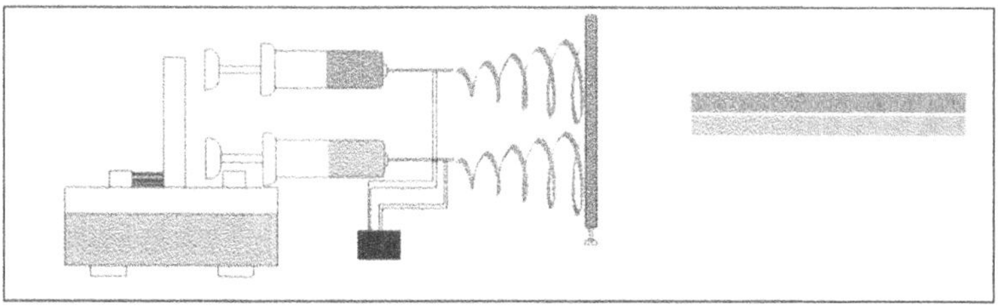

Figure 4.4 Side-by-side electrospinning setup [22].

a Janus structure because of the opposing forces, very few studies have been done on the subject.

4.6.1.2 *Coaxial Electrospinning*

A coaxially electrified jet is produced for coaxial electrospinning using a coaxial needle, which is constructed of two hollow needles stacked concentrically. Simply nestle a smaller (inner) needle inside of a bigger (outer) needle to create this. By connecting this to two syringe pumps, the flow rate of the solutions into the needles may then be independently regulated. As the core and shell solutions exit the coaxial needle, the shell solution will encircle the core solution with the help of an external electric field. Finally, a coaxial jet is produced by the resulting compound Taylor cone. Core-shell nanofibers with a variety of core and shell parts are shown to have evolved in Figure 4.5.

A lot of factors, such as the electrospinning parameters, the inner and outer solution characteristics, and others, have a big impact on how core-shell nanofibers develop. To guarantee a steady stream of jets, the viscosities of the inner and outer solutions must be suitable. It is also required to precisely manage the flow rates of the two solutions in order to guarantee that the inside solution is completely ringed by the outer solution. The diameter of the nanofibers and the thickness of the shell may both be changed by varying the flow rates [22].

For a variety of applications, coaxial electrospinning can produce core-shell nanofibers with more exact chemical compositions. Due to the capacity of the outer fluid to control the inner fluid, this technique also makes it possible to produce nanofibers from liquids that are challenging to spin. By separating the core from the resultant core-shell nanofibers, hollow nanofibers with variable wall thickness may be created. Multifunctional materials may be produced by combining a range of functional components into the two segments of coaxial electrospun fibres [12, 25, 33–45].

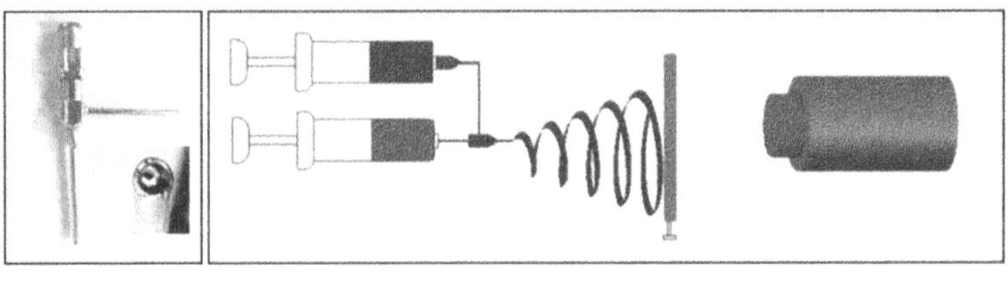

Figure 4.5 Coaxial spinneret and coaxial electrospinning setup [22].

4.6.1.3 Tri-Axial Electrospinning

Three concentric metal capillaries are shown in a trilayer spinneret with tri-axial electrospinning in Figure 4.6. Tri-axial electrospinning typically results in a three-layer structure that consists of a drug-filled core, a hydrophobic intermediate layer, and a hygroscopic outer layer.The immiscibility of the compartmentalized solutions or the simultaneous evaporation of solvents are two ways to prevent the merging of spinning solutions, which is necessary to produce multi-compartment fibers of outstanding quality. To achieve this, the spinneret must be created based on the particular application. In addition to producing the desired nanofiber formations, a spinneret with the right design may control the behavior of fluids in an electric field [46–49].

The pioneers in tri-axial electrospinning research were Lallave *et al.* [50]. In a tri-axial configuration with lignin, glycerine, and ethanol (in that sequence from the outermost to the innermost layer), they employed glycerine as a template fluid and ethanol as the sheath flow to stop the Taylor cone from forming.

4.6.1.4 Multichannel Electrospinning

Zhao *et al.*'s [51] innovative multifluid compound jet electrospinning method was proposed to quickly and easily create hierarchical multichannel microtubes that mimic biological structures. As seen in Figure 4.7,

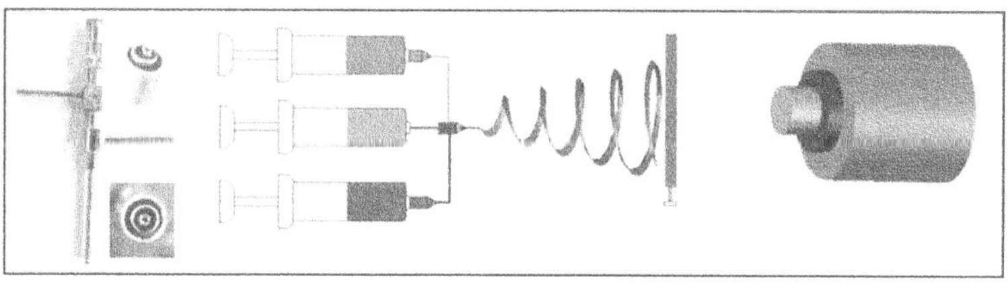

Figure 4.6 Tri-axial spinneret and tri-axial electrospinning setup [22].

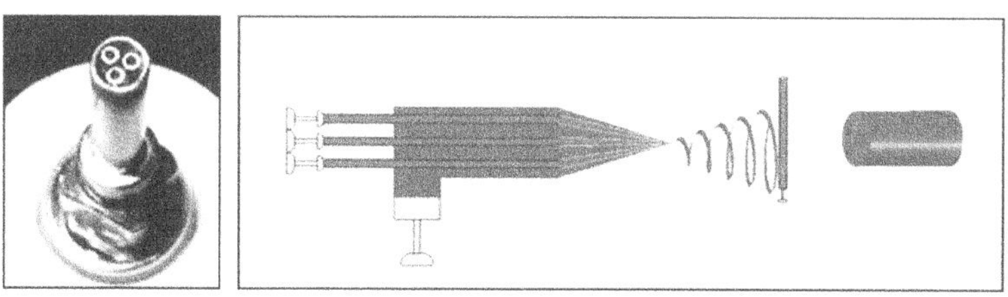

Figure 4.7 Multichannel spinneret and multichannel electrospinning setup [22].

this procedure involved inserting three metallic capillaries into a syringe in an equilateral triangle arrangement. Multiple inner axial paraffin oils suspended in a $Ti(OiPr)_4$ solution were employed in multi-axial electrospinning to generate the biomimetic system, which was then removed to assemble the multi-layered channel.

4.6.2 Needleless Electrospinning

Several cones are spun without the use of a needle or a tiny aperture in needleless electrospinning. In needleless electrospinning, the process of igniting the jet is a robotic process that takes place on a liquid surface without the assistance of capillary forces. According to this method, an external disturbance is used to focus the electric field on the liquid surface to an intensity high enough to form a Taylor cone [52–55].

4.6.3 Melt Electrospinning

Melt electrospinning allows for a complete avoidance of disposal, recycling, and environmental difficulties associated with the use of solvents. A capillary tube is used to contain the molten polymer, the electrically charged fluid, and the metal collector, and the entire procedure must take place in a vacuum. With very little variation in width, the fibers produced as a result are extremely uniform. Despite these advantages, there are drawbacks, including the need for specialised equipment, such as the viscosity and limited electrical conductivity of the molten polymer. Melt electrospinning has not been utilised as frequently as electrospinning from a polymer solution because of its high viscosity, high processing temperatures, and difficulty to create fibres with a diameter smaller than a nanometer [56–62].

4.6.4 Emulsion Electrospinning

There are two types of electrospinning that fall under the umbrella of emulsion electrospinning. In order to electrospin an emulsion that can reorganize into a core-shell fiber, as is the case with coaxial electrospinning, a single spinneret can first be used. As the emulsion reorganizes into a core-shell fiber, the second approach, which uses numerous spinnerets, can produce more jets at a faster rate. Using this kind of electrospinning, incompatible or low-affinity drug-polymer complexes, such as proteins put into a hydrophobic polymer for delayed release, may be loaded more densely. Emulsion electrospinning, as opposed to conventional mixing methods, does not require a common solvent for the drug and the polymer.

Emulsifiers such surfactants are typically added to encapsulate and preserve the drug phase [17, 63–68].

4.6.5 Solution Electrospinning

Solution electrospinning, the most often used method of electrospinning, is dissolving the polymer material to be spun into an acceptable solvent at an adequate concentration. For the purpose of producing nanofibers with the necessary properties, it is imperative to skilfully e use. These fibres can range in size from micro to nanometers, are porous and aligned, have a high surface to volume ratio, and have a big aspect ratio. Additionally, electrospinning enables the creation of fibres with an infinite variety of chemical compositions as well as the manufacturing of different morphologies by modifying the spinneret. Electrospun fibres can be employed in a variety of industries, including sensors, filtration, the energy sector, and medicine, to mention a few, due to the combination of all these features [69–71].

4.7 Applications of Electrospun Fibers

Due to the desirable properties of electrospun fibers, electrospinning is widely used in industry. These fibers feature a porous and aligned structure, a large aspect ratio, and a high surface-to-volume ratio. Their diameters can range from micro to nanometers. Additionally, by modifying the spinneret, electrospinning enables the creation of fibers with an incalculable range of chemical compositions as well as a variety of morphologies. Electrospun fibers have a wide range of applications in the disciplines of medical, filtration, energy, and sensors, to mention a few, due to the combination of all these features.

4.7.1 Biomedical Applications

The use of biocompatible polymers that have adhesion and biodegradability qualities is preferred in biomedical treatments. It is crucial to choose the best material for making these nanofibers since it affects their shape, compatibility, mechanical robustness, rate of disintegration, pattern of delivery, and ability to communicate with cells, which can result in a range of tissue responses.

For usage in biomedical applications, gelatin-chitosan nanofibers with an electrospun core-shell structure were produced by Jalaja *et al.* [34]. Gelatin can contain pharmaceuticals and other bioactive substances, whereas chitosan on the outer layer can mimic the extracellular matrix.

As shown in Figure 4.8, Singh *et al.*'s [45] unique coaxial airbrushing technique produced core-shell nanofibers for biological applications.

A coaxial needle and an air brush were used to produce a core-shell nanofiber. This nanofiber was produced using two different polymeric solutions that included biomolecules. Polycaprolactone (PCL) served as the foundation, and polyethylene oxide (PEO), poly-DL-lactide, and PCL

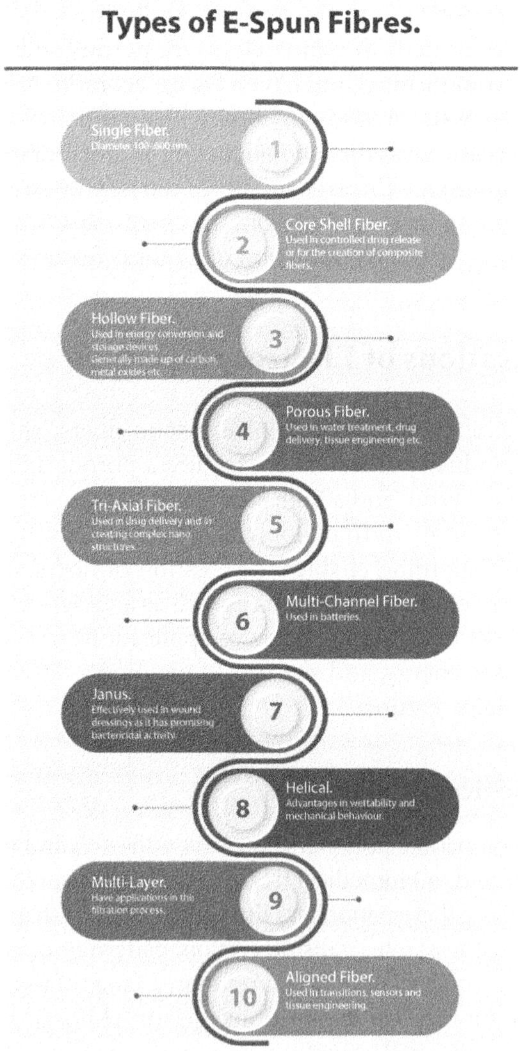

Figure 4.8 Different types of electrospun fibers.

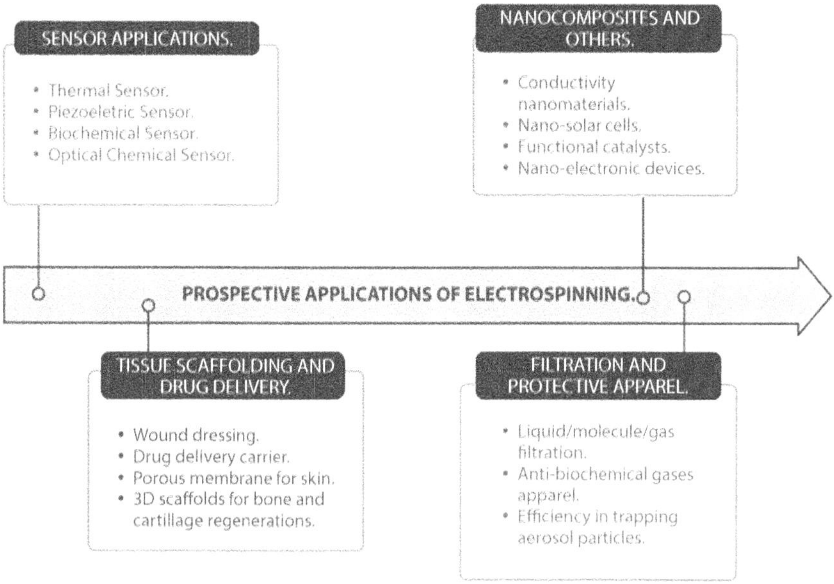

Figure 4.9 Applications of electrospinning.

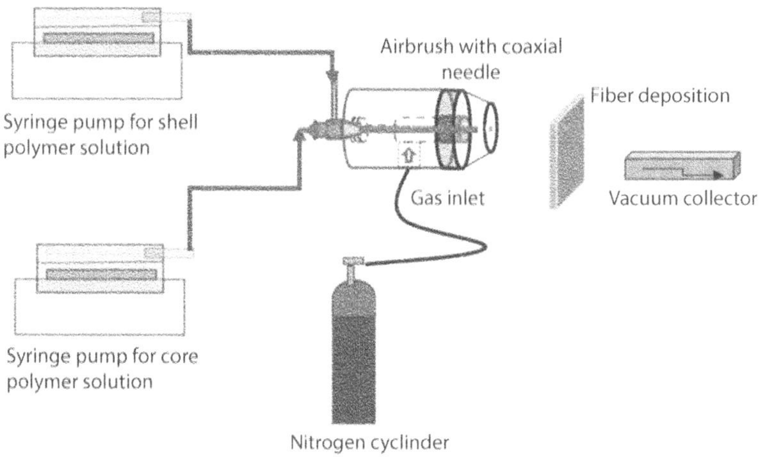

Figure 4.10 Experimental setup for core-shell fiber preparation via coaxial airbrushing methods [45].

made up the core (Figures 4.9 and 4.10). The extensive potential for coaxial electrospinning in the biomedical field is shown by the numerous papers and assessments that are accessible on the subject in the literature [45].

4.7.1.1 Drug Delivery Systems

The single-nozzle electrospinning technique results in fibres that can contain more medications at the fibre surface due to an exceptional surface area-to-volume ratio. When a considerable amount of the medication is introduced into the solution at the beginning of the operation, this often results in a fast release [38, 46, 72, 73]. As a result, several scientists have chosen to use core-shell nanofibers to control the pace of medication release.

4.7.1.2 Tissue Engineering

In order to create an extracellular matrix (ECM) and generate new tissues, tissue engineering depends on the interaction between cells and scaffolds. Inflammation, toxicity, and incompatibility caused by degenerating scaffolds were all problems that biocomposite scaffolds were created to address. Tissue engineering seeks to mimic the functions of the extracellular matrix (ECM) using electrospun nanofiber scaffolds. Core-shell is a versatile approach with fascinating capabilities for enclosing the active ingredient, such as growth factors and medicines in the fiber's core, which is beneficial for medication administration and tissue regeneration [33, 39, 40, 44, 65, 74–80].

4.7.1.3 Wound Healing

Hemostasis, inflammation, proliferation, and remodelling are the four separate phases of the multifaceted, intricate process of wound healing. This intricate process involves several mediators, including extracellular matrix components, cell growth factors, platelets, cytokines, and chemokines. Wound dressings must be versatile and able to deliver the medications required at each stage of healing in order to be effective.

Many advantages for treating wounds can be derived from electrospun nanofibers. The first reason is because these nanofibers' form and structure are similar to the extracellular matrix found in living things, which promotes the adhesion, migration, development, and differentiation of cells under the best possible conditions. Furthermore, electrospun nanofibers' blend of synthetic and natural polymers gives them a high level of biocompatibility and superior mechanical strength. By changing the electrospun nanofibers' fibre structure, it is possible to change how quickly and for how long medication is delivered. Because of this, they offer a great opportunity for developing improved bioactive dressings for wound healing [19, 32, 38, 40, 44, 45, 58, 79, 81–84].

4.8 Summary, Conclusion and Future Prospects

Despite having a lower mechanical strength than more common alternatives like carbon or glass fibers, the use of electrospun nanofibers is a viable choice for composite reinforcements. As a post-processing step, nanoparticles could be added to the nanofibers to improve their mechanical properties. The inability to raise scaffold thickness, control pore size, and create similar scaffolds is one of the key problems preventing electrospinning applications in tissue engineering from progressing.

It is critical to investigate the connection between drug-controlled release patterns and nanofiber topologies in more detail. Because of the structure of nanofibers, mathematical models of drug release from diverse nanofibers may be helpful for comprehending the drug transport mechanism and forecasting drug liberation kinetics.

Researchers have discovered the extraordinary characteristics of electrospun nanofibers, which make them ideal for sensing applications. Nano-fibrous membranes should have larger pores and more surface area if they are to become more sensitive. It is challenging to incorporate such nanoparticles into functional devices since it calls for materials with a particular orientation, size, and repeatability so they can be placed in their designated positions and directions.

References

1. Vollrath, F. and Knight, D.P., Liquid crystalline spinning of spider silk. *Nature*, 410, 541–548, 2001, [PubMed] [Google Scholar].
2. Heim, M., Keerl, D., Scheibel, T., Spider silk: From soluble protein to extraordinary fiber. *Angew. Chem., Int. Ed.*, 48, 3584–3596, 2009, [PubMed] [Google Scholar].
3. Andersson, M., Jia, Q., Abella, A., Lee, X.-Y., Landreh, M., Purhonen, P., Hebert, H., Tenje, M., Robinson, C.V., Meng, Q. *et al.*, Biomimetic spinning of artificial spider silk from a chimeric minispidroin. *Nat. Chem. Biol.*, 13, 262–264, 2017, [PubMed] [Google Scholar].
4. W.H. Carothers, Alkylene ester of polybasic acids. U.S. Pat. 2,012,267, 1935.
5. W.H. Carothers, Linear condensation polymers. U.S. Pat. 2,071,250, 1937.

6. McIntyre, J.E., *Synthetic fibres: Nylon, polyester, acrylic, polyolefin*, Woodhead, Cambridge, U.K., 2005, Chapter 1. [Google Scholar].
7. Gilbert, W., *De magnete*, Courier, New York, 1958, [Google Scholar].
8. Xue, J., Wu, T., Dai, Y., Xia, Y., Electrospinning and electrospun nanofibers: Methods, materials, and applications. *Chem. Rev.*, 119, 8, 5298–5415, 2019.
9. Neumann, C., Production, properties, and uses of the finest threads. *Proc. Phys. Soc. Lond.*, 489, 489–499, 1887.
10. J.F. Cooley, Apparatus for electrically dispersing fluids. US Pat 692, 631, 1900.
11. Soderlund, H., Kaariainen, L., Von Bonsdorff, C.H., Properties of Semliki Forest virus nucleocapsid. *Med. Biol.*, 53, 5, 412–417, 1975.
12. Yarin, A.L., Koombhongse, S., Reneker, D.H., Bending instability in electrospinning of nanofibers. *J. Appl. Phys.*, 89, 3018, 2001.
13. Reneker, D.H., Yarin, A.L., Fong, H., Koombhongse, S., Bending instability of electrically charged liquid jets of polymer solutions in electrospinning. *J. Appl. Phys.*, 87, 9I, 4531–4547, 2000.
14. Joshi, B., Samuel, E., Kim, Y., Yarin, A.L., Swihart, M.T., Yoon, S.S., Review of recent progress in electrospinning-derived freestanding and binder-free electrodes for supercapacitors. *Coord. Chem. Rev.*, 460, 21446, 2022.
15. Shi, X., Zhou, W., Ma, D., Ma, Q., Bridges, D., Ma, Y. *et al.*, Electrospinning of nanofibers and their applications for energy devices. *J. Nanomater.*, 2015, 140716, 2015.
16. Luo, C.J., Stoyanov, S.D., Stride, E., Pelan, E., Edirisinghe, M., Electrospinning versus fibre production methods: From specifics to technological convergence. *Chem. Soc. Rev.*, 41, 4708–4735, 2012, Available from: www.rsc.org/csr.
17. Yarin, A.L., Coaxial electrospinning and emulsion electrospinning of core-shell fibers. *Polym. Adv. Technol.*, 22, 3, 310–317, 2011.
18. Li, Y., Li, Q., Tan, Z., A review of electrospun nanofiber-based separators for rechargeable lithium-ion batteries. *J. Power Sources*, 443, August, 227262, 2019.
19. Fadil, F., Affandi, N.D.N., Misnon, M.I., Bonnia, N.N., Harun, A.M., Alam, M.K., Review on electrospun nanofiber-applied products. *Polym. (Basel)*, 13, 13, 1–31, 2021.
20. Angel, N., Li, S., Yan, F., Kong, L., Recent advances in electrospinning of nanofibers from bio-based carbohydrate polymers and their applications. *Trends Food Sci. Technol.*, 120, 308–324, 2022.
21. Kumar Sharma, G. and Rachel, J.N., Progress in electrospun polymer composite fibers for microwave absorption and electromagnetic interference shielding. *ACS Appl. Electron. Mater.*, 3, 11, 4657–4680, 2021.
22. Sharma, G.K. and James, N.R., Electrospinning: The technique and applications, in: *Recent Developments in Nanofibers Research*, IntechOpen, 2022, https://doi.org/10.5772/intechopen.105804.
23. Zhao, K., Wang, W., Yang, Y., Wang, K., Yu, D.G., From Taylor cone to solid nanofiber in tri-axial electrospinning: Size relationships. *Results Phys.*, 15, 102770, 2019.

24. Zargham, S., Bazgir, S., Tavakoli, A., Rashidi, A.S., Damerchely, R., The effect of flow rate on morphology and deposition area of electrospun nylon 6 nanofiber. *J. Eng. Fibers Fabr.*, 7, 4, 42–49, 2012.
25. Monfared, M., Taghizadeh, S., Zare-Hoseinabadi, A., Mousavi, S.M., Hashemi, S.A., Ranjbar, S. *et al.*, Emerging frontiers in drug release control by core–shell nanofibers: A review. *Drug Metab. Rev.*, 51, 4, 589–611, 2019.
26. Sharma, G.K. and James, N.R., Carbon black incorporated carbon nanofiber based polydimethylsiloxane composite for electromagnetic interference shielding. *Carbon Trends*, 8, 100177, 2022.
27. Yang, C., Jia, Z., Xu, Z., Wang, K., Guan, Z., Wang, L., Comparisons of fibers properties between vertical and horizontal type electrospinning systems. *Annu Rep - Conf Electr Insul Dielectr Phenomena: CEIDP*, pp. 204–207, 2009.
28. He, X.X., Zheng, J., Yu, G.F., You, M.H., Yu, M., Ning, X. *et al.*, Near-field electrospinning: Progress and applications. *J. Phys. Chem. C*, 121, 16, 8663–8678, 2017.
29. Rathore, P. and Schiffman, D.J., Beyond the single-nozzle: Coaxial electrospinning enables innovative nanofiber chemistries, geometries, and applications. *ACS Appl. Mater. Interfaces*, 13, 1, 48–66, 2020.
30. Sundarrajan, S., Tan, K.L., Lim, S.H., Ramakrishna, S., Electrospun nanofibers for air filtration applications. *Proc. Eng.*, 75, 159–163, 2014.
31. Tang, Zhong, B., Abd-El-Aziz SC, A.S., Electrospinning Series Editors: Titles in the Series, RSC Polymer Chemistry Series No.14, London, 2014.
32. Moradipour, P., Limoee, M., Janfaza, S., Behbood, L., Core-Shell nanofibers based on polycaprolactone/polyvinyl alcohol and polycaprolactone/collagen for biomedical applications. *J. Pharm. Innovation*, 3, 1–10, 2021.
33. Perez, R.A. and Kim, H.W., Core-shell designed scaffolds for drug delivery and tissue engineering. *Acta Biomater.*, 21, 2–19, 2015.
34. Jalaja, K., Naskar, D., Kundu, S.C., James, N.R., Potential of electrospun core–shell structured gelatin–chitosan nanofibers for biomedical applications. *Carbohydr. Polym.*, 136, 1098–1107, 2016.
35. Tong, Z., Liao, Z., Liu, Y., Ma, M., Bi, Y., Huang, W. *et al.*, Hierarchical Fe3O4/Fe@C@MoS2 core-shell nanofibers for efficient microwave absorption. *Carbon*, 179, 646–654, 2021.
36. Zupančič, Š., Core-shell nanofibers as drug delivery systems. *Acta Pharm.*, 69, 2, 131–153, 2019.
37. Neisiany, R.E., Saied Khorasani, N., Kong, J., Lee, Y., Javad, R. *et al.*, Core-shell nanofibers for developing self-healing materials: Recent progress and future directions. *Mater. Des. Process. Commun.*, 3, 90, 2021.
38. Kharaghani, D., Gitigard, P., Ohtani, H., Kim, K.O., Ullah, S., Saito, Y. *et al.*, Design and characterization of dual drug delivery based on *in-situ* assembled PVA/PAN core-shell nanofibers for wound dressing application. *Sci. Rep.*, 9, 1, 1–11, 2019.
39. Ezhilarasu, H., Sadiq, A., Ratheesh, G., Sridhar, S., Ramakrishna, S., Mohd, M.H. *et al.*, Functionalized core/shell nanofibers for the differentiation of

mesenchymal stem cells for vascular tissue engineering. *Nanomedicine*, 14, 2, 201–214, 2019.
40. Li, Z., Mei, S., Dong, Y., She, F., Li, P., Li, Y. *et al.*, Multi-functional core-shell nanofibers for wound healing. *Nanomaterials*, 11, 6, 1546, 2021.
41. Lee, S., Park, J., Kim, M.C., Kim, M., Park, P., Yoon, I.J. *et al.*, Polyvinylidene fluoride Core-Shell nanofiber membranes with highly conductive shells for electromagnetic interference shielding. *ACS Appl. Mater. Interfaces*, 13, 21, 25428–25437, 2021.
42. Elahi, M.F. and Lu, W., Core-shell fibers for biomedical applications-a review. *J. Bioeng. Biomed. Sci.*, 03, 01, 121, 2013.
43. Zhu, L.F., Zheng, Y., Fan, J., Yao, Y., Ahmad, Z., Chang, M.W., A novel core-shell nanofiber drug delivery system intended for the synergistic treatment of melanoma. *Eur. J. Pharm. Sci.*, 137, 105002, 2019.
44. Movahedi, M., Asefnejad, A., Rafienia, M., Khorasani, M.T., Potential of novel electrospun core-shell structured polyurethane/starch (hyaluronic acid) nanofibers for skin tissue engineering: *In vitro* and *in vivo* evaluation. *Int. J. Biol. Macromol.*, 146, 627–637, 2020.
45. Singh, R., Ahmed, F., Polley, P., Giri, J., Fabrication and characterization of core-shell nanofibers using a next-generation airbrush for biomedical applications. *ACS Appl. Mater. Interfaces*, 10, 49, 41924–41934, 2018.
46. Yu, D.G., Li, X.Y., Wang, X., Yang, J.H., Bligh, S.W.A., Williams, G.R., Nanofibers fabricated using Triaxial electrospinning as zero order drug delivery systems. *ACS Appl. Mater. Interfaces*, 7, 33, 18891–18897, 2015.
47. Ghosal, K., Augustine, R., Zaszczynska, A., Barman, M., Jain, A., Hasan, A. *et al.*, Novel drug delivery systems based on triaxial electrospinning based nanofibers. *React. Funct. Polym.*, 163, December 2020, 104895, 2021.
48. Nagiah, N., Murdock, C.J., Bhattacharjee, M., Nair, L., Laurencin, C.T., Development of Tripolymeric Triaxial electrospun fibrous matrices for dual drug delivery applications. *Sci. Rep.*, 10, 1, 1–11, 2020.
49. Han, D. and Steckl, A.J., Triaxial electrospun nanofiber membranes for controlled dual release of functional molecules. *ACS Appl. Mater. Interfaces*, 5, 16, 8241–8245, 2013.
50. Lallave, M., Bedia, J., Ruiz-Rosas, R., Rodríguez-Mirasol, J., Cordero, T., Otero, J.C. *et al.*, Filled and hollow carbon nanofibers by coaxial electrospinning of alcell lignin without binder polymers. *Adv. Mater.*, 19, 4292–4296, 2007.
51. Zhao, Y., Cao, X., Jiang, L., Bio-mimic multichannel microtubes by a facile method. *J. Am. Chem. Soc.*, 129, 4, 764–765, 2007.
52. Partheniadis, I., Nikolakakis, I., Laidmäe, I., Heinämäki, J., A mini-review: Needleless electrospinning of nanofibers for pharmaceutical and biomedical applications. *Processes, Basel, Switzerland*, 8, 6, 673, 2020.
53. Iranshahi, K., Schoeller, J., Luisier, N., Chung, M., Hashemizadeh, S., Fortunato, G. *et al.*, Improving needleless electrospinning throughput by tailoring polyurethane solution properties with Polysiloxane additives. *ACS Appl. Polym. Mater.*, 4, 3, 2205–2215, 2022.

54. Hwang M, O., Karenson, M., Elabd Y, A., High production rate of high purity, high fidelity nafion nanofibers via needleless electrospinning. *ACS Appl. Polym. Mater.*, 1, 10, 2731–2740, 2019.
55. Yang, Z., Zhang, X., Qin, Z., Li, H., Wang, J., Zeng, G. *et al.*, Airflow synergistic needleless electrospinning of instant noodle-like curly nanofibrous membranes for high-efficiency air filtration. *Small*, 18, 2107250, 2022.
56. Cao, L., Su, D., Su, Z., Chen, X., Fabrication of multiwalled carbon nanotube/polypropylene conductive fibrous membranes by melt electrospinning. *Ind. Eng. Chem. Res.*, 53, 6, 2308–2317, 2014.
57. Li, X., Zheng, Y., Mu, X., Xin, B., Lin, L., Investigation into jet motion and fiber properties induced by electric fields in melt electrospinning. *Ind. Eng. Chem. Res.*, 59, 5, 2163–2170, 2020.
58. Shao, Z., Chen, J., Jie, K.L., Wang, Q., Wang, X., Li, W. *et al.*, Directional transportation in a self-pumping dressing based on a melt electrospinning hydrophobic mesh. *ACS Biomater. Sci. Eng.*, 7, 12, 5918–5926, 2021.
59. Shabani, E., Azimi Yancheshme, A., Ronen, A., E. Gorga, R., Effect of the spin-line temperature profile on the translocation of the solidification point and jet thinning in unconfined melt electrospinning. *ACS Appl. Polym. Mater.*, 3, 1, 268–278, 2020.
60. Chen, F., Hochleitner, G., Woodfield, T., Groll, J., D. Dalton, P., G. Amsden, B., Additive manufacturing of a photo-cross-linkable polymer via direct melt electrospinning writing for producing high strength structures. *Biomacromolecules*, 17, 1, 208–214, 2015.
61. Xie, G., Wang, Y., Han, X., Gong, Y., Wang, J., Zhang, J. *et al.*, Pulsed electric fields on poly-l-(lactic acid) melt electrospun fibers. *Ind. Eng. Chem. Res.*, 55, 26, 7116–7123, 2016.
62. Dalton, D.P., Klinkhammer, K., Salber, J., Klee, D., Möller, M., Direct *in vitro* electrospinning with polymer melts. *Biomacromolecules*, 7, 3, 686–690, 2006.
63. Sharif, F., Arjmand, M., Moud, A.A., Sundararaj, U., Roberts, E.P.L., Segregated hybrid poly(methyl methacrylate)/graphene/magnetite nanocomposites for electromagnetic interference shielding. *ACS Appl. Mater. Interfaces*, 9, 16, 14171–14179, 2017.
64. Zhang, C. and Zhang, H., Formation and stability of core–shell nanofibers by electrospinning of gel-like corn oil-in-water emulsions stabilized by gelatin. *J. Agric. Food Chem.*, 66, 44, 11681–11690, 2018.
65. Pal, P., Kumar Srivas, P., Dadhich, P., Das, B., Maulik, D., Dhara, S., Nano-/microfibrous cotton-wool-like 3D scaffold with core–shell architecture by emulsion electrospinning for skin tissue regeneration. *ACS Biomater. Sci. Eng.*, 3, 12, 3563–3575, 2017.
66. Choi, S.H., Youn, D.Y., Mu Jo, S., Oh, S.G., Kim, I.D., Micelle-mediated synthesis of single-crystalline β(3C)-SiC fibers via emulsion electrospinning. *ACS Appl. Mater. Interfaces*, 3, 5, 1385–1389, 2011.
67. Zhao, P., Soin, N., Prashanthi, K., Chen, J., Dong, S., Zhou, E. *et al.*, Emulsion electrospinning of polytetrafluoroethylene (PTFE) Nanofibrous membranes

for high-performance triboelectric nanogenerators. *ACS Appl. Mater. Interfaces*, 10, 6, 5880–5891, 2018.
68. Qi, H., Hu, P., Xu, J., Wang, A., Encapsulation of drug reservoirs in fibers by emulsion electrospinning: Morphology characterization and preliminary release assessment. *Biomacromolecules*, 7, 8, 2327–2330, 2006.
69. Li, Y., Abedalwafa, M.A., Tang, L., Li, D., Wang, L., Electrospun nanofibers for sensors, in: *Electrospinning: Nanofabrication and Applications*, W. Andrew (Ed.), pp. 571–601, Elsevier, New York, United States, 2018.
70. Liang, J., Zhao, H., Yue, L., Fan, G., Li, T., Lu, S. *et al.*, Recent advances in electrospun nanofibers for supercapacitors. *J. Mater. Chem. A*, 8, 16747–16789, 2020.
71. Mamun, A., Blachowicz, T., Sabantina, L., Electrospun nanofiber mats for filtering applications technology, structure and materials. *Polym. (Basel)*, 13, 9, 1–14, 2021.
72. Davoodi, P., Gill, E.L., Wang, W., Shery Huang, Y.Y., Advances and innovations in electrospinning technology, in: *Biomedical Applications of Electrospinning and Electrospraying*, pp. 45–81, Woodhead Publishing, Elsevier, Sawston, Cambridge, 2021.
73. Bay Stie, M., Corezzi, M.D., Juncos Bombin, A., Ajalloueian, F., Attrill, E., Pagliara, S. *et al.*, Waterborne electrospinning of α-Lactalbumin generates tunable and biocompatible nanofibers for drug delivery. *ACS Appl. Nano Mater.*, 3, 2, 1910–1921, 2020.
74. Sarika, P.R., Cinthya, K., Jayakrishnan, A., Anilkumar, P.R., James, N.R., Modified gum arabic cross-linked gelatin scaffold for biomedical applications. *Mater. Sci. Eng.: C*, 43, 272–279, 2014.
75. Matthews, A.J., Wnek, E.G., Simpson, G.D., Bowlin, L.G., Electrospinning of collagen nanofibers. *Biomacromolecules*, 3, 2, 232–238, 2002.
76. Lim, D.J., Cross-linking agents for electrospinning-based bone tissue engineering. *Int. J. Mol. Sci.*, 23, 5444, 2022.
77. Nemati, S., Jeong, K.S., Shin, Y.M., Shin, H., Current progress in application of polymeric nanofibers to tissue engineering. *Nano Convergence*, 6, 1, 36, 2019.
78. Rahmati, M., Mills, D.K., Urbanska, A.M., Saeb, M.R., Venugopal, J.R., Ramakrishna, S. *et al.*, Electrospinning for tissue engineering applications. *Prog. Mater. Sci.*, 117, 100721, 2021.
79. Wnek, E.G., Carr, E.M., Simpson, G.D., Bowlin, L.G., Electrospinning of nanofiber fibrinogen structures. *Nano Lett.*, 3, 2, 213–216, 2002.
80. Jalaja, K., Sreehari, V.S., Kumar, P.R.A., Nirmala, R.J., Graphene oxide decorated electrospun gelatin nanofibers: Fabrication, properties and applications. *Mater. Sci. Eng.: C*, 64, 11–19, 2016.
81. Hoque, M.E., Nuge, T., Yeow, T.K., Nordin, N., Electrospun matrices from natural polymers for skin regeneration, in: *Nanostructured Polymer Composites for Biomedical Applications*, pp. 87–104, Elsevier, Amsterdam, Netherlands, 2019.

82. Zhong, W., Nanofibres for medical textiles, in: *Advances in Smart Medical Textiles*, pp. 57–70, Woodhead Publishing, Elsevier, Sawston, Cambridge, 2016.
83. Lv, P., Xu, W., Li, D., Feng, Q., Yao, Y., Pang, Z. *et al.*, Metal-based bacterial cellulose of sandwich nanomaterials for anti-oxidation electromagnetic interference shielding. *Mater. Des.*, 112, 374–382, 2016.
84. Rezvani Ghomi, E., Khosravi, F., Neisiany, R.E., Shakiba, M., Zare, M., Lakshminarayanan, R. *et al.*, Advances in electrospinning of aligned nanofiber scaffolds used for wound dressings. *Curr. Opin. Biomed. Eng.*, 22, 100393, 2022.

5

Development of Nanocellulose-Based Nanocomposites and Its Properties

Sunita Adak

Department of Food Science and Technology, School of Lifesciences, St. Joseph's University, Bangalore, Karnataka, India

Abstract

Advances in material science have led to the development of nanomaterials. Conversion of size from macro to nanoscale changes the bulk properties and instills certain unique features. Nanocelluloses are well suited for the fabrication of composites due to their variable morphology and size, mechanical and thermal characteristics, crystallinity, biodegradability, and renewability. The nanocellulose based nanocomposite is an emerging field of research and development due to its diversity, unique properties, adaptability, and huge scope of applicability. This chapter explores the uniqueness, diversity, and importance of nanocomposites. Their classification based on matrix and filler has been reviewed with an emphasis on the importance of molecular interactions. Conventional extraction techniques and sources of nanocellulose particles impart diversity in properties but add to cost and pollution. Alternate and emerging green technologies of extraction have been highlighted with information on the primary, secondary, and tertiary sources of cellulose. Since nanocomposite fabrication techniques influence nanoparticle distribution and interaction in the polymer, an elaborate understanding of the different fabrication methods is provided. Furthermore, a correlation between the factors governing the properties of designed nanocomposites has been highlighted with a few supporting research outputs. The chapter concludes with an insight into the underlying challenges in the commercialization of nanocellulose based nanocomposite materials and the probable area of research to overcome these challenges.

Keywords: Nanocomposite, nanocellulose, sources, extraction, fabrication, properties

Email: sunitaadak@gmail.com

Mainak Mukhopadhyay and Debalina Bhattacharya (eds.) Nanocellulose: A Biopolymer for Biomedical Applications, (113–140) © 2024 Scrivener Publishing LLC

5.1 Introduction

In order for humans to survive and flourish, one of the crucial aspects of our evolutionary process has been the creation of innovative materials. Extensive research and development at all scales, from macro to micro to nano, has resulted from the fabrication of materials with desired qualities to fulfill a variety of uses. The addition of inorganic and organic fillers in any polymer creates composite material having enhanced functional properties. Such polymer composites have gained attention in diverse fields. The progression of composite to nanocomposite introduced unique features in material properties which can be applied in sectors like sports goods, aviation, aerospace, automotive, biomedical, and food. Cellulose-based nanofillers have gained attention due to their low density, non abrasive, combustible, nontoxic, low cost, and biodegradable properties [1].

5.1.1 Nanocomposite

Composite materials are heterogeneous and anisotropic, consisting of a minimum of two or more materials combined together. These combined materials have different properties and impart new characteristics to the developed composite. Usually, one of these materials is the matrix, and the other is filler or reinforcement. Both maintain individual distributions with distinct boundaries and properties even when combined [2]. When one of the constituting materials in the composite possesses one, two, or all three dimensions on the nanometer scale (10^{-9} m), it is called a nanocomposite (Figure 5.1). Reduction in the size of the materials changes their property from that of bulk, instilling some unique properties in terms of

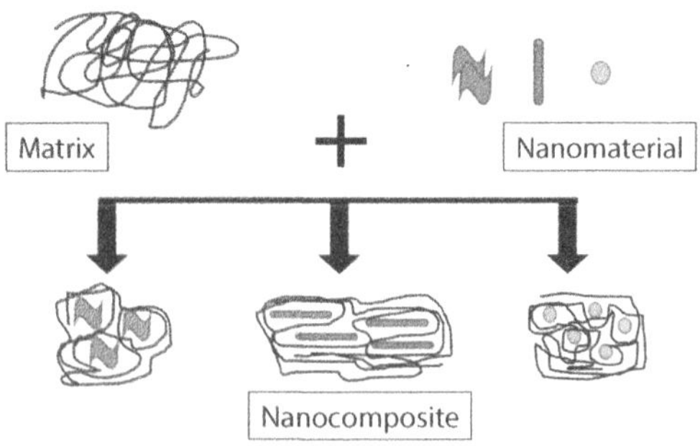

Figure 5.1 Composition of nanocomposite.

surface, magnetic, electrical, mechanical, catalytic, and optical properties. The nanomaterials used are in the form of nanoparticles, nanofibers, and nanoclays. Based on the dimension, shape, and aspect ratio of the nanomaterial used, the properties of the nanocomposites vary. As a result of these diverse properties, nanocomposites find wider applications in the fields of material science, automotive, aviation, clinical, health care, and food [3, 4].

The properties of a nanocomposite are equally affected by the matrix. Based on the matrix composition, nanocomposites are classified into:

i) Ceramic Matrix Nanocomposites (CMNC),
ii) Metal Matrix Nanocomposites (MMNC), and
iii) Polymer Matrix Nanocomposites (PMNC).

CMNCs are created using various methods that combine two or more ceramic materials with nanoreinforcements. Al_2O_3/SiO_2, SiO_2/Ni, Al_2O_3/TiO_2, Al_2O_3/SiC, and Al_2O_3/CNT are common combinations. Al_2O_3, SiN, SiC, etc. serve as matrices, and Fe, TiO_2, silica, clays, etc. serve as crystalline reinforcement. These nanocomposites are more resistant to wear and mechanical stress, and they remain stable when exposed to different temperatures and chemicals. Apart from diverse industrial applications, they are used in the biomedical and healthcare sectors. Similar to CMNCs, MMNCs are heterogeneous materials made up of ductile metals or alloys. Common combinations are Fe-Cr/Al_2O_3, Ni/Al_2O_3, Co/Cr, Fe/MgO, Al/CNT, and Mg/CNT. Most of the MMNC has Al, Mg, Pb, Sn, W, and Fe as the matrix materials, whereas the reinforcements are the same as those of the CMNC. These nanocomposites bear the characteristics of higher ductility, strength, Young's modulus, and toughness. Their field of application mainly encompasses electronics, automotive, defense, and aerospace. PMNCs are the most adaptable, with a wider range of combinations and applications. The various combinations are thermoplastic/thermoset, polymer/layered silicates, polyester/TiO_2, polymer/CNT, polymer/layered double hydroxides, etc. They possess better properties compared to CMNC and MMNC, like high strength and modulus, damage tolerance, flame resistance, barrier properties, and good damping. Each kind of nanocomposite is developed using specific processing techniques (Figure 5.2), which in turn affects the physical distribution, properties, and applications [3, 5]. The material properties of a nanocomposite are the result of the synergy between the nanomaterial, matrix, and processing technique.

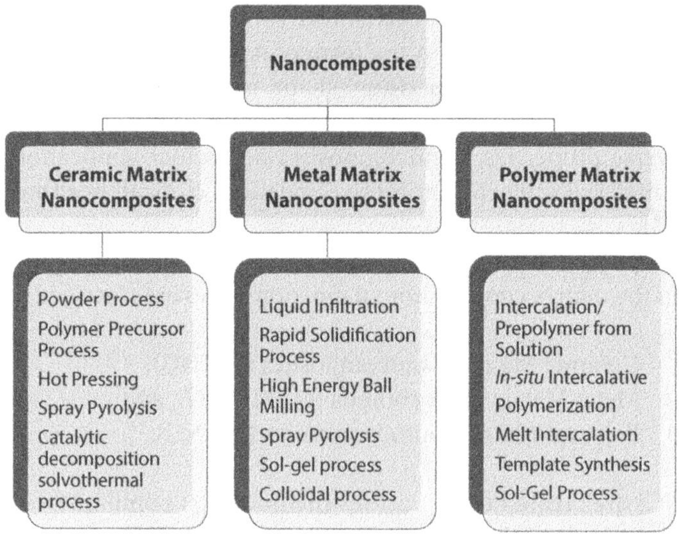

Figure 5.2 Classification of nanocomposites based on the matrices and processing techniques [5].

5.1.2 Nanofillers

Nanofillers or nanoreinforcements are added to materials to increase mechanical strength, stability, and resistance beyond what is naturally possible without significantly increasing weight. These fillers have dimensions in nanometers, either one, two, or all, which imparts two specific properties: a higher surface area and a quantum effect. The properties of nanocomposite are governed by the interfacial interactions between the matrix and nanofiller, with a larger interface allowing greater interaction. This further affects molecular arrangement, distribution, entanglement, and flexibility. Nanostructure brings in quantum effect-associated properties, which also dictate the nanocomposite properties [6].

Nanofillers come in various shapes and dimensions, ranging from spherical to sheets. There are three major shapes: sheets, fibers or tubes, and spheres. The sheets and fibers/tubes are 2-dimensional structures. Sheets have a thickness in the nanoscale (0.1–100 nm), and the other two dimensions are longer (>100 nm). Nanoclays, graphene, layered silicates, MXene, etc. are examples of nanosheets that create layered nanocomposites. Fibers/tubes have two dimensions at the nanoscale (diameter in the range of 0.1–100 nm), and the remaining dimension is longer (>100 nm). These structures are elongated structures with a fiber appearance like carbon nanotubes, carbon nanofibers, halloysite nanotubes, cellulose nanocrystals, etc. The last type has all dimensions at the nanoscale (diameter in the range of 0.1–100 nm) and is spherical with a low aspect ratio. Nanoparticles of this type include spherical silica, ceramic nanoparticles, semiconductor nanoclusters, fullerene, metal nanoparticles, and quantum dots [7, 8].

Nanofillers can also be divided into natural, synthetic, inorganic, and organic groups based on what they are made of and where they come from. Nanoclays, carbon nanoparticles, and metallic nanofillers come under inorganic, while nanocellulose, starch, chitin, and chitosan are organic nanofillers [9]. Inorganic nanofillers show diverse properties, which give a novel character to the developed nanocomposite. High surface properties, electrical conductivity, hardness, wear resistance, UV shielding property, lightweight, dielectric properties, photocatalytic activity, antimicrobial effect, flame retardation, wrinkle resistance, and thermal resistance are some of the common properties found in inorganic nanofillers [10]. Among all nanoclays, mineral clays are the most studied and utilized inorganic nanofillers. They are layered nanoparticles with sheets comprising hydrated silicates. Based on the difference in the arrangement of the sheets and their dimensions, they are further classified into montmorillonite/smectite, kaolinite, illite, and chlorite groups. Compared to nanoclay, metal oxides like silica, alumina, and titania generally come in spherical dimensions. Carbon nanoparticles come in a variety of shapes and sizes, including layered (graphene), nanotubes, nanofibers, and semicrystalline (carbon black). Concerns about health and environmental safety have led to the development of organic nanoparticles, which typically take the form of nanofibers. Nanocellulose, nanolignin, nanochitin, and chitosan are natural, biocompatible and possess properties suitable for their specific applications as an alternative to their inorganic counterparts [8].

5.2 Nanocellulose

Nanocellulose is an organic nanofiller isolated from natural fiber cellulose. Different classifications and nomenclatures of the nanocellulose are available. For each type, structure and properties vary based on the source and method used for isolation. Plant-based biomasses are the common sources, along with a few microorganisms that naturally produce cellulose. The mechanical, rheological, optical, and thermal properties of cellulose nanoparticles are excellent. Moreover, they are biocompatible, degradable, and can be easily modified/activated, making them suitable for various applications.

5.2.1 Types and Nomenclature

Cellulose with a nanoscale dimension in at least one dimension is termed nanocellulose. Nanocellulose has been synthesized from a variety of sources via physical, chemical, enzymatic, and biological processes. These bring differences in the morphology, size, structural arrangement, and properties of nanocellulose. There are three broad types of nanocellulose with specific nomenclature [11]:

i) Cellulose nanocrystals,
ii) Cellulose nanofibrils, and
iii) Bacterial nanocellulose.

Cellulose Nanocrystals (CNCs) are needle-shaped, highly crystalline, and less flexible. These have been known by various names, but the latest nomenclature given is nanocrystalline cellulose or cellulose nanowhiskers. Typically created through the acid hydrolysis of biomass-derived cellulose pulp. Acid hydrolysis dissolves amorphous regions while leaving most crystalline regions intact. CNCs show high crystallinity (54–88%) with a dimension of 5 nm × 20–100 nm and an aspect ratio of 11. Crystallinity and dimensions vary with the source of nanocellulose and the process of production [12, 13]. CNCs exhibit a wide range of properties, including high strength with a Young's modulus of 150 GPa, colloidal dispersion in aqueous, organic, or mixture solvents, and shear-thinning and liquid crystal functions at specific concentrations [13–15].

Cellulose nanofibrils (CNFs) are also known as nanofibrillated cellulose (NFC), nanofibrillar cellulose, and cellulose nanofibers. As the name suggests, they are fiber-like or rope-like in morphology, with both amorphous and crystalline regions. When compared to CNCs, biomass cellulose sources are subjected to mechanical treatment that disintegrates cellulose fibers into smaller dimensions (fibrillation) while retaining a significant number of amorphous regions. They have a rope-like structure, a diameter of 5-60 nm, a length of microns, an aspect ratio of 30-300, and a crystallinity of 59-64% [13, 18]. Various other chemical and enzymatic methods are also explored for the production of CNF. The properties of CNFs vary with the source and production process. The Young's modulus of CNF is approximately 30–180 GPa and it shows thixotropic and shear-thinning properties [14].

Bacterial nanocellulose (BNC) is produced and secreted by specific bacterial species belonging to the Rhizobium, Agrobacterium, Pseudomonas, Gluconacetobacter, and Sarcina genera. These bacteria are cultured, and the nanocelluloses are extracted. They are a pure form of nanocellulose compared to other types. BNCs have dimensions of 20-100 nm (diameter) micrometers (length), an aspect ratio of 50, and a crystallinity of 84-89%. All these properties vary depending on the culture conditions, media composition, and microbial source. Among the diverse functional properties of BNCs higher water holding capacity, Young's modulus of 78 GPa, low apparent density, interpenetrating structure, and high surface area are important [12, 13, 16].

Apart from the three major types of nanocellulose, other forms also have been developed and explored. Amorphous nanocellulose (ANC), Cellulose nanoyarn (CNY), and Cellulose nanoplatelets (CNP) are some other such types. Their properties and morphology difference arise due to variations in the preparation and processing techniques [12].

5.2.2 Sources and Preparation

The base material of nanocellulose is cellulose, which is one of the most abundant biomolecules on earth. The primary source of cellulose is plant biomass, which is more accurately called lignocellulosic biomass. Plant biomass is classified as wood and non-wood biomass. Wood lignocelluloses, derived from hardwood and softwood, account for the major share of cellulose pulp obtained from plant biomass. Secondary sources are agricultural wastes, which include seeds, stalks, bast, fruit, leaves, grass, and crane. These are better than primary sources in terms of renewability, biomass yield, and cellulose content. Some examples of secondary sources are rice straw, maize straw, corn cobs, wheat straw, coconut coir, barley straw, tea stalk, pineapple leaf, sisal leaf, flax bast, cotton linter, and bamboo. Byproducts of the food and beverage industry, municipal waste, and papermaking sludge are the different tertiary sources of cellulose. Tertiary sources such as bagasse, pomace, hull, husk, bran, and so on are both economical and environmentally friendly [17, 18].

Lignocellulosic biomass comprises hemicellulose (20–40%), lignin (15–25%), and cellulose (30–60%). All these components along with other minor constituents interact strongly, limiting the cellulose availability and purity. Thus, preparation of nanocellulose requires multiple consecutive steps. There are two major approaches of nanocellulose preparation:

i) Top-down (large to small), and
ii) Bottom-up (small to large).

The top-down process deals with the extraction of nanocellulose from the lignocellulosic biomass, which requires pre-treatment and disintegration of the biomass. The steps aim to purify the cellulose and further disintegrate cellulose fibers into nanosized particles. Both CNC and CNF are synthesized using a top-down approach. A reverse approach is the bottom-up process, where microorganisms are cultured to produce nanocellulose. Cultured bacteria utilize monomeric sugars to build up nanofibers. BNC synthesis is an example of a bottom-up approach [18, 19].

5.2.2.1 *Top-Down Process*

Preparation of nanocellulose (CNC, CNF) from biomass is a two-step process. The first step is pretreatment to remove unwanted constituents like hemicellulose, lignin, wax, and ashes and obtain pure cellulose fiber. The pure cellulose fiber is then fibrillated into nano-sized cellulose in the second step. Various pretreatment and extraction techniques have been discussed in this section.

Pretreatment serves two important purposes, one is the removal of impurities other than cellulose, and the other is the reduction in power consumption and cost of mechanical fibrillation. Several pretreatment processes serve different purposes, and the choice of pretreatment depends on the source of the biomaterial [20]. The organosolvent treatment removes waxes from the biomass, also known as dewaxing. This treatment also removes oils, phenolics, and pigments. Common organic solvents used are hexane-ethanol, benzene-methanol, ethanol-toluene, and benzene-toluene. In general, a mixture of solvents having polarity differences in varying ratios is used for this step to achieve optimum dewaxing. Other contributing factors are temperature and duration of treatment. Similarly, alkali-acid treatment is used to solubilize hemicellulose, pectin, and lignin from biomass before extraction. Alkalis like potassium hydroxide, sodium hydroxide, calcium hydroxide, sodium carbonate, and ammonia penetrate the crystalline region of cellulose and cause swelling, increase the surface area, and reduce the interactions. This allows better hydrolysis and delignification. Acid treatment involves using a strong acid, like hydrochloric acid, at a high temperature, which solubilizes hemicelluloses. To remove residual lignin, bleaching pretreatment is carried out using oxidizing agents including chlorine dioxide, hydrogen peroxide, ozone, and oxygen [13, 17]. The use of novel ionic liquids has also been investigated to optimize cellulose extraction. Enzymatic pretreatment has been used to avoid harsh chemicals, cellulose fiber structure damage, and to achieve specific target molecule/structure degradation. To modify the cellulose, cellobiohydrolases, endoglucanases, and hemicellulases are used individually or in combination. Though a prolonged duration of treatment is required compared to other pretreatment processes, the quality of the product is found to be more desirable [15, 20].

Extraction can be accomplished through a variety of conventional methods, including chemical and mechanical processes. Table 5.1 represents some of the extraction processes. The most widely used chemical extraction is acid hydrolysis (HCl, H_2SO_4, H_3PO_4, HBr). The acid dissolves the amorphous regions of cellulose, generating crystalline CNCs. Acid type, concentration, temperature, and duration of treatment affect the morphology, size, and properties of the CNC. Though a popular technique, it suffers from many drawbacks, like the generation of acidic wastewater, water wastage, corrosion, high cost, and pollution. Alternate chemical extractions involve oxidation using TEMPO (2,2,6,6-tetramethyl-1-piperidinyloxy) and ammonium persulfate (APS). TEMPO in combination with NaClO and NaBr under alkaline conditions produces CNC with better colloidal dispersal stability. While APS oxidation is not as toxic as TEMPO, it takes longer to generate the desired quantity and quality of the product [18, 20].

Ball milling, cryocrushing, and grinding are examples of traditional mechanical processes. The ball mill consists of a hollow rotating cylinder that is charged with cellulose solution and metal balls of a specific diameter. The rotation of the cylinder causes the balls to collide, which applies an impact force on the cellulose and disintegrates it into nanocellulose. The water content maintains the fiber structure, retaining the amorphous regions along with the crystalline regions. This is suitable for preparing CNFs. Grinding cellulose pulp between stones is another method for producing nanofibers. Stone grinders consist of two stones; one is fixed and the other rotates. The pulp passes between the two stones, and the sheer forces generated help induce fibrillation in CNFs. Similarly, cryocrushing requires impact forces to break the crystalline region, liberating nanofibers from cellulose. Before crushing, the cellulose is immersed in water and then frozen by exposure to liquid nitrogen. All these conventional techniques have high power consumption, low yield, low uniformity, are expensive, and generate heat [15, 16].

Other well-established eco-friendly processes are ultrasonication and high-pressure homogenization (HPH). Ultrasound waves are produced, and the cellulose is subjected to them. The high energy creates oscillations in the molecules and results in the formation, expansion, and implosion of air bubbles. The high pressure, energy, and temperature of implosion lead to structural disorientation in the cellulose and a reduction in the degree of polymerization. To produce CNCs and CNFs, ultrasonication is frequently

Table 5.1 Various nanocellulose extraction processes.

Biomass	Extraction process	Nanocellulose property	Reference
Pineapple leaf fiber	Steam explosion	CNF with 5-60 nm width, aspect ratio 50 and 35.9% crystallinity	Cherian *et al.* [21]
Jute fiber waste	Ball milling	Nanoparticle of 500 nm size distribution	Baheti *et al.* [22]
Cellulose powder	Ball milling in presence of ionic liquids	93.1% CNF yield with 10-25 nm diameter, 65.8% crystallinity and high thermal stability	Phanthong *et al.* [23]
Sugarcane bagasse	Acid hydrolysis	Spherical CNC with 111 nm average diameter and 76.01%. crystallinity	Wulandari *et al.* [24]
Microcrystalline cellulose	Acid hydrolysis + Ultrasonication	CNC with axiolitic shape, 89 nm hydrodynamic radius and crystallinity of 72.31%	Niu *et al.* [25]
Oil palm empty fruit bunch	Enzymatic hydrolysis + Sonication	CNF with diameter 30 nm and 100% yield	Aditiawati *et al.* [26]
Microcrystalline cellulose	Acid-microwave pretreatment + Enzymatic hydrolysis	84.4 wt.% CNC yield with high decomposition temperature	Qian *et al.* [27]
Hardwood (*Eucalyptus tereticornis* Sm. and *Casuarina equisetifolia* L.)	Steam explosion	NCF with average diameter of 27.801 nm - 28.690 nm and high aspect ratio	Raju *et al.* [28]

combined with other chemical or mechanical processes. HPH, on the other hand, is used to produce CNF by passing cellulosic pulp through a tiny nozzle at extremely high pressure. Shear and impact forces are created by the high velocity and pressure, reducing the size of the cellulose fiber to the nanometer range. HPH is a quick method that is scalable to industrial use. The associated drawbacks of higher power consumption and cost could be minimized by suitable pretreatments. Many new techniques in this field are emerging, such as hammer milling, microfluidization, microwave irradiation, electron beam irradiation, steam explosion, and electrospinning [16, 18].

5.2.2.2 Bottom-Up Process

BNC production by bacterial species is a bottom-up approach. BNC is chemically identical to biomass derived cellulose but is devoid of any hemicellulose, lignin, wax, etc. and have different macromolecular and physical characteristics. Bacteria produces cellulose by the following pathway:

- i) Glucose to Glucose-6-phospahte (phosphorylation by glucokinase),
- ii) Glucose-6-phosphate to Glucose-1-phosphate (isomerization by phosphoglucomutase),
- iii) Glucose-1-phosphate to UDP-Glucose (synthesis by UDPG-pyrophosphorylase),
- iv) UDP-Glucose to cellulose (polymerization by cellulose synthase).

Bacteria are cultured in media containing macro and micromolecules to produce BNC. Agricultural biomass or processing biowastes such as bagasse, molasses, pomace, corn steep liquor, and so on are used in industrial scale. Some of the media and culture techniques for BNC have been summarized in Table 5.2. The media composition and culture conditions affect the morphology, structural arrangement, and physicochemical character of BNCs. Cellulose-producing bacteria have been cultured by three methods: static, agitated, and bioreactor-based culturing. Static culture is carried out in media-filled containers at a specific temperature, pH, and time and produces BNC in the form of a hydrogel pellicle. Agitated cultures produce BNC with a lower degree of polymerization and crystallinity and with varied shapes: sphere, cocoon, pellet, or even irregular clumps. Mainly airlift or rotatory disc bioreactors have been employed for continuous bacterial culture for BNC production. When compared to static and

Table 5.2 BNC production from various microorganisms, biomass and culture mode.

Microorganism	Biomass	Culture mode	Yield (g/L)	Reference
A. xylinum BPR 2001	Molasses	Static/agitated	7.82	Kim *et al.* [32]
A. xylinum subsp. sucrofermentans BPR2001	Corn steep liquor	Airlift	3.8	Huang *et al.* [33]
Gluconacetobacter xylinus	Grape pomace extract, corn steep Liquor	Static	6.56	Cerrutti *et al.* [34]
Komagataeibacter xylinus BPR 2001	Molasses, corn steep liquor	Static	6.64	Rodrigues *et al.* [31]
Komagataeibacter saccharivorans MD1	Sugarcane molasses	Static	3.9	Abol-Fotouh *et al.* [35]
Kombucha starter culture	Beetroot peel	Static	11.57	Beluhan *et al.* [36]

agitated cultures, BNC produced by this process has lower crystallinity, elasticity, and degree of polymerization. Research on non-conventional media and media composition is being explored to optimize BNC production in terms of quality and quantity [29–31].

5.3 Nanocellulose-Based Nanocomposites

Because of its unique properties, nanocellulose - CNC, CNF, and BNC - has emerged as a promising nano-reinforcement material. They are natural, organic, and obtained from diverse, eco-friendly sources. Though using wood-based biomass as a source is not a green technology, using agro-industrial wastes as a source makes nanocelluloses sustainable and safe. Compared to conventional nanofillers (silica, metal oxides, glass fibers, nanotubes, etc.), they are non-toxic, biocompatible, cheap, recyclable, renewable, biodegradable, and environment-friendly [37].

5.3.1 Nanocellulose Properties

Based on the source, pretreatment and extraction method their morphology, size distribution, molecular arrangement and physicochemical properties vary. Owing to this huge diversity is available among nanocelluloses increasing their scope of applications in varied fields. Some of the comparable properties which make them good alternative to conventional nanofillers are [37, 38]:

- i) High aspect ratio,
- ii) High contact surface area,
- iii) High Young's modulus and tensile strength,
- iv) Crystallinity index,
- v) Flexibility,
- vi) High porosity,
- vii) High water uptake capacity,
- viii) Hygroscopic nature,
- ix) High dielectric properties,
- x) Good magnetic properties,
- xi) Good thermal properties (high thermal stability),
- xii) Light weight,
- xiii) Low density,
- xiv) Improved barrier properties, and
- xv) Optical transparency.

Apart from the above-mentioned properties, the presence of enormous hydroxyl groups in the nanocellulose structure allows extensive hydrogen bond formation and other interactions between the filler and matrix. This strengthens the nanocomposite material, as the nanocomposite properties depend on the nanofiller and matrix interface and interaction. Nanocelluloses in some cases show poor dispersion, which can be overcome by surface modifications. Surface modification and functionalization are possible due to the availability of hydroxyl groups. Some of the common chemical modifications are sulfate esters, ether linkage, carboxymethyl surfaces, urethane linkages, polymer grafting, phosphorylation, acetylation, and oligomeric silylated layer. Physical treatments (plasma, corona, flame, laser, ion beam, and hydrothermal treatments) and/or enzymatic treatments could also be used to functionalize the surface [17].

5.3.2 Nanocomposite Fabrication

Nanocelluloses are used as fillers in polymer-based composites made of both thermoplastics and thermosets. Polyolefins such as polypropylene (PP), high density polyethylene (HDPP), and polyolefin derivatives such as polystyrene (PS) were used to create conventional nanocomposites. Later, newer polymers have been explored, both synthetic and natural. Some of them are starch, soy protein, silk fibroin, polylactic acid (PLA), poly(hydroxyalkanoate) (PHA), cellulose acetate butyrate, regenerated cellulose, chitosan, poly(oxyethylene), poly(butylene adipate) (PBA), poly (ε-caprolactone) (PCL), poly(vinyl alcohol) (PVOH), and poly(vinyl chloride) (PVC). Likewise, thermoset epoxy and segmented polyurethane have been reinforced with nanocellulose [39]. A wide range of fabrication processes are available for these nanocomposites. The choice is determined by the type of filler, matrix, and desired final product properties. The desirable properties of these nanocomposites depend on the fabrication processes. Thus, optimization of the different production factors is important for achieving the required character in the final product. Solution casting, melt intercalation (mixing), impregnation, *in-situ* polymerization, and coating are common fabrication processes. Advanced methods such as electrospinning have also been investigated [17, 38]. This section discusses the various nanocellulose-based nanocomposite fabrication processes.

5.3.2.1 Solution Casting

Solution casting is an old technology developed to make plastic films. It has been widely used for the development of different composite materials, specifically bionanocomposites. This technique is simple, works at low to ambient temperatures, does not require specific equipment, and is efficient in both aqueous and non-aqueous solvents. The fabricated product, which is typically a film, has properties such as uniform thickness, optical purity, low haze, and isotropy. For polymers that are insoluble in water, the nanofiller dispersion solvent is exchanged with appropriate non-aqueous solvents. Followed by this, mixing with the polymer solutions is done to be cast into films. Though extensively used, they are mainly restricted to laboratory-scale fabrication due to the use of solvents, slow processing speeds, and additional cost incurred for solvent recovery [40].

This process requires dissolving the polymer in a solvent and the nanofiller too is dispersed either in the same solvent or different solvent. An additional step of sonication might be employed for even dispersion. The dissolved polymer and dispersed nanofiller is then mixed by stirring and casted on a glass plate or Teflon sheet. Some cases the nanofiller is directly added to the polymer solution and then casted. The casted solution is allowed to dry until the solvent evaporates to obtain the nanocomposite. To accelerate the process, the drying could be conducted under vacuum or at elevated temperatures [41].

Solution casting is used to create hydrophilic and hydrophobic polymer nanocomposites based on nanocellulose. In 2012, Savadekar and Mhaske [42] prepared thermoplastic starch (TPS)-NCF composite films. NCF was derived from cotton and dispersed in varying concentrations (0.1–1%) in distilled water by stirring at 1000 rpm. To the dispersed NCF, 5% TPS was added and stirred to complete homogeneity. They further heated the solution at 70° C for 30 minutes after adding glycerol and acetic acid for gelatinization. To obtain a uniform film, the solution was poured into an acrylic mold and allowed to dry at room temperature. Similarly, κ-carrageenan-NFC composite films were prepared using water as a solvent and glycerol as a plasticizer by Savadekar and coworkers [43]. Processing parameters maintained were continuous stirring at 1500 rpm followed by heating at 65°C. The films were cast on acrylic sheets and dried under circulating hot air at 45°C. Khoo *et al.* [44] developed nanocomposites of the hydrophobic polymer - poly(lactic acid) (PLA) using solution casting technique. In this case, they used chloroform as a solvent to dissolve PLA and CNC derived from microcrystalline cellulose by acid hydrolysis. The mixture was poured into glass petri dishes, and solvent evaporation was allowed for 24 hours at room temperature. Some researchers have made changes to the nanocomposite's composition and processing. A nanocomposite with wound healing capacity was created by combining chitosan, poly(vinyl pyrrolidone), and nanocellulose [45]. To improve the properties of natural rubber, incorporation of nanocellulose reinforcement was done using the solvent casting method with additional steps of ball milling and ultrasonication of the mixture of nanocellulose, latex, and crosslinker [46]. These modifications are aimed at achieving desirable dispersion of nanofillers, interaction with the matrix, homogeneity, and expected character in the nanocomposite for specific applications.

5.3.2.2 Melt Mixing

Melt mixing/intercalation/blending is another common and simple method of nanocomposite fabrication. Clay/polymer and semiconductor/polymer nanocomposites are the most commonly used. Nanoparticles are combined with molten thermoplastic polymers in this technique to allow polymer chains to intercalate between the nanofiller, improving polymer-reinforcement interactions. Nanofillers are mixed with molten polymers in the required ratio and extruded, often in the presence of an inert gas (argon, nitrogen, etc.). Alternatively, dry mixing of nanofillers and polymer could be done, followed by heating and mixing. In this case, high shear generated during mixing leads to intercalation [47, 48].

As solvents and chemicals are not required, the technique is environmentally friendly. It creates better interaction between the polymer and nanofiller compared to other techniques of fabrication. Melt mixing is more flexible, versatile, and compatible with other important industrial processes like extrusion and injection molding. The process can be modified by incorporating some additional techniques in case of incompatibility between polymer and filler, which makes it a flexible technique. All these account for the popularity of melt mixing in nanocomposite development, even at an industrial scale [17].

Recent melt mixing fabrications have investigated various methods to improve the dispersion of nanocelluloses in polymers, particularly hydrophobic ones. Re *et al.* [49] investigated the effect of wet feeding and PMMA latex on the properties of CNF and poly(-caprolactone) (PCL) nanocomposite fabrication using melt mixing. To improve upon the dispersion of NCF PMMA latex, which enhanced the mechanical properties. All of the constituents were dispersed in water before being dried under a fume hood to reduce the water content by 50%. This reduced mixture was subjected to melt blending by passing through a DSM twin-screw microcompounder at 120 °C at different feed rates and times. In another work, CNCs and reduced graphene oxide (rGO) reinforced a biodegradable PLA matrix through melt mixing and successfully produced nanocomposites with high tensile strength, non-toxicity, and biocompatibility [50]. Research on the development of an industrially feasible method for nanocomposites using melt mixing is in full swing. Polyamide 6 (PA 6)/CNF nanocomposites with a high CNF composition were processed by a high-speed kinetic mixer (GELIMAT®) followed by compression molding. Different ratios of CNF (5, 15, and 25 wt%) yielded products with high mechanical and thermal stability, making them suitable for high-performance applications [51]. Platnieks *et al.* [52] made a similar attempt to improve the efficiency of poly (butylene succinate) and CNF composites production. A masterbatch process strategy was proposed to reduce solvent usage, facilitate large-scale production, and improve the nanocomposite properties obtained through melt mixing. Additionally, combinations of various techniques are also being explored, like electrospinning with melt compounding [53].

5.3.2.3 In-Situ Polymerization

Solution casting gives good dispersion but only in thin films at small scales, whereas melt mixing can be scaled up to industrial scales but suffers from poor dispersion. Compared to these techniques, *in-situ* polymerization provides excellent dispersion of nano reinforcements in the bulk polymer. The uniform distribution of nanofillers allows for strong interaction between the matrix and filler (often in the form of covalent linkages), which results in a higher quality nanocomposite. Polymers unstable at higher temperatures and not suitable for melt mixing may be used for nanocomposite fabrication by this technique [54].

Nanofillers are dispersed in a single monomer or many monomers, and then the monomer is polymerized. The monomer is typically in a fluid or solution state, allowing for the dispersion of fillers and later polymerization to form the matrix. This leads to uniform distribution and uniform interaction between the polymer and filler. Curing agents, peroxides, and high temperatures are required for thermoset and thermoplastic polymerization [17].

A comparison of *in-situ* polymerization, melt mixing, and solution casting for CNC/nylon composites showed the best result in the case of *in-situ* polymerization. The nanocomposite had the best material property due to effective dispersal of the CNC, particle size, and morphology control [55]. Fujisawa and team [56] fabricated a PS/CNF nanocomposite with high thermostability, strength, and transparency. CNF was emulsified in a styrene solution by ultrasonication, followed by polymerization under stirring conditions at 70°C. Post-polymerization, the nanocomposite was precipitated by adding solvent (methanol and HCl), filtered, and vacuum dried. Similarly, PDA/NC nanocomposite was created by *in-situ* polymerization of dopamine (DA) monomer on the surface of nanocellulose. In this case, the precipitated nanocomposite was separated by centrifugation and freeze-dried. These nanocomposites had Cr ion chelating activity from waste water [57].

5.3.2.4 *Electrospinning*

As reinforcements, graphene, graphene oxide, metal (gold) nanoparticles, carbon nanotubes, hydroxyapatite, and other inorganic nanoparticles have been used to make a lot of electrospun nanocomposites. Research on nanocellulose-based nanocomposites with specific additives has been reported for biomedical applications [59]. Rahmat *et al.* [60] produced a nanocomposite of silane-grafted PLA and CNC using electrospinning that had enhanced mechanical strength. Silane-grafting was done to improve the crosslinking between CNC and PLA. The filtered suspension of CNC and PLA in DMF/chloroform was electrospun for 45 minutes with a horizontal setup at 15 kV, 1.5ml/h feed rate, room temperature, and 170mm distance. The fibers obtained were ethanol-treated and freeze-dried for storage. To achieve enhanced antimicrobial barrier properties in PLA-CNF composite, Vergara-Figueroa, and coworkers [61] carried out dual electrospinning techniques. Two different polymer blends (nZH-Cu-PLA and CNF-PLA) were injected into the electrospinning setup in a dual configuration with a rotary drum collector at the same time. Apart from the modification of composition, the electrospinning parameters can also be optimized to achieve desired properties in the nanocomposite. A CNC-based waterborne polyurethane (WPU)/poly(vinyl alcohol composite film having large porosity (62.61%) was developed by optimizing speed, spinning voltage, and curing distance [62].

5.3.2.5 *Other Fabrication Techniques*

Fabrication is also accomplished through the use of less common processes. Some examples include injection molding, sono-coprecipitation, and elastic kneading [63–65]. Among the most recent developments is the use of 3D printing technology to create nanocellulose-based nanocomposite materials. Nanocellulose is 3D printable, can combine with a variety of polymers, has excellent mechanical properties, and is renewable and biocompatible. Though applications in a variety of fields are anticipated, more research into cost reduction and making the 3D printing process more environmentally friendly is required [66].

5.4 Properties of Nanocellulose-Based Nanocomposites

Nanocellulose-based nanocomposites find varied applications in biomedicine, packaging, electronics, automobiles, aerospace, textiles, cosmetics, and the paper industry. Biomedical applications include tissue engineering, drug delivery, diagnostics, wound repair, and antimicrobial activity. They are suitable for food packaging due to their mechanical strength, barrier properties, non-toxicity, biocompatibility, and biodegradability.In the presence of suitable additives, they have been used for developing electrodes, electrolytes or separators, sensors, batteries, fuel cells, and supercapacitors. Moreover, they have been used for wastewater treatment, the removal and degradation of dyes, and the chelation of heavy metals [58, 67, 68]. All these applications are based on the unique set of properties of the developed nanocomposites. Some of these properties have been discussed in this section.

5.4.1 Mechanical Property

Mechanical properties of any material include strength, stiffness, ductility, brittleness, toughness, resilience, etc. In a nanocomposite, mechanical properties depend on the matrix, the nanofiller, and the interface between the nanofiller and the matrix. The mechanical properties measured are Young's modulus, tensile strength, and elongation at break. The blend of nanocellulose and polymer enhances the mechanical properties of the product. This is due to the inherent stiffness of the cellulose nanoparticles and the higher aspect ratio. A higher aspect ratio combined with a large hydrogen bonding capacity allows for extensive interactions between the polymer and nanocellulose, thereby strengthening the nanocomposite [38]. Likewise, better distribution of nanocellulose plays an important role. PS/CNF films prepared by *in-situ* polymerization with uniform distribution of components by sonication achieved higher Young's modulus (3.4 GPa), ultimate strength (50.1 MPa), and strain to failure (1.5%) compared to solvent-casted films [56]. The amount of nanocellulose in the blend has an impact on the mechanical properties. Mechanical strength gradually improves and reaches optimum as concentration increases. Beyond the optimum concentration, increasing the concentration does not improve mechanical properties because agglomeration between nanocellulose begins, affecting uniformity in size, distribution, and interactions, eventually leading to phase separation [43]. Surface modifications and the use of crosslinkers, which allow for better interactions, can provide additional strengthening effects [49].

5.4.2 Thermal Property

Packaging applications require good thermal properties in composite materials. It is equally important for the fabrication of other products aside from packaging materials. The common thermal properties like glass transition temperature (Tg), crystallization temperature (Tc), and melting temperature (Tm) are measured by differential scanning calorimetry (DSC), and decomposition temperature (Td), 10% weight loss temperature (T10), and maximum temperature (Tmax) are measured by thermogravimetric analysis (TGA). The thermal stability of TEMPO-CNF/polyethylene (PE) film was found to be better than that of only PE film. The storage moduli of nanocomposites were above 300° C, whereas for PE it was not measurable above 120° [64]. The ratio of nanocellulose in the nanocomposite imparted a similar effect on the thermal property as was observed in the case of the mechanical property. Khoo *et al.* [44] reported an increase in the T10 and Td of PLA with the increasing loading of CNC. The addition of a compatibilizer to PLA-CNF composites was found to increase the heat distortion temperature (HDT) by 16%. The main reason was proposed to be an improvement in the interaction between PLA and CNFs [69].

5.4.3 Barrier Property

Poor permeability to moisture, oxygen, carbon dioxide, flavors, etc. is required for food packaging materials. Although olefins are commonly used in this field, green alternatives are being investigated due to their non degradability. Because of their small size, higher aspect ratio, uniform distribution, and higher interactions, nanocellulose-based nanocomposites have good barrier properties. All these increase the tortuosity of the nanocomposites, decreasing their permeability. Though cellulose has a high moisture absorption capacity, a decrease in size and surface modification can reduce it significantly. Size reduction also reduces the gas permeability of nanocelluloses [70]. In a study, the PVA/nanocellulose nanocomposites had decreased moisture absorption capacity. The diffusion of water molecules was inhibited due to the extensive cross-linking and network formation between PVA and nanocellulose [71]. Barrier properties are equally important for wound dressing application. The water vapor transmission rate (WVTR) of poly(vinyl pyrrolidone) (PVP) decreased linearly to 1742.9 g/m2day when the nanocellulose was added up to 3 wt%, and then it slightly increased with an increase in the concentration of nanocellulose. The decrease occurred as a result of increased surface density and pore closure. While the oxygen permeability increased as the nanocellulose concentration increased [45].

5.4.4 Optical Property

The incorporation of nanofiller material alters the optical property of a matrix. This change might be desirable or undesirable based on the proposed application. Transmittance in the range of 200–1000 nm is measured and reported. The type of filler material can introduce absorption peaks at specific wavelengths. Carboxylated and sulfated nanocellulose show absorption peaks in the UV range, while the inclusion of ZnO to create a bionanocomposite added peaks at 390 and 366 nm [65]. Optical transparency is also affected by the dispersion and dense packing of the nanofiller, which might be due to the concentration of the nanofiller and/or the process of fabrication. This was observed in TPS/NCF films, where at low concentrations transparency increased, but at higher concentrations it decreased [42]. When a different fabrication process was used, Fujisawa *et al.* [56] found no significant difference in the transparency of PS/CNF nanocomposite film. However, Beck *et al.* [72] showed a significant effect of ultrasonication on the shift in reflectance pattern of NCC films.

5.4.5 Other Properties

Many other properties are important and are measured for characterization, like biodegradability, biocompatibility, and antimicrobial activity. Nanocellulose improves the biodegradability of nanocomposites, making them a more environmentally friendly alternative to traditional packaging materials. Abraham *et al.* [73] found an acceleration in the degradation rate of rubber with an increase in the concentration of nanocellulose. The presence of hydroxyl groups promotes microbial attack and, as a result, faster degradation. Biocompatibility and non-toxicity are important for biomedical applications. The cell viability test, cell culture, and MTT assay are used to assess it. Pal *et al.* [50] observed all nanocomposites of PLA/CNC and PLA/graphene modified CNC to be nontoxic and cytocompatible to HEK293 cells, making them suitable for tissue engineering. Another important property to be checked for biomedical applications is antimicrobial activity. High antimicrobial activity was observed in cellulose nanocrystal/ZnO bionanocomposites, ZnO/Cellulose nanocomposites, and chitosan-PVP-nanocellulose composites, indicating their potential use as wound dressings [45, 65, 74].

5.5 Conclusion

The advent of nanotechnology and environmental awareness has put an impetus on the development of green technology and sustainable materials. As the demand for organic, natural nano reinforcements in composite fabrication has increased, nanocellulose has proven to be a suitable candidate. Argo-industrial residues have been successfully utilized to extract nanocelluloses. Though the conventional methods of extraction are not eco-friendly and cost-effective, recent attempts at developing greener processes like enzymatic hydrolysis, ultrasonication, etc. can mitigate the undesirable effects. The source of cellulose and processing techniques impact the morphology, shape, mechanical, and physicochemical properties of the nanocellulose, bringing variety and flexibility in terms of application. Nanocomposite fabrication employs both traditional and cutting-edge techniques to create materials with enhanced desirable properties. Nanocellulose-based nanocomposites have gained importance and are an emerging material of interest in the fields of biomedical, packaging, electronics, optics, and remediation. The current challenge lies in the development of eco-friendly extraction and fabrication processes with cost efficiency and industrial scalability. Equally essential aspects are achieving uniform distribution of nanocelluloses, preventing agglomeration, compatibility with a wide range of matrixes, water resistance, and superior mechanical stability, all of which drastically affect the material quality. To commercialize green nanocomposites, thorough research, and development are required, which can bring in a multitude of possibilities.

References

1. Kalia, S., Dufresne, A., Cherian, B.M., Kaith, B.S., Avérous, L., Njuguna, J., Nassiopoulos, E., Cellulose-based bio- and nanocomposites: A review. *Int. J. Polym. Sci.*, 2011, 837875, 35, 2011.
2. Amir, S.M.M., Sultan, M.T.H., Jawaid, M., Ariffin, A.H., Mohd, S., Salleh, K.A.M., Ishak, M.R., Md Shah, A.U., Nondestructive testing method for Kevlar and natural fiber and their hybrid composites, in: *Woodhead Publishing Series in Composites Science and Engineering, Durability and Life Prediction in Biocomposites, Fibre-Reinforced Composites and Hybrid Composites*, M. Jawaid, M. Thariq, N. Saba (Eds.), pp. 367–388, Woodhead Publishing, United Kingdom, 2019.
3. Omanović-Mikličanin, E., Badnjević, A., Kazlagić, A., Hajlovac, M., Nanocomposites: A brief review. *Health Technol.*, 10, 51–59, 2020.

4. Shameem, M.M., Sasikanth, S.M., Annamalai, R., Raman, R.G., A brief review on polymer nanocomposites and its applications. *Mater. Today: Proc.*, 45, 2, 2536–2539, 2021.
5. Camargo, P.H.C., Satyanarayana, K.G., Wypych, F., Nanocomposites: Synthesis, structure, properties and new application opportunities. *Mater. Res.*, 12, 1, 1–39, 2009.
6. Hussain, F., Hojjati, M., Okamoto, M., Gorga, R.E., Review paper: Polymer-matrix nanocomposites, processing, manufacturing, and application: An overview. *J. Compos. Mater.*, 40, 17, 1511–1575, 2006.
7. Fu, S., Sun, Z., Huang, P., Li, Y., Hu, N., Some basic aspects of polymer nanocomposites: A critical review. *Nano Mater. Sci.*, 1, 2–30, 2019.
8. Rallini, M. and Kenny, J.M., Nanofillers in polymers, in: *Modification of Polymer Properties*, C.F. Jasso-Gastinel and J.M. Kenny (Eds.), pp. 47–86, Elsevier, United Kingdom, 2017.
9. Shankar, S. and Rhim, J.-W., Bionanocomposite films for food packaging applications, in: *Reference Module in Food Science*, pp. 1–10, 2018.
10. Jamróz, E., Kulawik, P., Kopel, P., The effect of nanofillers on the functional properties of biopolymer-based films: A review. *Polym.*, 11, 675, 2019.
11. Nasir, M., Hashim, R., Sulaiman, O., Asim, M., Nanocellulose: Preparation methods and applications, in: *Woodhead Publishing Series in Composites Science and Engineering Cellulose-Reinforced Nanofibre Composites*, M. Jawaid, S. Boufi, A.H.P.S. Khalil (Eds.), pp. 261–276, Woodhead Publishing, United Kingdom, 2017.
12. Trache, D., Tarchoun, A.F., Derradji, M., Hamidon, T.S., Masruchin, N., Brosse, N., Hussin, M.H., Nanocellulose: From fundamentals to advanced applications. *Front. Chem.*, 8, 392, 2020.
13. Kaur, P., Sharma, N., Munagala, M., Rajkhowa, R., Aallardyce, B., Shastri, Y., Agrawal, R., Nanocellulose: Resources, physio-chemical properties, current uses and future applications. *Front. Nanotechnol.*, 3, 747329, 2021.
14. Randhawa, A., Dutta, S.D., Ganguly, K., Patil, T.V., Patel, D.K., Lim, K.-T., A review of properties of nanocellulose, its synthesis, and potential in biomedical applications. *Appl. Sci.*, 12, 7090, 2022.
15. Ullah, M.W., Manan, S., Ul-Islam, M., Vasilyevich, R.V., Thomas, S., Yang, G., Introduction to nanocellulose, in: *Nanocellulose: Synthesis, Structure, Properties and Applications*, G. Yang, M.W. Ullah, Z. Shi (Eds.), pp. 1–50, World Scientific Publishing Europe Ltd, London, 2021.
16. Thomas, B., Raj, M.C., Athira, K.B., Rubiyah, M.H., Joy, J., Moores, A., Drisko, G.L., Sanchez, C., Nanocellulose, a versatile green platform: From biosources to materials and their applications. *Chem. Rev.*, 118, 11575–11625, 2018.
17. Norizan, M.N., Shazleen, S.S., Alias, A.H., Sabaruddin, F.A., Asyraf, M.R.M., Zainudin, E.S., Abdullah, N., Samsudin, M.S., Kamarudin, S.H., Norrrahim, M.N.F., Nanocellulose-based nanocomposites for sustainable applications: A review. *Nanomaterials*, 12, 3483, 2022.

18. Pradhan, D., Jaiswal, A.K., Jaiswal, S., Emerging technologies for the production of nanocellulose from lignocellulosic biomass. *Carbohydr. Polym.*, 285, 119258, 2022.
19. Wang, Q., Yao, Q., Liu, J., Sun, J., Zhu, Q., Chen, H., Processing nanocellulose to bulk materials: A review. *Cellulose*, 26, 7585–7617, 2019.
20. Gupta, G.K. and Shukla, P., Lignocellulosic biomass for the synthesis of nanocellulose and its eco-friendly advanced applications. *Front. Chem.*, 8, 601256, 2020.
21. Cherian, B.M., Leão, A.L., Ferreira de Souza, S., Thomas, S., Pothan, L.A., Kottaisamy, M., Isolation of nanocellulose from pineapple leaf fibres by steam explosion. *Carbohydr. Polym.*, 81, 720–725, 2010.
22. Baheti, V.K., Abbasi, R., Militky, J., Ball milling of jute fibre wastes to prepare nanocellulose. *World J. Eng.*, 9, 1, 45–50, 2012.
23. Phanthong, P., Karnjanakom, S., Reubroycharoen, P., Hao, X., Abudula, A., Guan, G., A facile one-step way for extraction of nanocellulose with high yield by ball milling with ionic liquid. *Cellulose*, 24, 2083–2093, 2017.
24. Wulandari, W.T., Rochliadi, A., Arcana, I.M., Nanocellulose prepared by acid hydrolysis of isolated cellulose from sugarcane bagasse. *IOP Conf. Ser.: Mater. Sci. Eng.*, 107, 012045, 2016.
25. Niu, F., Mengya., L., Huang, Q., Zhang, X., Pan, W., Yang, J., Li, J., The characteristic and dispersion stability of nanocellulose produced by mixed acid hydrolysis and ultrasonic assistance. *Carbohydr. Polym.*, 165, 197–204, 2017.
26. Aditiawati, P., Dungani, R., Amelia, C., Enzymatic production of cellulose nanofibers from oil palm empty fruit bunch (efb) with crude cellulase of *Trichoderma* sp. *Mater. Res. Express*, 5, 034005, 2018.
27. Qian, M., Lei, H., Villota, E., Zhao, Y., Wang, C., Huo, E., Zhang, Q., Mateo, W., Lin, X., High yield production of nanocrystalline cellulose by microwave-assisted dilute-acid pretreatment combined with enzymatic hydrolysis. *Chem. Eng. Process.: Process Intensif.*, 160, 108292, 2021.
28. Raju, V., Revathiswaran, R., Subramanian, K.S., Parthiban, K.T., Chandrakumar, K., Anoop, E.V., Chirayil, C.J., Isolation and characterization of nanocellulose from selected hardwoods, viz., *Eucalyptus tereticornis* Sm. and *Casuarina equisetifolia* L., by steam explosion method. *Sci. Rep.*, 13, 1, 1199, 2023.
29. Corujo, V.F., Cerrutti, P., Foresti, M.L., Vázquez, A., Production of bacterial nanocellulose from non-conventional fermentation media, in: *Multifunctional Polymeric Nanocomposites Based on Cellulosic Reinforcements*, D. Puglia, E. Fortunati, J.M. Kenny (Eds.), pp. 39–59, Elsevier, United Kingdom, 2016.
30. Barja, F., Bacterial nanocellulose production and biomedical applications. *J. Biomed. Res.*, 35, 4, 310–317, 2021.
31. Rodrigues, A.C., Fontão, A.I., Coelho, A., Leal, M., Soares da Silva, F.A.G., Wan, Y., Dourado, F., Gama, M., Response surface statistical optimization of bacterial nanocellulose fermentation in static culture using a low-cost medium. *N. Biotechnol.*, 49, 19–27, 2019.

32. Kim, S., Li, H., Oh, I., Kee, C., Kim, M., Effect of viscosity-inducing factors on oxygen transfer in production culture of bacterial cellulose. *Korean J. Chem. Eng.*, 29, 792–797, 2012.
33. Huang, L., Chen, X., Nguyen, T.X., Tang, H., Zhang, L., Yang, G., Nanocellulose 3D-networks as controlled-release drug carriers. *J. Mater. Chem. B*, 1, 2976–2984, 2013.
34. Cerrutti, P., Roldán, P., García, R.M., Galvagno, M.A., Vázquez, A., Foresti, M.L., Production of bacterial nanocellulose from wine industry residues: Importance of fermentation time on pellicle characteristics. *J. Appl. Polym. Sci.*, 133, 14, 9, 2016.
35. Abol-Fotouh, D., Hassan, M.A., Shokry, H., Roig, A., Azab, M.S., Kashyout, A.E.B., Bacterial nanocellulose from agro-industrial wastes: Low-cost and enhanced production by *Komagataeibacter saccharivorans* MD1. *Sci. Rep.*, 10, 1, 3491, 2020.
36. Beluhan, S., Herceg, F., Leboš Pavunc, A., Djaković, S., Preparation and structural properties of bacterial nanocellulose obtained from beetroot peel medium. *Energies*, 15, 9374, 2022.
37. Singla, R., Guliani, A., Kumari, A., Yadav, S.K., Nanocellulose and nanocomposites, in: *Nanoscale Materials in Targeted Drug Delivery, Theragnosis and Tissue Regeneration*, S. Yadav (Ed.), Springer, Singapore, 2016, https://doi.org/10.1007/978-981-10-0818-4_5.
38. Mondal, S., A review on nanocellulose polymer nanocomposites. *Polym. Plast. Technol. Eng.*, 57, 13, 1377–1391, 2017.
39. Saba, N. and Jawaid, M., Recent advances in nanocellulose-based polymer nanocomposites, in: *Woodhead Publishing Series in Composites Science and Engineering, Cellulose-Reinforced Nanofibre Composites*, M. Jawaid, S. Boufi, A.H.P.S. Khalil (Eds.), pp. 89–112, Woodhead Publishing, United Kingdom, 2017.
40. Mathew, A.P. and Oksman, K., Processing of bionanocomposites: Solution casting, in: *Materials and Energy Handbook of Green Materials*, K. Oksman, A.P. Mathew, A. Bismarck, O. Rojas, M. Sain (Eds.), pp. 35–52, World Scientific, Singapore, 2014.
41. Das, R., Pattanayak, A.J., Swain, S.K., Polymer nanocomposites for sensor devices, in: *Woodhead Publishing Series in Composites Science and Engineering, Polymer-based Nanocomposites for Energy and Environmental Applications*, M. Jawaid and M.M. Khan (Eds.), pp. 205–218, Woodhead Publishing, United Kingdom, 2018.
42. Savadekar, N.R. and Mhaske, S.T., Synthesis of nano cellulose fibers and effect on thermoplastics starch based films. *Carbohydr. Polym.*, 89, 146–151, 2012.
43. Savadekar, N.R., Karande, V.S., Vigneshwaran, N., Bharimalla, A.K., Mhaske, S.T., Preparation of nano cellulose fibers and its application in kappa-carrageenan based film. *Int. J. Biol. Macromol.*, 51, 1008–1013, 2012.

44. Khoo, R.Z., Ismail, H., Chow, W.S., Thermal and morphological properties of poly (lactic acid)/nanocellulose nanocomposites. *Proc. Chem.*, 19, 788–794, 2016.
45. Poonguzhali, R., Khaleel Bashab, S., Kumari, S.V., Synthesis and characterization of chitosan-PVP-nanocellulosecomposites for *in-vitro* wound dressing application. *Int. J. Biol. Macromol.*, 105, 111–120, 2017.
46. Thomas, M.G., Abraham, E., Jyotishkumar, P., Maria, H.J., Pothen, L.A., Thomas, S., Nanocelluloses from jute fibers and their nanocomposites with natural rubber: Preparation and characterization. *Int. J. Biol. Macromol.*, 81, 768–77, 2015.
47. Rane, A.V., Kanny, K., Abitha, V.K., Thomas, S., Methods for synthesis of nanoparticles and fabrication of nanocomposites, in: *Micro and Nano Technologies, Synthesis of Inorganic Nanomaterials*, S.M. Bhagyaraj, O.S. Oluwafemi, N. Kalarikkal, S. Thomas (Eds.), pp. 121–139, Woodhead Publishing, United Kingdom, 2018.
48. Kango, S., Kalia, S., Thakur, P., Kumari, B., Pathania, D., Semiconductor–polymer hybrid materials. *Adv. Polym. Sci.*, 267, 283–312, 2015.
49. Re, G.L., Engström, J., Wu, Q., Malmström, E., Gedde, U.W., Olsson, R.T., Berglund, L., Improved cellulose nanofibril dispersion in melt-processed polycaprolactone nanocomposites by a latex-mediated interphase and wet feeding as LDPE alternative. *Appl. Nano Mater.*, 1, 2669–2677, 2018.
50. Pal, N., Banerjee, S., Roy, P., Pal, K., Melt-blending of unmodified and modified cellulose nanocrystals with reduced graphene oxide into PLA matrix for biomedical application. *Polym. Adv. Technol.*, 30, 12, 3049–3060, 2019.
51. Sridhara, P.K. and Vilaseca, F., High performance PA 6/cellulose nanocomposites in the interest of industrial scale melt processing. *Polym.*, 13, 9, 1495, 2021.
52. Platnieks, O., Sereda, A., Gaidukovs, S., Thakur, V.K., Barkane, A., Gaidukova, G., Filipova, I., Ogurcovs, A., Fridrihsone, V., Adding value to poly (butylene succinate) and nanofibrillated cellulose-based sustainable nanocomposites by applying masterbatch process. *Ind. Crops Prod.*, 169, 113669, 2021.
53. Martínez-Sanz, M., López-Rubio, A., Villano, M., Oliveira, C., Majone, M., Reis, M., Lagaron, J.M., Production of bacterial nanobiocomposites of polyhydroxyalkanoates derived from waste and bacterial nanocellulose by the electrospinning enabling melt compounding method. *J. Appl. Polym. Sci.*, 133, 14, 2015.
54. Grothe, J., Kaskel, S., Leuteritz, A., Nanocomposites and hybrid materials, in: *Polymer Science: A Comprehensive Reference*, K. Matyjaszewski and M. Möller (Eds.), pp. 172–209, Elsevier, United Kingdom, 2012.
55. Motaung, T.E., Mochane, M.J., Linganiso, Z.L., Mashigo, A.P., *In-situ* polymerization of nylon-cellulose nano composite. *Polym. Sci.*, 3, 1–8, 2017.
56. Fujisawa, S., Togawa, E., Kuroda, K., Facile route to transparent, strong, and thermally stable nanocellulose/polymer nanocomposites from an aqueous pickering emulsion. *Biomacromolecules*, 18, 1, 266–271, 2017.

57. Alzahrani, F.M., Amari, A., Katubi, K.M., Alsaiari, N.S., Tahoon, M.A., The synthesis of nanocellulose-based nanocomposites for the effective removal of hexavalent chromium ions from aqueous solution. *Open Chem.*, 20, 970–983, 2022.
58. Kausar, A., Polymeric nanocomposite via electrospinning: Assessment of morphology, physical properties and applications. *J. Plast. Film Sheeting*, 37, 1, 70–92, 2021.
59. Teixeira, M.A., Paiva, M.C., Amorim, M.T.P., Felgueiras, A.H.P., Electrospun nanocomposites containing cellulose and its derivatives modified with specialized biomolecules for an enhanced wound healing. *Nanomaterials*, 10, 3, 557, 2020.
60. Rahmat, M., Karrabi, M., Ghasemi, I., Zandi, M., Azizi, H., Silane crosslinking of electrospun poly (lactic acid)/nanocrystalline cellulose bionanocomposite. *Mater. Sci. Eng. C Mater. Biol. Appl.*, 68, 397–405, 2016.
61. Vergara-Figueroa, J., Alejandro-Martin, S., Cerda-Leal, F., Gacitúa, W., Dual electrospinning of a nanocomposites biofilm: Potential use as an antimicrobial barrier. *Mater. Today Commun.*, 25, 101671, 2020.
62. Ge, L., Yin, J., Yan, D., Hong, W., Jiao, T., Construction of nanocrystalline cellulose-based composite fiber films with excellent porosity performances via an electrospinning strategy. *ACS Omega*, 6, 4958–4967, 2021.
63. Wei, L., Luo, S., Mcdonald, A., Agarwal, U., Hirth, K., Matuana, L., Sabo, R., Stark, N., Preparation and characterization of the nanocomposites from chemically modified nanocellulose and poly(lactic acid). *J. Renew. Mater.*, 5, 5, 410–422, 2017.
64. Noguchi, T., Niihara, K., Iwamoto, R., Matsuda, G., Endo, M., Isogai, A., Nanocellulose/polyethylene nanocomposite sheets prepared from an oven-dried nanocellulose by elastic kneading. *Compos. Sci. Technol.*, 207, 108734, 2021.
65. Dawwam, G.E., Al-Shemy, M.T., El-Demerdash, A.S., Green synthesis of cellulose nanocrystal/ZnO bio-nanocomposites exerting antibacterial activity and downregulating virulence toxigenic genes of food-poisoning bacteria. *Sci. Rep.*, 12, 1, 16848, 2022.
66. Ee, L.Y. and Li, S., Recent advances of 3D printing nanocellulose: Structure, preparation and application prospects. *Nanoscale Adv.*, 3, 1167–1208, 2021.
67. Ganguly, K. and Lim, K.-T., Nanocellulose-based polymer nanohybrids and nanocomposite applications, in: *Micro and Nano Technologies, Multifunctional Hybrid Nanomaterials for Sustainable Agri-Food and Ecosystems*, K.A. Abd-Elsalam (Ed.), pp. 485–504, Elsevier, Netherlands, 2020.
68. Xu, T., Du, H., Liu, H., Liu, W., Zhang, X., Si, C., Liu, P., Zhang, K., Advanced nanocellulose-based composites for flexible functional energy storage devices. *Adv. Mater.*, 33, 2101368, 2021.
69. Ghasemi, S., Behrooz, R., Ghasemi, I., Yassar, R.S., Long, F., Development of nanocellulose-reinforced PLA nanocomposite by using maleated PLA (PLA-g-MA). *J. Thermoplast. Compos. Mater.*, 31, 1–12, 2017.

70. Dufresne, A., Nanocellulose: A new ageless bionanomaterial. *Mater. Today*, 16, 6, 220–227, 2013.
71. Mandal, A. and Chakrabarty, D., Studies on the mechanical, thermal, morphological and barrier properties of nanocomposites based on poly(vinyl alcohol) and nanocellulose from sugarcane bagasse. *J. Ind. Eng. Chem.*, 20, 462–473, 2014.
72. Beck, S., Bouchard, J., Berry, R., Controlling the reflection wavelength of iridescent solid films of nanocrystalline cellulose. *Biomacromolecules*, 12, 1, 167–172, 2011.
73. Abraham, E., Elbi, P.A., Deepa, B., Jyotishkumar, P., Pothen, L.A., Narine, S.S., Thomas, S., X-ray diffraction and biodegradation analysis of green composites of natural rubber/nanocellulose. *Polym. Degrad. Stab.*, 97, 2378–2387, 2012.
74. Lefatshe, K., Muiva, C.M., Kebaabetswe, L.P., Extraction of nanocellulose and *in-situ* casting of zno/cellulose nanocomposite with enhanced photocatalytic and antibacterial activity. *Carbohydr. Polym.*, 164, 301–308, 2017.

6

Surface Functionalization Process: Its Advantages and Disadvantages

Pranabi Maji[1]* and Dibyarupa Pal[2]

[1]*Department of Physics, JIS University, Kolkata, India*
[2]*Department of Biosciences, JIS University, Kolkata, India*

Abstract
Cellulose, the most abundant natural polymer on Earth, is an excellent performer for the fabrication of sustainable materials due to its low cost, easy availability, durability, strength, nontoxicity, renewability, and biocompatibility. However, due to poor solubility, an imbalance in hydrophilic and hydrophobic constituents, and poor surface energy between matrix and fiber, large-scale exploitation is hindered. Extracting cellulose at the nanoscale dimension by various chemical, enzymatic, and physical approaches can scale up the process. Hence, various surface modification processes are applied through the expansion and creation of new surface molecules to reinforce its features. In this context, the chapter highlights the advantages and disadvantages of the different surface modification processes that have enhanced the properties of cellulose.

Keywords: Nanocellulose, surface modification, advantages, disadvantages

6.1 Chemical Approach

The nanocellulose's high surface-to-volume ratio facilitates surface functionalization or other -OH chemical operations via various chemistries. To balance the hydrophilic–hydrophobic balance and enhance interfacial properties, this chemical modification is necessary. Therefore, various chemical modification techniques have been developed for their

**Corresponding author*: pranabi.ism@gmail.com

Mainak Mukhopadhyay and Debalina Bhattacharya (eds.) Nanocellulose: A Biopolymer for Biomedical Applications, (141–158) © 2024 Scrivener Publishing LLC

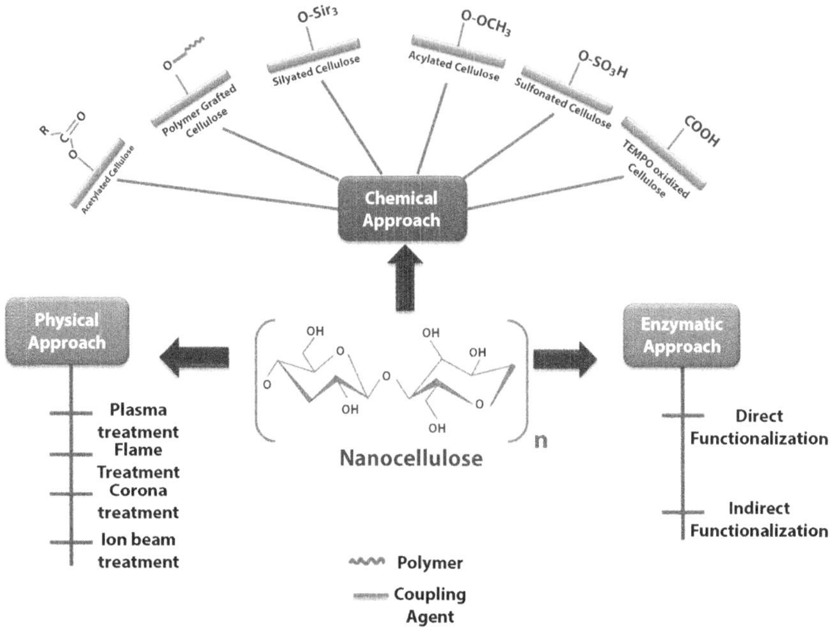

Figure 6.1 Schematic diagram for surface modification of nanocellulose.

sophisticated application, but some of the most important ones, which we have mentioned in this chapter, are acetylation/esterification, polymer grafting, silylation, acylation/alkanoylation, sulfonation, and TEMPO-mediated oxidation (Figure 6.1).

6.1.1 Acetylation/Esterification

Acetylation is the process used to introduce -$COCH_3$ onto the surface of nanocellulose. It is the most extensively used esterification reaction process, as this is the simplest technique to reduce the hydroxyl density on the surface of nanocellulose. There are two routes for acetylation: one homogeneous and one heterogeneous, in which a small amount of either sulfuric acid or perchloric acid is used as a catalyst along with a mixture of dry acetic acid and acetic anhydride (Ac2O) as an acyl donor. The homogeneous route is more advantageous in terms of excellent product recovery, reproducibility, and better control over the degree of substitution (DS), i.e., the average number of replaced hydroxyl groups per repeating unit. Depending on the reaction conditions, the theoretical maximum possible DS value is 3.0. Acetylated nanocellulose with DS value 1 disperses better in polar solvents; however, DS values closer to 3.0 disperse better in nonpolar solvents. Modified nanocellulose with a greater DS value can interact as a filler in the hydrophobic matrices to increase its mechanical strength. Thus, modifying the DS value by varying the ratio of acetic anhydride and acetic acid is an essential factor for numerous applications of nanocellulose. The surface functionalization by acetylation process was simplified by introducing the "one-pot reaction," a novel single-step reaction method, by Barun and Dorgan [1]. The authors were able to simultaneously hydrolyze and functionalize the surface of nanocellulose with the Fischer esterification process using a mixture of acetic acid and HCl. This reaction method offers the freedom to control the length of the cellulose by using different types of organic solvents. CNFs with poly(lactic acid) (PLA) filler content are reported to have a significant reduction in oxygen and water vapor transfer rate, which is critical for food packaging applications [2–4]. But only 0.3% of celluloses are soluble in the conventional solvent; others are not soluble because of their strong hydrogen bonding network and their ordered structure. So, for the last decade, the search for effective and eco-friendly green solvents (ionic solvents) has been a major focus. The main advantage of this process is that the ionic green solvent can be reprocessed for the next cycle of functionalization. After surface functionalization, the materials are more porous and maintain their nanoscale dimension [5].

6.1.2 Polymer Grafting

Nanocellulose polymer grafting is a versatile and effective process for surface functionalization. This reaction occurs when a small organic molecule or a long macromolecule reacts with the reactive functional group present on the surface of nanocellulose. Achieving a highly stable suspension of nanocellulose in a polar solvent is the major goal of this approach. This process can be classified by two techniques: "grafting from" and "grafting onto". In the "grafting from" technique, nanocellulose is functionalized by an initiator to initiate the polymerization on its surface. The main advantage of this technique is that it produces high-density nanocellulose, but the main disadvantage is that the resulting materials are difficult to characterize and determine their molecular weight. The "grafting onto" technique uses a polymer that has a reactive end group and a covalent bond with the coupling agent on the surface of nanocellulose. This technique gets more scientific and technological interest due to its simplicity and limiting grafting of the functional group [6]. Lizundia *et al.* (2016) [7] describe a technique for successfully grafting L-Lactide monomer with the catalyst Tin (II) ethyl hexanoate ($Sn(Oct)_2$). Similarly, Chuensangjun *et al.* (2019) [8] reported the synthesis of 2,2,6,6-tetramethylpiperidine-1-oxyl (TEMPO) oxidized cellulose nanofibers (TOCNs), which were used to create TOCN-graft-poly(lactic acid) (PLA) nanocomposites with high crystallinity using ring-opening polymerized L-Lactide in a polar solvent. In both studies, they discovered that the CNC-to-L-Lactide monomer ratio had a significant impact on the optical, thermal, and mechanical properties of the materials, as well as their grafting effectiveness. A team of researchers reported that It is feasible to graft natural rosin, a product from pine resin, onto the surface of CNC [9].

6.1.3 Silylation

Another commonly used surface modification technique is silylation, where silyl groups (R_3Si) are added to the CNC's surface. At room temperature, the silyl groups have a strong affinity with the hydroxyl group present on the surface of nanocellulose and are considered effective silane coupling agents. Hexamethyldisilazane (HDMS), 3-methacryloxy-propyltrimethoxysilane (MPS), 3-isocyanatepropyltriethoxysilane (IPTS), N-(-aminoethyl)--aminopropyl-trimethoxysilane (AEAPTMS), 3-glycidoxypropyltrimethoxysilane (GPTMS), and 3-aminopropyl-Tri (APTES) serve as silane coupling agents for stable surface functionalization (see below). This silane coupling agent reduces the hydroxyl group on the nanocellulose surface and changes the hydrophilic character of the material into a hydrophobic state. These also enhance the interfacial strength of the natural cellulose polymer. Indarti *et al.* (2020) [10] reported the effect of temperature on the silylated nanocellulose using oil palm empty fruit bunch (OPEFB) pulp and 3-aminopropyltriethoxysilane (APS). The FTIR data analysis revealed that the optimum temperature for modification is 80°C. Chandrashekar *et al.* (2022) [11] synthesized biobased polyurethane nanocellulose nanocomposites from cottonseed oil. The polyurethane matrix was then filled with silylated nanocellulose derived from pineapple leaves for Rhodamine B dye adsorption from a water medium. They found that the incorporation of silylated nanocellulose increased the adsorption efficiency. However, the solvent exchange technique and the requirement on organic solvents of this conventional functionalization process are one of its key drawbacks. It has been demonstrated that silylated nanocellulose with DS values ranging from 0.6 to 1 increases dispersibility in low-polarity organic solvents while retaining morphological integrity.

6.1.4 Acylation/Alkanoylation

In this modification process, an acyl group is introduced to the chemical structure of nanocellulose. Acyl chlorides, a regularly used acylation agent for cellulose, are considered undesirable owing to their negative impact on the characteristics of the final cellulose products since strong acid (HCl) is needed in this procedure [12]. Therefore, alkenyl succinic, maleic, or phthalic anhydrides are used to introduce acylation in cellulose. The main advantage of this technique is that the cellulose whiskers exhibit improved dispersivity in solvents with low polarities, such as dimethyl sulfoxide, dichloromethane, 1,4-dioxane, alcohols, and chloroform following modification. The properties may be further enhanced by the addition of heat. With the incremental DS value, trifluoroacetic anhydride and various fatty acids acylated nanocellulose can destroy the influence of moisture absorption on its surface [13]. Moreover, the crystallinity and zeta potential value may be enhanced using this modification technique. Sun *et al.* (2023) [14] modified the tunicate nanocelluloses (TNCs) with acetic anhydride and (3-aminopropyl) triethoxysilane, respectively, and used them to fabricate composite filter media with glass fiber substrate. This modified cellulose increased the filtration efficiency and decreased the air resistance. *Escherichia coli* (*E. coli*) and *Staphylococcus aureus* (*S. aureus*) bacteria were also found to be resistant to the antibacterial properties of acetylated nanocellulose.

6.1.5 Sulfonation

It is another functionalization technique to fill the surface of nanocellulose with anionic surface charge. It is a two-step modification technique. First, cellulose is oxidized using sodium periodate, and then oxidized nanocellulose is sulfonated to produce sodium bisulfite. This modification technique improves the water retentivity and solubility of nanocellulose. The sulfonated nanocellulose shows better divalent cation adsorption compared to unmodified nanocellulose with a high pH value. Adsorption affinity is primarily determined by the electronegativity of the materials. Since the reaction is sensitive to experimental conditions, including temperature, duration, and concentration of acid, it is difficult to control the quantity of sulfate grafting onto sulfuric acid-hydrolyzed nanocellulose. For environmental implications, green sulfonation of nanocellulose is an environmentally friendly approach.

6.1.6 TEMPO-Mediated Oxidation

This is an important and common functionalization technique. The main advantage of this technique is that it increases the stability of nanocellulose in aqueous suspension by imbuing it with negative charges on its surface. 2,2,6,6-Tetramethyl piperidine-1-oxyl (TEMPO) is used in a standard TEMPO modification process in the presence of NaBr and NaClO. This approach involves dissolving a catalytic amount of TEMPO and NaBr in polysaccharide solution at a pH range of 10–11, followed by the addition of a primary oxidant, such as NaClO solution, to begin the oxidation process. This technique reduces the crystallinity of nanocellulose. Half of the hydroxymethyl group in nanocellulose can be modified because of its higher degree of crystallinity. Because chemicals are used, this modification technique is regarded as expensive and non-ecofriendly [6]. Using TEMPO/NaBr/NaclO oxidation, Fang *et al.* (2014) [15] were able to construct a highly optically transparent and highly hazing solar cell by weakening the hydrogen bonds of the nanocellulose and introducing a carboxyl group, which caused swelling and led to the formation of a highly dense film with excellent optical properties. A novel method for cellulose solubilization was described by Wang *et al.* (2022) [16], combining 2,2,6,6-tetramethylpiperidine-1-oxyl radical (TEMPO)-mediated oxidation with a zincate–NaOH system. $Zn(OH)_4^{2-}$ was used as a stabilizing agent to reduce agglomeration and increase cellulose dissolution. Barbash *et al.* (2022) [17] reported that from organosolv reed pulp (ORP), TEMPO-mediated nanocellulose followed by ultrasonication treatment increased density, tensile strength, carboxyl group content, and optical transparency but decreased crystallinity.

6.2 Enzymatic Approach

Several enzymatic approaches involve direct functionalization methods like phosphorylation via hexokinases, polymerization via laccases, esterification via lipases, and indirect enzymatic approaches like oxidation and grafting to produce new cellulose derivatives with enhanced properties (Figure 6.1).

6.2.1 Direct Enzymatic Approach

6.2.1.1 Phosphorylation via Hexokinases

Hexokinases catalyze the transfer of the phosphate group to the regioselective C-6 group where cellulose phosphate esters are produced. The benefit of this regiospecific modification ensures more efficient cellulose processing under various conditions. Tzanov *et al.* (2002) [18] suggested using baker's yeast hexokinase enzyme to phosphorylate cotton cellulose's primary hydroxyl groups. Phosphorylated cellulose has the advantage of selective adsorption of different metal ions like Ag+, Cu2+, and Fe3+ from industrial effluents with a high sorption capacity. Furthermore, phosphorylated cellulose has significantly higher thermal stability and flame resistance than native cellulose. Phosphorylated cellulose has significant biomedical applications, like providing a suitable matrix for bone regeneration compared to native cellulose [19]. The advantages of enzymatic hydrolysis are the nonproduction of toxic components in the process, which saves the cost of utilizing detoxification inhibitors. Besides, with high substrate and enzyme concentrations, a higher yield is expected. The high cost of ATP limits scaling up, and a high and dry biomass concentration is required to initiate enzymatic hydrolysis [20].

6.2.1.2 Cellulose Acylation/Esterification by Hydrolases

Cellulose esters are an important group of biopolymers that are made by reacting the primary hydroxyl group of small sugar molecules with acyl groups. The proteases of Subtilisin Carlsberg (*Bacillus licheniformis*) and *Aspergillus niger* were used to make cellulose propionate and acrylate [21]. Gremos *et al.* (2011) [22] made the polysaccharide chains accessible by unwrapping the cellulose structure before pretreatment with different hydrolases. The enzyme treatment improves the fiber's flexibility, tensile strength, and smoothness, as well as its water absorption capacity and fibrillation [23]. The mixture of cellulase and xylanase also amended pulp properties like reduced fiber length, better tear strength, clarity, and defibrillation [24]. The disadvantage is that excessive enzyme use can reduce pulp strength and wear out the fiber surface [25].

6.2.1.3 Esterification via Lipases

Catalyzation by laccases, an oxidative enzyme, involves grafting reactions with lignin and surface modification of dodecyl gallate, a hydrophobic component present in unbleached cellulose [26, 27]. Several other reactions involving cellulose hydrophobization involve the coupling of short nonpolar chains on cellulose surfaces [28, 29]. During enzymatic reactions, complex intermediate polymers are formed, and the medium's viscosity rises, impairing bioreactor operation. Another disadvantage caused by the accumulation of by-products such as lignin is enzyme deactivation.

6.2.2 Indirect Enzymatic Approach

6.2.2.1 Surface Modification via Xyloglucan Endotransglucosylase

Xyloglucan is a main part of hemicellulose. It binds to cellulose without making a chemical bond, and endotransglucosylase or endohydrolase activities change the backbone of xyloglucan. Cross-linking of xyloglucan between cellulose microfibrils results in cellulose/xyloglucan composites, which increase the stiffness of the cellulose matrix. Further cellulose/xyloglucan composites interacting with xyloglucan endotransglucosylase enhance plant freezing tolerance during freeze–thaw cycles [30].

6.2.2.2 Grafting of Active Molecules

Grafting is most often used because it exposes the most reactive binding sites, which increases the efficiency of sorption. It also has ion-exchange properties and makes more functional groups available.

6.3 Physical Techniques

The physical methods include plasma, flame, corona, and ion beam treatments, which are used for surface modification of nanocellulose (Figure 6.1).

6.3.1 Plasma Treatment

This technique utilizes ionized gasses like free ions, electrons, and radicals generated from micro- and radiowaves or an extremely hot filament discharge. This technique involves the direct or indirect introduction of new functional groups on the cellulose surface. The advantage of the plasma technique is that no aqueous or chemical solvent is required, which minimizes the cost of chemicals and water. As a result, it is a more energy-efficient technique than other surface modification methods. Plasma application is limited to 100 nm or less from the surface; thus, no significant damage to cellulose occurs. The limitations faced are the extreme high temperatures of this treatment, which damage cellulose fibers but can be overcome by the application of the cold plasma technique. Thus, technical knowledge to choose the right kind of plasma application is highly recommended for the operation. Moreover, this technology can

Table 6.1 Summary of advantages and disadvantages of different surface modification techniques for nanocellulose.

Modification approach	Modification techniques	Advantages	Disadvantages	References
Chemical Approach	Acetylation/ Esterification	• Simple and effective technique • Reduce the hydroxyl density on to the surface of nanocellulose by the introduction of acetyl group • Excellent product recovery and better control over degree of substitution (DS) • Disperse in both polar and nonpolar solvent depending on DS value	• Very difficult to control the esterification degree • Only 0.3% celluloses are soluble in conventional solvent due to their ordered structure and strong hydrogen bonding network	[1–5]
	Polymer grafting	• Limiting grafting of functional group [6] • Achieving a highly stable suspension of nanocellulose in polar solvent[6] • Produce high density Nanocellulose [6]	• Difficulties in characterizing • Determining the molecular weight of the resulting materials with high density [6]	[6]

(Continued)

Table 6.1 Summary of advantages and disadvantages of different surface modification techniques for nanocellulose. (*Continued*)

Modification approach	Modification techniques	Advantages	Disadvantages	References
	Silylation	• Reduce some of the hydroxyl group into silent group carbomethyl group • Improve interfacial strength • Increased dispersibility in low polarities organic solvents (DS value 0.6 to 1) • Preserves its morphological integrity	• Complicated solvent exchange technique • Requirement of organic solvents	[10,11]
	Acylation/ Alkanoylation	• 1 stage Modification • Reduce some of the hydroxyl group into carboxyl group • The cellulose whiskers exhibit improved dispersivity in the solvents of low polarity • Destroy the influence of moisture absorption on the surface of nanocellulose • Crystallinity and zeta potential value enhanced	• Negative impact on the characteristics of the final cellulose products due to the use of strong acid HCl	[12–14]

(*Continued*)

Table 6.1 Summary of advantages and disadvantages of different surface modification techniques for nanocellulose. (*Continued*)

Modification approach	Modification techniques	Advantages	Disadvantages	References
	Sulfonation	• 2 stage modification • Reduce the hydroxyl group into sulfonic group • Increase water retentivity and solubility	• The reaction is sensitive to experimental condition, including as temperature, duration, and concentration of acid • Difficult to exact control over the quantity of sulfate grafting on to sulfuric acid-hydrolyzed nanocellulose	[6]
	TEMPO mediated oxidation	• 2,2,6,6-tetramethyl piperidine-1-oxyl (TEMPO) is the agent • Reduce some of the hydroxyl group into sodium carboxylate group • Intermediate stage for many modification • Improves stability of nanocellulose aqueous suspension	• Expensive and non-ecofriendly • Reduces the crystallinity of nanocellulose	[15–17]

(*Continued*)

Table 6.1 Summary of advantages and disadvantages of different surface modification techniques for nanocellulose. (*Continued*)

Modification approach	Modification techniques	Advantages	Disadvantages	References
Enzymatic Approach	Direct Functionalization	• Stereospecific and region-specific modifications • Enhanced hydrophobicity, thermal stability and flame-resistance properties • No toxic components produced	• High cost for enzyme preparation and purification • Complex intermediate polymers produced • Increases in medium viscosity • Enzyme deactivation	[18–29]
	Indirect Functionalization	• Reactive binding sites exposed • Generation of diverse substrates	• Poor operational stability • Sensitive to environmental conditions • Reusing and recycling problems	[30]
Physical Approach	Plasma treatment	• No aqueous or chemical solvent is required • Energy saving technique • No degradation of cellulose • Compounded with grafting techniques	• Cold plasma preferred • Technical knowledge necessary for applications	[31]

(*Continued*)

Table 6.1 Summary of advantages and disadvantages of different surface modification techniques for nanocellulose. (*Continued*)

Modification approach	Modification techniques	Advantages	Disadvantages	References
	Flame treatment	• Better adhesion with polyethylene • Large surface area covered	• Constant monitoring regarding flame height, flame chemistry and treatment time	[6, 33]
	Corona treatment	• Improves the surface energy of matrix-fiber • Increases bonding with different coatings and adhesives • Improved physico-mechanical properties	• Electrical conductivity decreases • Melt viscosity decreases formation of low molecular weight moieties post treatment	[32]
	Ion beam treatment	• Specific reaction with metals and polymers	• Difficulty to control over experimental parameters such as, the types of ion used, temperature for the ion treatment and beam frequency utilized	[33]

be synergistically used with grafting techniques, which enhance the tensile strength, durability, and stability of cellulose yarn further [31].

6.3.2 Flame Treatment

Flame treatment is a high-speed process that surface-functionalized textiles, glass, paperboard, and polymers. This treatment gives better adhesion than other treatment procedures by generating free radicals. Constant monitoring of flame chemistry, time of treatment, and distance between the surface to be modified and the flame are all limitations encountered.

6.3.3 Corona Treatment

This method makes use of low-temperature visible electrical discharge between the cellulose surface and the liquid adhesive. Corona treatment improves the surface energy of the treated surface, which enhances bonding with matrix fiber but decreases the electrical conductivity property. The advantage of this technique is that it enhances the mechanical properties and joint strength of the fabric [32]. Simultaneously, it decreases the melt viscosity of the treated fiber, rendering the chain flexible.

6.3.4 Ion Beam Treatment

This technique is applied to materials like metals and polymers. Ion beam treatment causes significant structural changes to the cellulose surface as intramolecular and intermolecular hydrogen bonds are affected. Depending on the intensity of the ionic beams applied, the degree of crystallinity of cellulose is affected. Major limitations faced are the types of ions used, the temperature for the ion treatment, and the beam frequency utilized because crossing the optimized parameters might cause substantial damage to cellulose.

6.4 Conclusion

This chapter discusses various surface modification approaches that improve the physicomechanical properties of cellulose such as durability, biocompatibility, surface hydrophobicity, and the availability of bioactive sites for functional modifications. Chemical processes such as acetylation and silylation achieve significant features such as moisture absorption reduction, fiber reinforcement for structural applications, and increased adsorption. TEMPO-assisted oxidation prevents the formation of strong interfibril bonds, whereas various enzymatic approaches improve cellulose fiber-matrix bonding. Physical methods enhance surface modification properties by introducing new functional groups. Thus, for commercial applications of nanocellulose, these surface treatments of natural fibers are a primary prerequisite.

References

1. Braun, B. and Dorgan, J.R., Single-step method for the isolation and surface functionalization of cellulosic nanowhiskers. *Biomacromolecules*, 10, 334, 2009.
2. Tavassoli, M., Khezerlou, A., Bangar, S.P., Bakhshizadeh, M., Haghi, P.B., Moghaddam, N., Ehsani, A., Functionality developments of Pickering emulsion in food packaging: Principles, applications, and future perspectives. *Trends Food Sci. Technol.*, 132, 171–187, 2023.
3. Wu, C.S., Wu, D.Y., Wang, S.S., Preparation and characterization of polylactic acid/bamboo fiber composites. *ACS Appl. Bio Mater.*, 5, 1038–1046, 2022.
4. Trifol, J., Plackett, D., Sillard, C., Hassager, O., Daugaard, A.E., Bras, J., Szabo, P.J., A comparison of partially acetylated nanocellulose, nanocrystalline cellulose, and nanoclay as fillers for high-performance polylactide nanocomposites. *Appl. Polym. Sci.*, 133, 2016.
5. Azimi, B., Maleki, H., Gigante, V. *et al.*, Cellulose-based fiber spinning processes using ionic liquids. *Cellulose*, 29, 3079–3129, 2022.
6. Ghasemlou, M., Daver, F., Ivanova, P.E., Habibi, Y., Adhikari, B., Surface modifications of nanocellulose: From synthesis to high-performance nanocomposites. *Prog. Polym. Sci.*, 119, 101418, 2021.
7. Lizundia, E., Fortunati, E., Dominici, F., Luis, J., Manuel, L., Armentano, I., Torre, L., Kenny, J.M., PLLA-grafted cellulose nanocrystals: Role of the CNC content and grafting on the PLA bionanocomposite film properties. *Carbohydr. Polym.*, 142, 105–13, 2016.
8. Chuensangjun, C., Kanomata, K., Kitaoka, T. *et al.*, Surface-modified cellulose nanofibers-graft-poly(lactic acid)s made by ring-opening polymerization of l-lactide. *J. Polym. Environ.*, 27, 847–861, 2019.
9. Oliveira, D., Castro, D., Bras, J., Gandini, A., Belgacem, N., Surface grafting of cellulose nanocrystals with natural antimicrobial rosin mixture using a green process. *Carbohydr. Polym.*, 137, 1, 2016.
10. Indarti, E., Marwan, Wanrosli, W.D., Effect of temperature on the silylation of nanocrystalline cellulose from oil palm empty fruit bunch with 3-aminopropyltriethoxysilane. *IOP Conf. Ser.: Earth Environ. Sci.*, 425, 012065, 2020.

11. Chandrashekar, A., Vargheese, S., Vijayan, J.G. *et al.*, Highly efficient removal of Rhodamine B dye using nanocomposites made from cotton seed oil-based polyurethane and silylated nanocellulose. *J. Polym. Environ.*, 30, 4999–5011, 2022.
12. Huang, F.Y., Thermal properties and thermal degradation of cellulose tri-stearate (CTs). *Polymers*, 4, 1012–1024, 2012.
13. Huang, F., Wu, X., Yu, Y., Lu, Y., Chen, Q., Acylation of cellulose nanocrystals with acids/trifluoroacetic anhydride and properties of films from esters of CNCs. *Carbohydr. Polym.*, 155, 525, 2017.
14. Sun, Z., Huang, Z., Guo, L., Hu, S., Wang, H., Meng, L., Tang, M., Qi, H., Acetylated tunicate nanocellulose-based high-efficient air filter media with antibacterial property. *J. Membr. Sci.*, 669, 121307, 2023.
15. Fang, Z., Zhu, H., Yuan, Y., Ha, D., Zhu, S., Preston, C., Chen, Q., Li, Y., Han, X., Lee, S., Chen, G., Li, T., Munday, J., Huang, J., Hu, L., Novel nanostructured paper with ultrahigh transparency and ultrahigh haze for solar cells. *Nano Lett.*, 12, 765–773, 2014.
16. Wang, F.P., Li, J.J., Zhang, Y., Sun, M.Y., Li, L., Wahid, F., Xie, Y.Y., Jia, S., Li, W., Zhong, C., TEMPO-mediated ,oxidation promotes cellulose dissolution in a zincate–NaOH system at suprazero temperatures. *ACS Sustain. Chem. Eng.*, 10, 7374–7384, 2022.
17. Barbash, V.A., Yashchenko, O.V., Gondovska, A.S. *et al.*, Preparation and characterization of nanocellulose obtained by TEMPO-mediated oxidation of organosolv pulp from reed stalks. *Appl. Nanosci.*, 12, 835–848, 2022.
18. Tzanov, T., Stamenova, M., Cavaco-Paulo, A., Phosphorylation of cotton cellulose with baker's yeast hexokinase. *Macromol. Rapid Commun.*, 23, 16, 2002.
19. Fricain, J.C., Granja, P.L., Barbosa, M.A., Jéso, B.D., Barthe, N., Baquey, C., Cellulose phosphates as biomaterials. *In vivo* biocompatibility studies. *Biomaterials*, 23, 971–980, 2002.
20. Battista, F. and Bolzonella, D., Some critical aspects of the enzymatic hydrolysis at high dry-matter content: A review. *Biofuels Bioprod. Biorefin.*, 12, 711–723, 2018.
21. Xie, J. and Hsieh, Y.-L., Modification of cellulose solids by enzyme-catalyzed transesterification with vinyl esters in anhydrous organic solvents, in: *Biocatalysis in Polymer Science*, vol. 840, pp. 217–230, 2002.
22. Gremos, S., Zarafeta, D., Kekos, D., Kolisis, F., Direct enzymatic acylation of cellulose pretreated in BMIMCl ionic liquid. *Bioresour. Technol.*, 102, 1378–82, 2011.
23. Lin, X., Wu, Z., Zhang, C., Liu, S., Nie, S., Enzymatic pulping of lignocellulosic biomass. *Ind. Crops Prod.*, 120, 16–24, 2018.
24. Kumar, N.V., Rani, M.E., Gunaseeli, R., Kannan, N.D., Paper pulp modification and deinking efficiency of cellulase-xylanase complex from *Escherichia coli* SD5. *Int. J. Biol. Macromol.*, 111, 289–295, 2018.
25. William, K. and Thomas, J., Enzyme processes for pulp and paper: A review of recent developments. *ACS Symp. Ser.*, 845, 210–239, 2003.

26. Saastamoinen, P., Mattinen, M.-L., Hippi, U., Nousiainen, P., Sipilä, J., Lille, M., Suurnäkki, A., Pere, J., Laccase aided modification of nanofibrillated cellulose with dodecyl gallate. *Bioresources*, 7, 5749–5770, 2012.
27. Cusola, O., Valls, C., Vidal, T., Roncero, B.M., Application of surface enzyme treatments using laccase and a hydrophobic compound to paper-based media. *Bioresour. Technol.*, 131, 521–526, 2013.
28. Takahashi, D., Johnson, K.L., Hao, P. *et al.*, Cell wall modification by the xyloglucan endotransglucosylase/hydrolase XTH19 influences freezing tolerance after cold and sub-zero acclimation. *Plant Cell Environ.*, 44, 915–930, 2021.
29. Habibi, Y., Key advances in the chemical modification of nanocelluloses. *Chem. Soc. Rev.*, 43, 1519–1542, 2014.
30. Wolf, R.A., Girourard, N., Xu, S., Meredith, J.C., Shofner, M.L., Cross, L., Mintz, E., Schueneman, G.T., Adhesion improvements in cellulose nanocrystal interfaces. *Plast. Eng.*, 69, 32–37, 2013.
31. Mather, R.R., 13 - Surface modification of textiles by plasma treatments, in: *Woodhead Publishing Series in Textiles, Surface Modification of Textiles*, Q. Wei (Ed.), pp. 296–317, Woodhead Publishing, New York, 2009.
32. Ragoubi, M., George, B., Molina, S., Bienaime, D., Merlin, A., Hiver, J.M., Dahoun, A., Effect of corona discharge treatment on mechanical and thermal properties of composites based on miscanthus fibres and polylactic acid or polypropylene matrix. *Composites (Part A)*, 43, 675–685, 2012.
33. Pakharenko, V., Pervaiz, M., Pande, H., Sain, M., Sain, M., Chemical and physical techniques for surface modification of nanocellulose reinforcements. *Interface/Interphase in Polymer Nanocomposites*, 283–310, 2018.

7

Applications of Nanocellulose in Tissue Engineering and Tissue Grafting

Sneha Dey, Paramjeet Singh, Sampriti Sarkar and Avik Acharya Chowdhury*

Department of Biosciences, JIS University, Agarpara, Kolkata, West Bengal, India

Abstract
In recent times, the development of effective systems for tissue grafting and tissue engineering through the use of inert and biocompatible nanomaterials has become an important topic of research. Nanocellulose has shown promise in the manufacturing of contact lenses, protective barriers, and frameworks for the engineering of blood vasculature, bone, cartilage, the dura mater, the liver, and the tissue of the sensory organs. Due to its unique physical features, low cytotoxicity, unique surface area, and other biological aspects, nanocellulose is one of the materials that have attracted a lot of attention. Massive cellulose units can disintegrate into nanocellulose, a form of cellulose that can only be seen at the nanoscale, by rupturing the intramolecular and intermolecular hydrogen connections between the chains. The idea that nanocellulose has the potential be employed as a transplant material for achieving the goal of the replacement of tissues was motivated by a number of advantages of the material, including its capacity to absorb water, adequate flexibility and durability, customizable shape, biocompatibility, a porous, nanofibrous structure, and an adaptable form. Hence, the current literature aims to summarize the administration of nanocellulose in the field of tissue engineering and grafting, emphasizing the advantages and future developments and related materials for tissue engineering.

Keywords: Tissue engineering, tissue repair, nanocellulose, bacterial nanocellulose, nanofibrillated cellulose, animal nanocellulose, algal nanocellulose, tissue grafting

**Corresponding author*: avikacharya@jisuniversity.ac.in

Mainak Mukhopadhyay and Debalina Bhattacharya (eds.) Nanocellulose: A Biopolymer for Biomedical Applications, (159–192) © 2024 Scrivener Publishing LLC

7.1 Introduction

As the world works to become more sustainable, there is an increasing demand for biorenewable natural raw materials with outstanding mechanical qualities and cost-effective manufacture. The requirements must be accomplished using economically and environmentally acceptable resources [1]. Recent nanotechnology breakthroughs have included cellulose, one of the Earth's most prevalent naturally occurring compounds, in the unique structure of nanocellulose. Anselme Payen, a French scientist, became the first person to discover cellulose in 1838. The most prevalent biopolymer is cellulose $(C_6H_{10}O_5)_n$, which is packed into microfibrils by intramolecular hydrogen bonds and intermolecular Van der Waals forces, is made up of a structure composed of several hundred to many thousands of glucan chains connected by β-1,4-glycosidic bonds with cellobiose residues serving as the repeating unit at varying extents of polymerization. Nanocellulose is a nanosized form of material taken from plants, animals, tunicates, or microbes [2, 3].

A vast range of bionanomaterials can be created from the various components of the enormous diversity of plants that are gathered worldwide. Research in material science and engineering is very interested in the production of nanoparticles and nanostructured materials from agricultural waste. The three primary plant components—cellulose (nearly 50%), hemicellulose (in a range of 20–40%), and lignin (20–30%)—have been separated using a variety of techniques to generate nanocellulose, nano-hemicelluloses, and nano-lignin with an extensive range of variable features [4].

Nanomaterials, due to their exclusive properties and structures, can mimic naturally occurring nanoscale components of the extracellular matrix (ECM), can traverse cellular membranes, and can deliver biologically active substances directly. Nanoparticles can also aid in protein absorption by cells. According to the aforementioned factors and some physicochemical features, such as wettability or electrical charge, a range of industries can manufacture and use nanoparticles with natural properties or functionalized with additional suitable attributes, which include scaffolds for the purpose of tissue engineering [5], medication transport [6], bioimaging [7], vascular grafts, wound dressing [8], gene therapy [9], and others.

The biocompatibility and bioactivity of numerous forms and variations of cellulose are being studied extensively in the scientific community with diverse cell types [10]. In contrast to uncharged nanocellulose, both types of charge (i.e., positive or negative) significantly enhanced cell attachment and proliferation while reducing immune cell activation [11, 12]. As opposed to cationic nanocellulose, which was produced by adding ammonium groups to nanocellulose, anionic nanocellulose frequently offered adequate support to enhance cell attachment and growth [13]. Anionic nanocellulose, on the other hand, triggered a more significant immunological response in cells than that of cationic nanocellulose [14].

Tissue engineering is a novel discipline of bioengineering that involves the methodologies and concepts of engineering, cell biology, and material science to generate tissue succedaneum that architecturally and physiologically mimics natural tissues [15].

Regenerated cellulose has been employed to make 3-D tissue with blood vessels and in combination with cellulose acetate to engineer heart tissue [16, 17]. Nonwoven cellulose fibers and injectable cellulose-based hydrogels have both been used successfully as *in vitro* and *in vivo* chondrocyte carriers for cartilage regeneration [18, 19]. The biocompatibility of viscose cellulose sponges with bone tissue and their potential as scaffolds for cartilage regeneration have both been investigated [20, 21].

Functional tissue replacements necessitate precise spatiotemporal management of biological activities and real-time monitoring. Growth factors, chemokines, inhibitors, genes, and other bioactive substances must be delivered and transported locally in a controlled manner in order to exert control over and evaluate the created tissues. Due to this requirement, NP-based systems have been used in scaffolds for tissue engineering for the administration of various factors that promote growth, the provision of contrast for imaging, and also for regulating the scaffold's properties [5].

7.2 Nanocellulose and Its Properties

Nanocellulose, as a naturally occurring nanosized substance, has several advantages over other materials. It possesses a distinctive shape and dimensions of geometry, a high surface area, high crystallinity, accessibility, biocompatibility, recyclability, mechanical reinforcement, barrier properties, orientation, mechanical stability, alignment, liquid crystalline action, rheological qualities, and is non-toxic. Nanocellulose is also a material with high availability and renewability that is quite inexpensive. These extensive, distinctive properties have resulted in the establishment of a diverse array of applications, from rheological modifiers and composite reinforcement to biological functionalization such as tissue grafting and tissue engineering [22]. These distinct features fall into three categories: physical traits, surface characteristics, and biological traits.

7.2.1 Physical Properties

Both the structured (crystalline) and unorganized (amorphous) portions of the nanoparticle can be used to describe the physical characteristics of nanocellulose. The presence of cellulosic chains in unstructured regions promotes bulk material suppleness and plasticity, whereas the presence of these chains in ordered parts promotes stiffness and elasticity. One theory is that a combination rule between the modulus of the structured domains and the ill-defined component will produce the modulus of various forms of nanocellulose. Because CNF and BC fibrils exhibit both crystalline and non-crystalline forms, their stiffness and modulus should be lower than those of CNC [23].

Especially for the characterization of nanocellulose, choosing appropriate measurement techniques is crucial. For instance, Park *et al.* looked into how four measuring methodologies affected the evaluation of the cellulose crystallinity index [24].

The elastic characteristics of CNC were calculated using theoretical calculations and indirect experimental observations utilizing several pieces of equipment, namely, the atomic force microscope (AFM), analysis on the diffraction of X-rays (XRD), inelastic X-ray scattering, and Raman scattering [25]. The X-ray diffractogram (XRD) height approach, which is the most straightforward and popular methodology, was utilized, and it was found that it yielded greater values than other methods. They put forth alternate X-ray diffractogram and nuclear magnetic resonance methods that assess the contribution of both crystalline and amorphous regions to the overall spectrum, allowing for a more precise measurement [24]. A wide range of results were obtained, which suggests that the Young's modulus of crystalline cellulose should lie in a range of 100–200 GPa, which turns out to be stronger than steel (200–220 GPa). In a recent study, the resilient modulus of crystalline cellulose was investigated using atomistic simulations that combined complementary techniques based on nanoscale indentation with the traditional uniform deformation methodology. It was discovered that the elastic modulus is 139.53.5 GPa [26].

The elastic modulus of cellulose microfibrils was determined using AFM tips and a three-point bending experiment. The size of cellulose microfibrils has been found to have a significant impact on their mechanical properties, with the horizontal modulus of pulp CNF predicted to be 8,112 GPa [27]. Recently, a Raman spectroscopic method that included determining the local molecular deformation of BC through a shift in the center location of the 1095 cm^{-1} Raman band was used to determine the modulus of BC to be 114 GPa [28].

Nanocellulose has the potential to be employed as a load-bearing component for a variety of host materials due to its outstanding mechanical properties. High-modulus nanocellulose demonstrates the potential for nano-reinforcement due to homogeneous distribution and interfacial interaction adhesion, allowing adequate stress transmission from the host element (matrix) to the reinforcing material (nanocellulose) [23].

7.2.2 Surface Chemistry

1,4-Anhydro-D-glucopyranose units make the substantial molecular weight homopolysaccharide structure of cellulose. These units assume a chair shape, with the hydroxyl group in an equatorial position that does not exactly lie in the plane of the structure and the subsequent glucose residues spinning through a 180° orientation with regard to the molecular axis [29]. The creation of fibrils and semi-crystalline packaging, which has a substantial impact on the physiological features of this highly cohesive material, depends on these hydroxyl groups' capacity to establish hydrogen bonds [30].

Surface functionalization broadens the application possibilities of nanocellulose [31]. Self-modulation chemistry and the functional alteration of pre-existing OH groups in the nanocellulose structure serve to illustrate chemical functionalization. Nanocellulose's hydroxyl groups are very reactive and simple to access for the necessary changes. When cellulose chains are arranged unidirectionally parallel within elementary fibrils during synthesis and deposition, crystals with hydroxyl functionality on one end—referred to as the non-reducing end—and hemiacetal functionality on the other—referred to as the reducing end—are formed [32].

A number of scientific papers have been published in the last 15 years that have concentrated on the alterations of chemical components of nanocellulose to modify its surface characteristics or balance its hydrophobic/hydrophilic ratio [33]. Because of its poor miscibility in both organic solvents (non-polar) and polymer matrix, pure nanocellulose without surface modification is frequently restricted to fields involving hydrophilic or polar media. This aspect is a significant restriction, especially in the case studies regarding polymer science, where nanoparticles are commonly employed as reinforcing additives. As a consequence, investigation in surface modification of nanocellulose was driven by the field of polymer science to achieve hydrophobization and improvised the mixing of nanocrystals with polymer matrix [34]. The hydrophilic nature of nanocellulose can be adjusted by its hydrophobic properties, and vice versa, potentially improving its congruence, miscibility, and relevant field of performance [35].

The reaction kinetics of the hydroxyl groups can be modified by changing the reaction conditions. The surface alterations have considerably boosted nanocellulose's potential for a number of industrial applications [36]. The characteristics of nanocellulose are frequently altered through etherification, amidation, amination, polymer grafting, oxidation, hydrolysis, sialylation, esterification, carbamation, cross-linking, sulfonation, phosphorylation, surfactant modulation, and fluorescent labeling [37–39].

Other than reactive groups, surface charges—most notably the negatively charged sulfate esters (-OSO_3-) on CNC—are a significant factor in the nanocellulose surface properties. Surface hydroxyls and an H_2SO_4 molecule undergo condensation esterification (also known as sulfation) during sulfuric acid hydrolysis, with the assistance of another H_2SO_4 molecule acting as a condensation agent. As a result of the hydrolysis by H_2SO_4, the CNC is strongly negatively charged and forms an evenly disseminated aqueous colloidal suspension. The amount of surface charge from sulfate groups in CNC is affected by the temperature and duration of H_2SO_4 hydrolysis. Furthermore, surface OSO_3 groups with negative charges improve CNC's solvent stability, which makes it suitable to be applied in biomedical fields such as protein or enzyme electrostatic adsorption [40].

7.2.3 Biological Characteristics

7.2.3.1 Biological and Hematological Compatibility

Biocompatibility is an important need for biomedical materials. It is defined as the capability of a foreign element (implanted inside a host) to coexist in harmony with tissue without generating negative alterations [41]. Different studies [42, 43] evaluating cellulose biocompatibility produce varying results due to the implications of different techniques. Early evidence suggests that cellulose possesses high biocompatibility and has the record of eliciting a mild response to foreign implants *in vivo*, sometimes. Because the human body lacks cellulolytic enzymes, cellulose cannot be easily broken down, resulting in some incompatibility.

Another aspect regarding biocompatibility is referred to as hemocompatibility (blood compatibility), which is particularly important for biomaterials that come into contact with blood as well as artificial organs like arteries and hearts. The hemocompatibility and distinctive biological activity of TEMPO-oxidized cellulose nanofibers appear to be promising aspects here [44].

With regards to collagen in particular, BC has a distinct nanofibril network topology that resembles the extracellular matrix's unique features. Due to the hydrophilicity of cellulose, the substance aggressively binds water molecules tightly and exhibits hydrogel-like behavior. Unlike many polymeric materials, BC has biocompatibility and the ability to get integrated into the body. It possesses distinctive mechanical qualities that are comparable to those of soft tissues. The ability of the nanofibrils to regulate cells makes BC a highly desirable scaffold for tissue engineering [45]. For their respective tissues of origin, cellulose, and collagen is frequently the main structure providing mechanical support.

Similar in size to BC fibers, collagen fibers are about 100 nm in diameter [46]. While BC has the distinguishing advantage of being a polysaccharide and having the advantage of being less immune-stimulatory, proteins are more likely to elicit immunologic reactions. Using the dorsal skinfold chamber model, a recent study discovered a good amount of tissue integration and production of extracellular matrix around the BC implant [47]. The complement activation indices (sC3a and sC5b–9) for BC were substantially greater in a current *in vitro* study comparing the blood compatibility of vascular grafts composed of BC with the materials typically used for grafts [48].

7.2.3.2 In Vivo Biodegradability

Non-biodegradable, biocompatible materials are found to be appropriate for some purposes (such as synthetic heart valves), whereas bioresorbable materials that promote tissue regeneration are mostly preferred for some other applications (such as artificial bone grafts) [45]. Hydrolysis degrades cellulose. However, due to its compact structure, it is particularly resistant to deterioration. Stomach acid and human GI tract hydrolyzing enzymes, such as amylase, are unable to significantly alter cellulose. The extent of cellulose disintegration, absorption, and immunological response can all be impacted by its form, which includes its crystallinity, hydration, and swelling. There are no biocompatible processes for the large-scale degradation of cellulose within the human body. Though certainly speculative, nonenzymatic, completely spontaneous hydrolysis of the chains of cellulose might explain the gradual degradation of unmodified BC in the body [49]. Regenerated cellulose made by deacetylating cellulose acetate did not regress significantly over a 6-week trial period. In contrast, equivalent samples of unstructured regenerated cellulose were disintegrated and assimilated to the tune of up to 75% (w/w) during the same study period. A different study claimed that in aquatic conditions, CNC was more recyclable than fullerenes and carbon nanotubes without an *in vivo* evaluation of biodegradability [50]. Pre-irradiation and sodium periodate oxidation processes on BC membranes have been shown to cause two primary phases of *in vitro* decomposition of oxidized BC: (1) a rapid disintegration of approximately 70–80% of the total sample in the initial stage, and (2) a gradual degradation of 5–10% that eventually leaves a small amount of non-resorbable material. In addition, *in vivo* degradation tests on male New Zealand white rabbits revealed that oxidized BC membranes degraded significantly all the time, with the fastest degradation taking place in the first 2–4 weeks [51].

7.3 Classification of Nanocellulose and Their Synthesis

Nanocellulose is categorized based on its creation and production method. The categories include (1) cellulose nanocrystals (CNCs), (2) cellulose nanofibrils (CNFs), and (3) bacterial cellulose (BC) [55].

7.3.1 Cellulose Nanocrystals (CNCs)

CNCs are derived from cellulose microcrystals through a process of acid hydrolysis at high temperatures. The production of CNCs involves breaking down cellulosic particles with strong acids like concentrated sulfuric acid, resulting in the creation of crystalline surfaces from specific unstructured regions of cellulose. This hydrolytic process produces a highly purified form of CNC with a few sulfate impurities [56, 57]. The process of acid hydrolysis leads to the penetration of hydronium ions into the cellulose chains within the amorphous domains. This causes the cleavage of glycosidic bonds, freeing individual crystallites, which are collectively referred to as cellulose nanocrystals (CNCs) [58].

CNC is composed of needle- or rod-shaped nanoparticles. Each rod is considered a single cellulosic crystal without defects. The structure of the nanocrystal varies based on factors like the cellulose source, ionic strength, and hydrolysis conditions. The kind of acid used for the hydrolysis process determines, to a great extent, the surface chemistry of CNC. The common use of sulfuric acid leads to the generation of sulfate esters on the CNC surface, which enhances the stability of suspensions but decreases thermal stability. If observed under polarized light using crossed-Nicols, CNC suspensions display the formation of birefringent domains. Other methods have been proposed to produce CNC as well [59, 60]. Other techniques for CNC production include ultrasonication-assisted $FeCl_3$-catalyszed hydrolysis, periodate oxidation, ammonium persulfate oxidation, ionic liquid, hydrolysis with gaseous acid, TEMPO oxidation, enzymatic hydrolysis, liquefaction, subcritical water, and acid deep eutectic solvents [59, 61, 62].

Cellulose nanocrystals (CNCs) are cylindrical, elongated, rigid, and less flexible, with a typical width ranging from 4 to 70 nm, and a characteristic length nearly from 100 to 6000 nm, along with a crystallinity of 54–88% [63].

7.3.2 Cellulose Nanofibrils (CNFs)

Cellulose nanofibrils are structurally elongated fibers with a specific diameter in the nanoscale range. They are produced through the application of high pressure that crushes cellulose into pulp, resulting in tightly twisted fibril networks. Cellulose nanofibril structures contain both amorphous and crystalline regions, and they are very small in both length and diameter [64].

Cellulose nanofibrils can be produced through a number of methods: (a) mechanical methods: homogenization and crushing; (b) chemical methods, such as TEMPO (2,2,6,6-Tetramethylpiperidinyloxy) oxidation; and (c) a combination of both mechanical and chemical methods [65–67].

TEMPO-mediated oxidation is often used for the production of CNFs (TEMPO CNF), where carboxyl groups are selectively added at the C6 position of glucose units and facilitate the isolation of nanofibers. Endoglucanases are fascinating substances for enzymatic nanocellulose production. Their reaction principle is that they cause fibers to swell and randomly break the internal linkages of the glycan chains, leading to the exposure of the non-reducing end of the cellooligosaccharides. Endoglucanases also help in reducing energy consumption during the later mechanical treatment [68, 69].

Ultrafine grinding is another commonly used method for producing cellulose nanofibrils (Enz-CNFs). This process involves passing pulp slurry through a grinder equipped with static and rotating grinding stones. The shear forces generated during grinding help break down the hydrogen bonds and cell wall structure of the cellulose, resulting in the isolation of most of the pulp into nanoscale dimensions [70].

A microfluidizer is employed with a similar mechanism to a high-pressure homogenizer. The high shear forces generated by the microfluidizer disrupt the cellulose fibers by disrupting the hydrogen bonds between them, which leads to the generation of CNF. The physical features of the CNF can be controlled by adjusting the flow rate, pressure, and number of cycles through the microfluidizer. Utilization of a microfluidizer allows the generation of CNF, demonstrating a considerable degree of uniformity and homogeneity, making them useful for an extensive set of applications such as reinforcing materials, filters, and biomedical materials [71, 72].

Cellulose nanofibrils (CNFs) are 3–15 nm in diameter, possessing 1–3 μm in length, and they possess good colloidal stability. These nanofibrils exhibit high mechanical strength (2–6 GPa) and comparatively lower thermal expansion, thermal stability, and chemical functionality from the surface hydroxyls. A high optical transmittance and gas barrier have been found to be characteristic properties of the CNF films. Chemically modified cellulose nanofibrils of smaller diameters can also be produced on an industrial scale. The length of chemically modified CNFs varies, with Enz-CNFs having a length of 1450 nm, TEMPO-CNFs having a length of 370 nm, and Holo-CNFs showing an exceptionally long length of 2060 nm [25, 73, 74].

7.3.3 Bacterial Cellulose (BC)

BC is highly pure, making it a superior source for the production of CNFs. Unlike the cellulose obtained from plants, BC has a higher degree of crystallinity, making it easier to process into nanofibrils. Additionally, the bacterial cellulose structure is highly ordered and uniform, leading to CNFs with consistent physical and mechanical properties. BC-CNFs exhibit unique characteristic features like enhanced water-holding capacity, high transparency, and high mechanical strength, which makes them ideal for a wide range of applications, including tissue engineering, food packaging, and filtration media [75]. Bacterial cellulose is synthesized by bacteria through the secretion of cellulose from the cytoplasm to the outside surface of the cell, where it forms a thin, continuous sheet. Unlike plant cellulose, which is formed by the polymerization of individual glucose units, bacterial cellulose is synthesized as a single, continuous sheet without the need for the polymerization of smaller units. This distinguishes it and ensures that it is free of lignin and other contaminants that are common in plant-based celluloses [76]. In contrast, bacterial cellulose is formed without the need for the polymerization of small molecules, whereas cellulose nanocrystals and nanofibrils are both created through the polymerization of small particles [66].

Bacterial cellulose (BC) is produced by microorganisms, including *Gluconacetobacter xylinus*, *Agrobacterium*, *Acanthamoeba*, *Acetobacter*, *Achromobacter*, *Alcaligenes*, *Rhizobium*, and algal species (*Cladophora*, *Microdiction*, *Rhizoclonium*, and *Chaetomorpha*) [77, 78]. The production process can be either agitated or static fermentation, and the type of BC produced is dependent on the morphologies and properties desired. Static fermentation leads to the establishment of gelatinous pellicles at the air–liquid interface of the culture medium, while agitated fermentation produces irregular pellets that remain suspended in the culture medium. Although agitated fermentation is easier to scale up for industrial production, static fermentation results in greater genetic stability among these specific bacterial species. Unlike cellulose nanocrystals and cellulose nanofibrils, bacterial cellulose is produced without the polymerization of small molecules [79, 80].

Due to its good biocompatibility and fine fiber network, great tensile strength, and excellent water holding capacity, the BC film is considered to be an efficient scaffold material. In addition to having a tensile strength of 200 to 300 MPa and an elongation of 1.5 to 2.0%, the elastic modulus of BC films ranges from 15 to 35 GPa [19, 81].

7.4 Composites Based on Nanocellulose

Nanocellulose is utilized for making composites because it has excellent reinforcing properties. Composites are composed of two or more reinforcing and matrix phases that have been blended to create a single new entity. These phases, which are different in terms of physicochemical properties, have a clearly defined interface. A composite is referred to as a bio-composite if a biopolymer was used in its production. In such nanocellulose-based composites, the polymer acts as the matrix phase, whereas nanocellulose acts as the reinforcing phase. The reinforcement phase, as opposed to the matrix phase, accounts for a larger amount of the composite [82].

Diverse composites have been manufactured in different ways, including in sheets, paper, and film. Fascinatingly, the substantial surface area that nanocellulose filler adds leads to a noticeable improvement in the characteristic features of the composite with only a small amount of nanocellulose [83].

7.4.1 Nanocellulose and Graphene Composite

A transparent, thin sheet of graphene that is both thermally and electrically conductive functions as a basic building block of all carbon allotropes, such as graphite, fullerenes, and carbon nanotubes. Incorporating cellulose nanoparticles enhances the dispersion of graphene nanoparticles in the aqueous environment and prevents aggregation without chemically functionalizing the material. Nanocellulose/graphene composites are made via a water-based dispersion process. There are a number of methods for producing composite materials, including filtration, filtering with hot pressing, freeze-casting, and freeze-drying. Incorporating graphene into the bacterial process of producing nanocellulose and also putting graphene atop a layer of nanocellulose are other fabrication techniques [84–88]. Nanocellulose/graphene composites can be created using several types of nanocellulose and graphene, and these composites can be enhanced with a variety of substances, such as carbides, oxides, ceramic nanoparticles, enzymes, and polymers, to change their properties for specific applications [89, 90]. After the addition of the graphene oxide to the electrospun cellulose acetate nanofibrous scaffold, increased osteogenic development is achieved by mesenchymal stem cells isolated from the human umbilical cord [91–93]. Additionally, it enhances the potentiality of these cells to proliferate and their adherence properties. The mechanical and antibacterial qualities of nanocellulose/graphene composites demonstrate great ability for the development of antibacterial fabrics and wound dressings. Increased antibacterial activity against *Bacillus cereus* and *Bacillus subtilis* is produced by an antibacterial textile made by electrospinning a mixture of cellulose acetate, graphene oxide sheets, and TiO_2 [94–97].

7.4.2 Nanocellulose and Polyvinyl Alcohol Composite

The development of green bio-nanomaterials works on the approach of combining nanocellulose with polyvinyl alcohol (PVA). Researchers extracted nanocrystalline cellulose (NCC) from *Acacia mangium* and mixed it with PVA film to generate CNC-PVA (Cellulose Nanocrystals/Polyvinyl Alcohol) film via solution casting. In comparison to pure PVA film, the film's tensile strength increased by 30%, 35%, 38%, and 50% after adding 2%, 5%, 7%, and 10% of NCC. Researchers have looked into the thermodynamic, mechanical, and swelling properties of PVA/CNC composites using various membrane ratios. According to their findings, the tensile strength as well as the elastic modulus of the composite membrane are directly proportional to the concentration of CNC in conditions of elevated relative humidity. The membrane's thermal stability, however, is slightly decreased by using CNC [98, 99].

A different team of researchers combined poly (D,L-lactide-co-glycolide) nanoparticles to create a CNC/PVA nanocomposite. The incorporation of CNC enhanced the membrane's elongation properties and

Young's modulus, making this nanocomposite a promising drug delivery method. A different study looked into an approach to bone tissue generation that suggests the use of nano-hydroxyapatite and CNF as fillers in an electrospun PVA nanofiber-based composite scaffold. Fillers influenced the *in vitro* disintegration of phosphate-buffered saline and improved the scaffold's cellular activity. The findings backed up the use of a fibrous scaffold for engineering tissues [100, 101].

7.4.3 Nanocellulose and Chitosan Composite

Bacterial cellulose (BC) is a flexible substance that has demonstrated its capability for several biomedical and aesthetic applications, which are complemented by its mechanical strength, form, biocompatibility, non-toxicity, and chemical controllability. Numerous biomedical products can be produced using composites made from bacterial nanocellulose (BNC) and other components, including synthetic and natural polymers [102, 103]. Both *in situ* and *ex situ* approaches are frequently used in the generation of BNC composites. In order to facilitate BC synthesis, the *in situ* approach adds reinforcing elements into the growth media; these components eventually end up as components of the hydrogel produced [104–106]. Composites are created via the *ex vivo* method by impregnating or inserting reinforcement into a synthetic polymer. Both *in situ* and *ex situ* procedures were employed to generate chitosan (Ch) and BC composites; the existence of bonds like the O-H bond and functional N-H groups, along with the structural resemblance, induce strong cooperation between Chitosan and BC, and the composite compound subsequently demonstrates significantly enhanced biological and physico-mechanical properties [107–109]. Gelatin's amazing biological features have shown immense potential in the application of scaffolding materials like BC. Considering the *in situ* and *ex situ* techniques, both are used to create BC-gelatin composites. The composites demonstrated better cell proliferation and adhesion capabilities and have been used in biomedical applications. Several polymers (collagen, Novo-aloe vera, and alginate) can be combined with BC to increase its medicinal potential. These polymeric materials are combined to create composites that have qualities that are antifungal, antimicrobial, and promote tissue regeneration. Scientists created a BC-Ag composite through a technique that included immersing BC sheets in an $AgNO_3$ solution. With the help of UV absorption and X-ray diffraction spectral peaks, nanoparticles that had attached to the BC surface were investigated. Gram-positive and Gram-negative bacteria were very sensitive to the BC-Ag composite's extremely powerful antibacterial properties. The study concluded that adding metals and metal oxides to BC composites results in the addition of electrical conductivity and antibacterial characteristics [110–113].

7.5 Applications of Nanocellulose in Tissue Engineering and Tissue Grafting

7.5.1 Nanocellulose for Engineering Skin and Tissues

Materials made of nanocellulose provide many benefits when used as tissue engineering scaffolds. They exhibit excellent mechanical qualities, great optical transparency, high water absorption, and high water retention. Their development follows a wide spectrum of methodologies, which include electrospinning, 3D printing, solvent casting, and freeze-drying. For simple, customized techniques for the engineering of tissues for damaged tissues or organs, cellulose materials are available [114, 115]. They can be made to have the desired surface functional groups, pore size, hierarchical structure, and mechanical properties. Hence, to engineer blood vessels, skin, and organs, cellulose-based materials hold great promise. High biocompatibility is an essential condition that scaffolds must fulfill to imitate the native conditions in the tissue and the extracellular matrix (ECM). Additionally, it is to promote division, development, and proliferation of the cells [116].

Consequently, bacterial cellulose (BC) offers hope for tissue engineering. The development of an excellent network resembling a collagen network was visible in BC pellicles. It utilized parafilm particles that were sterilized and were of three distinct sizes (90–150 m, 150–300 m, and 300–500 m) to create scaffolds with varied diameters of pores. Alginate and other biomolecules were employed for the preparation of sponges for the regeneration of tissues in the oral cavity. Additionally, tubular template materials like PDMS (polydimethylsiloxane) can also be used for the formation of tubes [117].

Cellulose has several benefits. Enhanced cellular compatibility, mechanical strength, and absorption of water are all characteristics of the cellulose composite. Cellulose-based products also demonstrate a substantial level of stability in water and phosphate buffered saline (PBS) buffers. BC promoted multipotent mesenchymal stem cell (MSC) expansion. MSC adhesion, proliferation, and differentiation were all enhanced by BC scaffolds. Thus, musculoskeletal tissue can be created using BC/MSC hydrogel.

The composites of polyvinyl alcohol (PVA) and BC have the ability to function as effective biocompatible materials for replacing soft tissues. Functioning as a bioscaffold, nanocellulose has the capability to encourage cell binding and multiplication, exhibiting mechanical features and biocompatibility in resemblance with natural tissues. Various distinct types of tissues have been cultured on nanocellulosic materials [118, 119].

Application of electrospun nanofibers in combination with nanocellulose is effective for supporting the growth of human cells on scaffolds. Electrospinning involves the stretching or melting of a polymer solution using electric force, which results in the generation of a small stream that then solidifies into continuous fibers ranging from nano- to micro-sized [120, 121]. This process creates porous, fibrous scaffolds, which have been found to be considerably compatible with cells, promoting adhesion and proliferation. These fibers possess a large surface area, thus allowing ideal cell growth. It has also been demonstrated in a different study that the electrospun membranes made of polylactide-polyglycolide (PLGA) and nanocellulose membranes are particularly preferred for promoting cell adhesion, growth, proliferation, and distribution of fibroblasts. Studies have shown that nanocellulose-loaded PLGA membranes have properties that are similar to those of the native skin. Additionally, freeze-drying, also known as lyophilization, leads to the sublimation of the solvent and the generation of dry materials, which are utilized for the creation of scaffolds that maintain their initial structure and shape after rehydration. Employment of a suspension of brown algae nanofibrillar cellulose (BANFC) in combination with chitin, which is quaternized and organic rectorite (BANFC/QCR), can also create sponges with high porosity, water absorption, and antibacterial properties that aid in skin healing. It demonstrates the potential to accelerate the healing of the skin by permitting the generation of collagen and promoting neovascularization. Usage of nanocellulose in the scaffolds of polyvinyl alcohol (PVA) created through the process of freeze-drying improves mechanical, thermal, and swelling properties, and the uniform porous structure and hydrophilicity contribute to a suitable environment for the growth of the cell and for them to function efficiently [122–124].

7.5.2 Nanocellulose for Repairing Blood Vessels

Tissue-engineered blood vessels (TEBVs) employ bacterial cellulose (BC). BC-based TEBV is promising for caliber vascular grafts to rebuild tissues associated with vascular abnormalities. BC was used to produce artificial blood vessels that can be used in microsurgery. Bacterially synthesized cellulose is referred to as BASYC®, the abbreviation for the vessels. In their wet form, the prepared vessels have demonstrated good mechanical strength and high water retention. Additionally, they demonstrated minimal interior surface roughness and completely "vitalized" the rat. In microsurgery, BASYC® was suggested as a synthetic blood vessel. For the fabrication of both cartilage tissues and temporary skin substitutes, BC has been well described as a scaffold [125]. BC was used for seeding *in vitro* a wide spectrum of human cell lines, including osteoblasts, keratinocytes, endothelial cells, and a number of stem cell lines, which after a period of 2 weeks of culture on the bacterial cellulose pellicles showed the capability of promoting cellular proliferation of smooth muscle cells and demonstrated great biocompatibility as well [123, 125, 126].

Artificial tubes that are utilized to replace tiny vascular grafts were developed in part because of their robustness and blood biocompatibility. It features a considerable potential for holding water and a smooth inner tube surface. Covalent bonds are employed to enable nanocellulose to adhere to the fibrin matrix, after which it is sufficiently reinforced with nanoparticles. Novel polyurethane bonded with nanocellulose has significant relevance in vascular replacement due to its enhanced elastic qualities, excellent blood clotting capacity, and outstanding physical and mechanical properties. In reference to the engineering of vascular tissues, scaffolds that are electrospun have linked, porous structures that are critical for cell proliferation and nutrient exchange [127]. They also have an increased surface area. Electrospinning and deacetylation are used to create cellulose fiber and cellulose acetate scaffolds that contain microcrystalline and nanocrystalline cellulose [128]. Because of the cellulose particles, the scaffolds have a multiscaled structure, which therefore mimics the extracellular matrix in nature. Together, microcrystalline cellulose and nanocrystalline cellulose promote the adhesion of cells, while the former assists in cell anchoring. The morphology and viability have been greatly enhanced by the improvisation of the microcrystalline cellulose and nanocrystalline. By combining more nanofibrillar cellulose (NFC) with poly(lactic acid) and poly(butylene succinate), it is possible to create electron-spinning scaffolds with sustainable degradation. Nanofibrillar cellulose scaffolds show improved tensile strength, elastic modulus, fiber structure, and biocompatibility, all of which are extremely beneficial [129].

Vascular scaffolds made of nanocellulose should be very strong to withstand the mechanical strain that develops in the various sites in the tissues of the blood vessels. To grow blood vessels with a smaller diameter, scaffolds containing cellulose acetate propionate (CAP) and nanocrystalline cellulose are employed. Even at 0.2 weight percent, nanocrystalline cellulose significantly increases the directional stiffness and strength of the scaffolds. Strong nanocrystalline cellulose hydrogen bonds give the scaffolds noticeable external strength and pore-filled nanostructures, which, in turn, helps them resist the pressure in physiological conditions and mirror the characteristics seen naturally in blood vessels and the extracellular matrix. The NCC-CAP solution is considerably affected due to the application of a tiny magnetic field force. It causes NCC to be correctly ordered in the CAP matrix. The orientation and oozing of nanoparticles have been improved after applying a modified magnetic field, which, in turn, also decreased the concentration of nanocrystalline cellulose to obtain satisfactory qualities [125, 128, 129].

7.5.3 Nanocellulose in Neural Tissue Engineering

It is highly challenging for neurons to heal after damage since neuronal tissue cannot mend itself. Axon development should be directed, the axon channel should be modulated, and specific neural stimulation and activity should be promoted using neural tissue engineering scaffolds. Neurons modify cellular activity with relatively modest electrochemical impulses. These impulses can be transmitted via the materials, and they can also produce electrical stimuli for the establishment of new neurons. Specifically, promotion of cell activity by these scaffolds with outstanding capabilities is necessary for engineering of the brain tissue. These signals can be transmitted, and electrical stimulation for neuronal development can be provided by the conductive materials [130].

An important feature of the scaffolds is their potential to induce correct cellular activity. The electrospun nanocellulose-based scaffolds demonstrate adjustable pore count, mechanical characteristics, positioning, and adaptability and are a good choice for the development and engineering of neural tissue. Using a revolving disc that serves as an electrospinning collector, electrospun cellulose/nanocrystalline cellulose nanocomposite nanofibers (ECCNN) are created that are uniaxially aligned. The tensile strength and elastic modulus of ECCNN are raised in the direction of fiber orientation by including 20 wt% of nanocrystalline cellulose, and the thermal stability is also greatly improved. Along with rapid proliferation on the ECCNN surface, the cells may also do so inside it. Nanofibers with ECCNN alignment control cellular organization and benefit brain tissue engineering by acting as scaffolds [131, 132].

7.5.4 Nanocellulose in Bone Tissue Engineering

The options to be considered for the replacement of cartilage include nanocellulose, which is highly durable, is compatible with tissues, and causes less calcification. As a crosslinker, bacterial cellulose/polyacrylamide gels are employed. These gels have the ability to provide tensile strength and length similar to the ligaments. Nanocellulose proves to show great promise as a scaffold for chondroblasts and osteoblasts. Nanocellulose can be utilized for the regeneration and restoration of osteoblast cells. Bone regeneration may be facilitated by membranes made of bacterial cellulose and hydroxyapatite. It promotes bone cell development and proliferation and increases the activity of the ALP (alkaline phosphatase) enzyme, promoting osteoblast function, and nodule formation and growth in bone [133, 134].

Researchers also looked at the sustained release of vascular endothelial growth factor (VEGF) and bone morphogenic protein (BMP) from hydrogels made of nanocellulose. The newly created hydrogel scaffolds demonstrated improved cell adhesion and proliferation characteristics. An increased ability for the creation of new bones in orthopedic deformities by BMP2 and VEGF-loaded scaffolds was observed in both Rat bone marrow stem cells (RBMSCs) that were treated with stem cells and the ones that were not treated with the stem cells. Higher expression of ALP observed in BMP2 and VEGF-loaded scaffolds compared to the control indicated ALP's increased bone regeneration capability. Human bone marrow-derived mesenchymal stem cells (hBMSCs) responded more favorably to CNC-incorporated alginate/gelatine scaffolds than pure alginate/gelatin scaffolds [135, 136].

The presence of CNC in the polymer matrices had a significant impact on cellular activity. The mechanical strength and swelling properties of CNC-incorporated alginate/gelatin scaffolds were found to be higher compared to those of pure alginate/gelatin scaffolds. In comparison to scaffolds made of pure chitosan, it has been observed that chitosan/CNC (CH/CNC) scaffolds facilitate comparatively enhanced cellular activity and hence increased potential for osteogenicity. The creation of carboxymethyl cellulose/hydroxyapatite (CMC/HA) nanocomposites for drug delivery and engineering of bone tissue was another significant development. To track the drug's transport into the cells and tissues, carbon dots were coupled with the CMC/HA nanocomposites [133, 136].

The applications of nanocellulose have been outlined in Figure 7.1. Table 7.1 summarizes the synthesis and applications of the various types of nanocellulose nanomaterials.

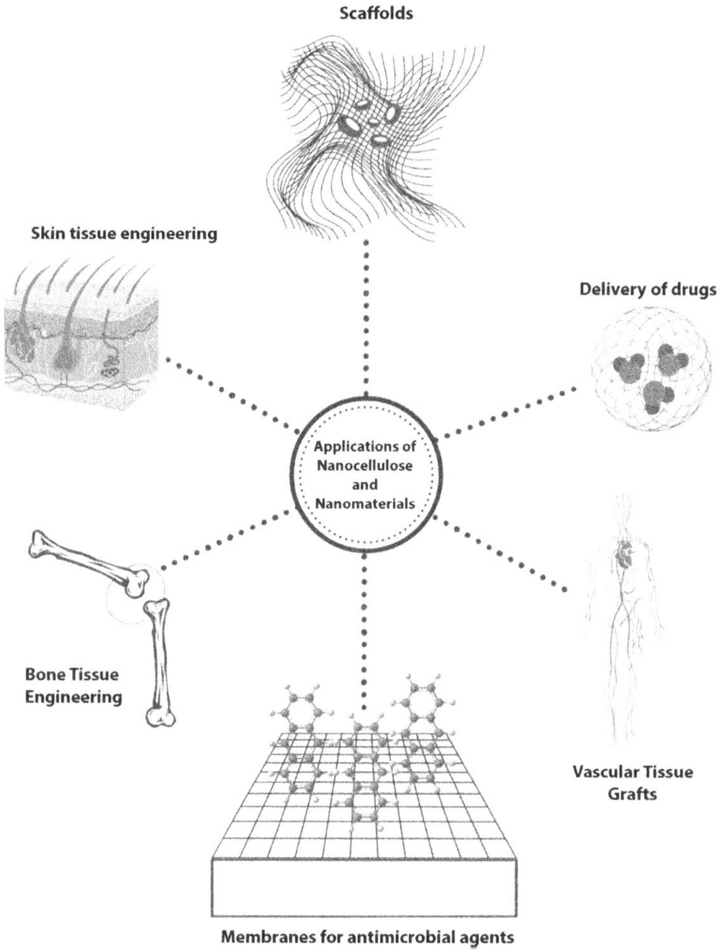

Figure 7.1 Instances of applications of nanocellulose in tissue engineering and repair.

7.6 Limitations of Nanocellulose

There are several limitations to using nanocellulose for applications in the engineering of tissues, which include the following:

7.6.1 Production Cost

Nanocellulose production remains a challenging and expensive process, mainly because of the need for specialized equipment and its high energy consumption. Furthermore, the building materials used during the production of nanocellulose are frequently derived from natural sources,

Table 7.1 Summary of Features and application of different types of nanocellulose.

Type of nanocellulose	Methods of preparation	Prime features	Application in tissue engineering	Ref
Cellulose nano crystals	Acid hydrolysis (Sulfuric acid); Tempo-Oxidation; Ultrasonication-assisted $Fecl_3$ catalyzed hydrolysis; Periodate oxidation; Ammonium persulfate oxidation; Ionic liquid hydrolysis with gaseous acid; Liquefaction; Subcritical water and acid dip eutectic solvents.	Cylindrical; Elongated; Rigid; Less flexible; Width: 4-70nm; Length: 100-6000nm; Crystallinity: 54-88%.	Scaffold preparation: Bone tissue engineering in form of Chitosan/CNC scaffold; CNC-alginate gelatin scaffold; Drug delivery: Carboxy methyl cellulose/ Hydroxy appetite; Neural tissue engineering.	[59, 61–63, 131–133, 136]
Cellulose nanofibrils	Homogenization; Crushing; Tempo-Oxidation; Ultrafine grinding.	Diameter: 3-15nm; Length: 1-5 Micrometer; Colloidal stability; Low thermal expansion; High mechanical strength (2-6 GPa); Thermal stability.	Skin tissue engineering; Brain tissue engineering.	[25, 65–67, 73, 129, 131, 132]
Bacterial Cellulose	Agitated fermentation; Static fermentation.	Fine-fiber network; High tensile strength; Good biocompatibility; Water holding capacity.	Tissue and skin engineering; Blood vessel repair and engineering; vascular grafts; Neural tissue engineering; Ligament and tendon repair; Bone osteoblast restoration.	[19, 81, 116, 123, 125, 133]

which can be scarce and contribute to the high production costs. The high cost of nanocellulose limits its widespread utilization in tissue engineering and other applications [137].

7.6.2 Mechanical Properties

While nanocellulose has exceptional mechanical features such as high tensile strength and stiffness, it lacks flexibility and toughness, making it a less ideal material for a few of the tissue engineering applications that require materials that can bend and stretch without breaking. As such, further research is needed for the enhancement of the mechanical properties of nanocellulose or to find ways to combine it with other materials to enhance its performance [138, 139].

7.6.3 Scalability

Producing nanocellulose on a large scale can be a challenge due to the need for specialized equipment and the substantial energy consumption required in the production process. This makes it difficult to use nanocellulose in large-scale tissue engineering projects that require large amounts of the material. The scalability of nanocellulose production seems to be a serious obstacle to its widespread utilization in tissue engineering, along with several other applications [140].

The following factors can make large-scale nanocellulose production difficult:

Technical barriers: The production of nanocellulose requires specialized processes and equipment, and scaling up production to a commercial level can be technically challenging and require significant investment. Raw material availability: Nanocellulose is produced from cellulose, which is a common renewable resource. However, sourcing cellulose on a large scale can be difficult and may require significant land use, which can have environmental impacts. Quality control: Ensuring consistent quality and properties of nanocellulose on a large scale can be difficult, as the production process can be influenced by a number of factors like temperature, pH, and the cellulose source. Cost: The production of nanocellulose can be expensive, especially at a large scale, due to the requirement of a high amount of energy and resources, along with specialized equipment in the production process [141, 142].

7.6.4 Compatibility with Cells

Although nanocellulose has shown promise as a support structure for cell growth and proliferation, it may not be compatible with all cell types. More research is needed to determine nanocellulose's suitability, enhance its applicability in tissue engineering, and develop methods to improve cells' compatibility with the material. Because of its substantial aspect ratio and nanoscale dimensions, which might result in cytotoxicity and other biological reactions including inflammation, nanocellulose may not be totally compatible with cells [143].

Nanocellulose's surface charge and hydrophilicity can also influence cell attachment and proliferation. These factors may limit nanocellulose's use as a scaffold material in tissue engineering [52, 53].

Additionally, the physical and mechanical properties of nanocellulose can influence cell behavior and tissue development. For example, the mechanical stiffness of nanocellulose can affect cell differentiation and tissue morphogenesis. The high tensile strength of nanocellulose can also lead to cellular stress and reduced cell viability [52, 54].

7.7 Future Prospects

The discovery and development of nanocellulose-based nanoparticles and nanomaterials have been a breakthrough, finding great applications across the techniques of tissue engineering and grafting. It is a highly effective material for biomedical applications due to its great biocompatibility and hemocompatibility. Membranes made of cellulose have been shown to support the cultivation of mesenchymal stem cells at a high level of viability, with over 95% of cells remaining alive. Because of their poor biodegradability, cellulose scaffolds are well suited for tissue engineering applications. Cellulose nanomaterials display good mechanical properties and can also be utilized in enhancing the mechanical strength of nanomaterials. While nanocellulose has great potential in tissue engineering, its compatibility with cells is still a matter of concern.

Further research is needed to optimize the use of nanocellulose in this field and to determine the best ways to modify its properties to improve its production, characteristics, and biocompatibility. For instance, several studies have demonstrated that chemically functionalized nanocellulose's surface characteristics can enhance cell adhesion and proliferation. Other research has found that combining nanocellulose with hydrogels or polymers improves its biocompatibility and mechanical qualities, making it more appropriate for usage as a scaffold material, such as:

Polymers: Nanocellulose can be blended with polymers such as polyvinyl alcohol (PVA), polyethylene glycol (PEG), or polylactic acid (PLA) to improve its mechanical properties and biocompatibility. Hydrogels: Hydrogels can be combined with nanocellulose to improve their cell compatibility and create 3D structures that have the potential to promote cellular growth and proliferation. Minerals: Minerals such as hydroxyapatite can be incorporated into nanocellulose to improve its osteoconductive properties, making it an appropriate substance for applications involving bone tissue engineering; Growth factors: growth factors such as platelet-derived growth factor (PDGF) or transforming growth factor beta (TGF-beta) can be incorporated into nanocellulose for the enhancement of its cell-stimulating properties and promotion of tissue regeneration; Biomaterials: Other biomaterials, such as collagen, can be combined with nanocellulose for the creation of hybrid materials demonstrating enhanced mechanical properties and cell compatibility.

The future goals concerning research and studies with nanocellulose should focus on the minimization or nullification of the toxicity of the material for the human population.

7.8 Conclusion

Different types of nanocellulose have recently received a lot of attention due to their ease of construction, biodegradability, and biocompatibility. A variety of sources can provide nanocellulose and allow for the optimization of its properties to improve the characteristics of the native polymers. The book chapter has focused on the rationale behind the utilization of nanocellulosic nanomaterials and nanoparticles for tissue grafting and tissue engineering and the mechanisms and applications of these nanomaterials in various biomedical defects. Nanocellulose finds several applications in the field of biomedical engineering, and a wide range of methods have been analyzed and employed by the scaffolds for tissue grafting and tissue engineering purposes. However, additional research is required to mitigate and correct the negative impacts that nanoparticles have on humans. Nanocellulose promises to be a material that can be applied in the delivery of different molecules, and the antimicrobial properties of this material allow it to be used for biomedical purposes. More studies will open up new avenues for research and for understanding the utilization of nanocellulose-based nanoparticles and nanomaterials for more advanced applications. Future research will, however, concentrate on the creation of unique hybrids with antibacterial capabilities that do not compromise their original traits.

References

1. Li, Y., Wang, B., Ma, M., Wang, B., Review of recent development on preparation, properties, and applications of cellulose-based functional materials. *Int. J. Polym. Sci.*, 2018, 1–18, 2018.
2. Holban, A.M. and Grumezescu, A., *Materials for biomedical engineering: Nanomaterials-based drug delivery*, pp. 131–164, Elsevier, Amsterdam, 2019.

3. Zabed, H., Sahu, J.N., Boyce, A.N., Faruq, G., Fuel ethanol production from lignocellulosic biomass: An overview on feedstocks and technological approaches. *Renewable Sustainable Energy Rev.*, 66, 751–774, 2016.
4. Barhoum, Jeevanandam, J., Rastogi, A., Samyn, P., Boluk, Y., Dufresne, A., Danquah, M.K., Bechelany, M., Plant celluloses, hemicelluloses, lignins, and volatile oils for the synthesis of nanoparticles and nanostructured materials. *Nanoscale*, 12, 22845–22890, 2020.
5. Fathi-Achachelouei, M., Knopf-Marques, H., da Silva, C.E.R., Barthes, J., Bat, E., Tezcaner, A., Vrana, N.E., Use of nanoparticles in tissue engineering and regenerative medicine. *Front. Bioeng. Biotechnol.*, 7, 113, 2019.
6. Yang, Y., Wang, S., Wang, Y., Wang, X., Wang, Q., Chen, M., Advances in self-assembled chitosan nanomaterials for drug delivery. *Biotechnol. Adv.*, 32, 7, 1301–1316, 2014.
7. Bhunia, S.K., Saha, A., Maity, A.R., Ray, S.C., Jana, N.R., Carbon nanoparticle-based fluorescent bioimaging probes. *Sci. Rep.*, 3, 1473, 2013. Halib, N. and Ahmad, I., *Nanocellulose: Insight into health and medical applications*, pp. 1–19, Springer International Publishing, Cham, 2018.
8. Keles, E., Song, Y., Du, D., Dong, W.J., Lin, Y., Recent progress in nanomaterials for gene delivery applications. *Biomater. Sci.*, 4, 1291–1309, 2016.
9. Mikulikova, R., Svorcik, V., Nahlik, J., Sopuch, T., Havelka, P., Cytocompatibility of surface ground PE doped with calcium salt of 6-carboxycellulose. *Cellulose*, 15, 473–479, 2008.
10. Jonsson, M., Brackmann, C., Puchades, M., Brattås, K., Ewing, A., Gatenholm, P., Enejder, A., Neuronal networks on nanocellulose scaffolds. *Tissue Eng. Part C Methods*, 21, 1162–1170, 2015.
11. Lopes, V.R., Sanchez-Martinez, C., Strømme, M., Ferraz, N., *In vitro* biological responses to nanofibrillated cellulose by human dermal, lung and immune cells: Surface chemistry aspect. *Part. Fibre Toxicol.*, 14, 1, 2017.
12. Skogberg, A., Mäki, A.J., Mettänen, M., Lahtinen, P., Kallio, P., Cellulose nanofiber alignment using evaporation-induced droplet-casting, and cell alignment on aligned nanocellulose surfaces. *Biomacromolecules*, 18, 3936–3953, 2017.
13. Hua, K., Ålander, E., Lindström, T., Mihranyan, A., Strømme, M., Ferraz, N., Surface chemistry of nanocellulose fibers directs monocyte/macrophage response. *Biomacromolecules*, 16, 2787–2795, 2015.
14. Gomes, M.E., Rodrigues, M.T., Domingues, R.M.A., Reis, R.L., Tissue engineering and regenerative medicine: New trends and directions-a year in review. *Tissue Eng. Part B, Rev.*, 23, 3, 211–224, 2017.
15. Ko, I.K. and Iwata, H., An approach to constructing three-dimensional tissue. *Ann. N.Y. Acad. Sci.*, 944, 443–455, 2001.
16. Entcheva, E., Bien, H., Yin, L., Chung, C., Farrell, M., Kostov, Y., Functional cardiac cell constructs on cellulose-based scaffolding. *Biomaterials*, 25, 5753–5762, 2004.

17. Müller, F.A., Müller, L., Hofmann, I., Greil, P., Wenzel, M.M., Staudenmaier, R., Cellulose- based scaffold materials for cartilage tissue engineering. *Biomaterials*, 27, 3955–3963, 2006.
18. Vinatier, C., Gauthier, O., Fatimi, A., Merceron, C., Masson, M., Moreau, A., Moreau, F., Fellah, B., Weiss, P., Guicheux, J., An injectable cellulose-based hydrogel for the transfer of autologous nasal chondrocytes in articular cartilage defects. *Biotechnol. Bioeng.*, 102, 1259–1267, 1998.
19. Pulkkinen, H., Tiitu, V., Lammentausta, E., Laasanen, M.S., Hämäläinen, E.R., Kiviranta, I., Lammi, M.J., Cellulose sponge as a scaffold for cartilage tissue engineering. *Biomed. Mater. Eng.*, 16, S29–S35, 2006.
20. Gatenholm, P. and Klemm, D., Bacterial nanocellulose as a renewable material for biomedical applications. *MRS Bull.*, 35, 208–213, 2010.
21. Lin, N. and Dufresne, A., Nanocellulose in biomedicine: Current status and future prospect. *Eur. Polym. J.*, 59, 302–325, 2014.
22. Salimi, S., Sotudeh-Gharebagh, R., Zarghami, R., Chan, S.Y., Yuen, K.H., Production of nanocellulose and its applications in drug delivery: A critical review. *ACS Sustainable Chem. Eng.*, 7, 19, 15800–15827, 2019.
23. Moon, R.J., Martini, A., Nairn, J., Simonsen, J., Youngblood, J., Cellulose nanomaterials review: Structure, properties and nanocomposites. *Chem. Soc. Rev.*, 40, 3941–3994, 2011.
24. Wu, X., Moon, R.J., Martini, A., Crystalline cellulose elastic modulus predicted by atomistic models of uniform deformation and nanoscale indentation. *Cellulose*, 20, 43–55, 2013.
25. Cheng, Q., Wang, S., Harper, D.P., Effects of process and source on elastic modulus of single cellulose fibrils evaluated by atomic force microscopy. *Compos. Part A*, 40, 583–8, 2009.
26. Hsieh, Y.C., Yano, H., Nogi, M., Eichhorn, S.J., An estimation of the Young's modulus of bacterial cellulose filaments. *Cellulose*, 15, 507–13, 2008.
27. Habibi, Y., Lucia, L.A., Rojas, O.J., Cellulose nanocrystals: Chemistry, self-assembly, and applications. *Chem. Rev.*, 110, 6, 3479–3500, 2010.
28. French, A.D., Bertoniere, N.R., Brown, R.M., Chanzy, H., Gray, D., Hattori, K., Glasser, W., *Encyclopedia of Polymer Science & Technology*, J.I. Kroschwitz (ed.), vol. 5, pp. 473–507, John Wiley & Sons, Inc., New York, 2003.
29. Plackett, D., Letchford, K., Jackson, J., Burt, H., A review of nanocellulose as a novel vehicle for drug delivery. *Nord. Pulp Pap. Res. J.*, 29, 105–118, 2014.
30. Eyley, S. and Thielemans, W., Surface modification of cellulose nanocrystals. *Nanoscale*, 6, 7764–7779, 2014.
31. Huang, J., Dufresne, A., Lin, N., *Nanocellulose: From fundamentals to advanced materials*, pp. 115–153, Wiley-VCH Verlag GmbH & Co. KGaA, Weinheim, 2019.
32. Kovalenko, A., Predictive multiscale modeling of nanocellulose based materials and systems. *IOP Conf. Ser. Mater. Sci. Eng.*, 64, 1, 012040, 2014.
33. Abdul Khalil, H.P.S., Bhat, A.H., Ireana Yusra, A.F., Green composites from sustainable cellulose nanofibrils: A review. *Carbohydr. Polym.*, 87, 963–979, 2012.

34. Newman, R.H. and Davidson, T.C., Molecular conformations at the cellulose–water interface. *Cellulose*, 11, 23–32, 2004.
35. Khine, Y.Y. and Stenzel, M.H., Surface modified cellulose nanomaterials: A source of non- spherical nanoparticles for drug delivery. *Mater. Horiz.*, 7, 1727–1758, 2020.
36. Pérez, S. and Samain, D., Structure and engineering of celluloses, in: *Advances in Carbohydrate Chemistry and Biochemistry*, vol. 64, pp. 25–116, 2010.
37. Natterodt, J., Petri-Fink, A., Weder, C., Zoppe, J., Cellulose nanocrystals: Surface modification, applications and opportunities at interfaces. *Chim. Int. J. Chem.*, 71, 376–383, 2017.
38. Lin, N. and Dufresne, A., Surface chemistry, morphological analysis and properties of cellulose nanocrystals with gradiented sulfation degrees. *Nanoscale*, 6, 5384–93, 2014.
39. Dugan, J.M., Gough, J.E., Eichhorn, S.J., Bacterial cellulose scaffolds and cellulose nanowhiskers for tissue engineering. *Nanomedicine*, 8, 287–98, 2013.
40. Miyamoto, T., Takahashi, S., Ito, H., Inagaki, H., Noishiki, Y., Tissue biocompatibility of cellulose and its derivatives. *J. Biomed. Mater. Res.*, 23, 125–33, 1989.
41. Märtson, M., Viljanto, J., Hurme, T., Laippala, P., Saukko, P., Is cellulose sponge degradable or stable as implantation material? An *in vivo* subcutaneous study in the rat. *Biomaterials*, 20, 1989–95, 1999.
42. Shimotoyodome, A., Suzuki, J., Kumamoto, Y., Hase, T., Isogai, A., Regulation of postprandial blood metabolic variables by TEMPO oxidized cellulose nanofibers. *Biomacromolecules*, 12, 3812–8, 2011.
43. Petersen, N. and Gatenholm, P., Bacterial cellulose-based materials and medical devices: Current state and perspectives. *Appl. Microbiol. Biotechnol.*, 91, 1277–86, 2011.
44. Bäckdahl, H., Helenius, G., Bodin, A., Nannmark, U., Johansson, B.R., Risberg, B., Gatenholm, P., Mechanical properties of bacterial cellulose and interactions with smooth muscle cells. *Biomaterials*, 27, 2141–2149, 2006.
45. Esguerra, M., Fink, H., Laschke, M.W., Delbro, D., Jeppsson, A., Gatenholm, P., Menger, M.G., Risberg, B., Intravital fluorescent microscopic evaluation of bacterial cellulose as scaffold for vascular grafts. *J. Biomed. Mater. Res. Part A*, 93A, 140–149, 2010.
46. Fink, H., Hong, J., Drotz, K., Risberg, B., Sanchez, J., Sellborn, A., An *in vitro* study of blood compatibility of vascular grafts made of bacterial cellulose in comparison with conventionally-used graft materials. *J. Biomed. Mater. Res. Part A*, 97A, 52–58, 2011.
47. Czaja, W.K., Young, D.J., Kawecki, M., Browm, R.M., The future prospects of microbial cellulose in biomedical applications. *Biomacromolecules*, 8, 1–12, 2007.
48. Kümmerer, K., Menz, J., Schubert, T., Thielemans, W., Biodegradability of organic nanoparticles in the aqueous environment. *Chemosphere*, 82, 1387–92, 2011.

49. Czaja, W., Kyryliouk, D., DePaula, C.A., Buechter, D.D., Oxidation of γ-irradiated microbial cellulose results in bioresorbable, highly conformable biomaterial. *J. Appl. Polym. Sci.*, 131, 39995, 2014.
50. Clift, M.J., Foster, E.J., Vanhecke, D., Studer, D., Wick, P., Gehr, P., Rothen-Rutishauser, B., Weder, C., Investigating the interaction of cellulose nanofibers derived from cotton with a sophisticated 3D human lung cell coculture. *Biomacromolecules*, 12, 10, 3666–3673, 2011.
51. Vartiainen, J., Pöhler, T., Sirola, K., Pylkkänen, L., Alenius, H., Hokkinen, J., Tapper, U., Lahtinen, P., Kapanen, A., Putkisto, K., Hiekkataipale, P., Eronen, P., Ruokolainen, J., Laukkanen, A., Health and environmental safety aspects of friction grinding and spray drying of microfibrillated cellulose. *Cellulose*, 18, 775–86, 2011.
52. Pereira, M.M., Raposo, N.R., Brayner, R., Teixeira, E.M., Oliveira, V., Quintão, C.C., Camargo, L.S., Mattoso, L.H., Brandão, H.M., Cytotoxicity and expression of genes involved in the cellular stress response and apoptosis in mammalian fibroblast exposed to cotton cellulose nanofibers. *Nanotechnology*, 24, 075103, 2013.
53. Mahfoudhi, N. and Boufi, S., Nanocellulose as a novel nanostructured adsorbent for environmental remediation: A review. *Cellulose*, 24, 1171–1197, 2017.
54. Reshmy, R., Philip, E., Thomas, D., Madhavan, A., Sindhu, R., Binod, P., Varjani, S., Awasthi, M.K., Pandey, A., Bacterial nanocellulose: Engineering, production, and applications. *Bioengineered*, 12, 2, 11463, 2021.
55. Lu, P. and Hsieh, Y.L., Preparation and characterization of cellulose nanocrystals from rice straw. *Carbohydr. Polym.*, 87, 1, 564–573, 2012.
56. Dufresne, A., Nanocellulose: A new ageless bionanomaterial. *Mater. Today*, 16, 6, 220–227, 2013.
57. Dufresne, A., Nanocellulose processing properties and potential applications. *Curr. For. Rep.*, 5, 76–89, 2019.
58. Lin, K.H., Enomae, T., Chang, F.C., Cellulose nanocrystal isolation from hardwood pulp using various hydrolysis conditions. *Molecules*, 24, 3724, 2019.
59. Sirviö, J.A., Visanko, M., Liimatainen, H., Acidic deep eutectic solvents as hydrolytic media for cellulose nanocrystals production. *Biomacromolecules*, 17, 3025–32, 2016.
60. Kargarzadeh, H., Huang, J., Lin, N., Ahmad, I., Mariano, M., Dufresne, A., Thomas, S., Gałęski, A., Recent developments in nanocellulose-based biodegradable polymers, thermoplastic polymers, and porous nanocomposites. *Prog. Polym. Sci.*, 87, 197–227, 2018.
61. Naz, S., Ali, J.S., Zia, M., Nanocellulose isolation characterization and applications: A journey from non-remedial to biomedical claims. *Biodes. Manuf.*, 2, 187–212, 2019.
62. Jiang, F. and Hsieh, Y.L., Chemically and mechanically isolated nanocellulose and their self-assembled structures. *Carbohydr. Polym.*, 95, 1, 32–40, 2013.

63. Lavoine, N., Desloges, I., Dufresne, A., Bras, J., Microfibrillated cellulose - its barrier properties and applications in cellulosic materials: A review. *Carbohydr. Polym.*, 90, 2, 735–764, 2012.
64. Jacob, S., R, R., Antony, S., Madhavan, A., Sindhu, R., Kumar Awasthi, M., Kuddus, M., Pillai, S., Varjani, S., Pandey, A., Binod, P., Nanocellulose in tissue engineering and bioremediation: Mechanism of action. *Bioengineered*, 13, 5, 12823–12833, 2022.
65. Khalil, H.A., Bhat, A.H., Yusra, A.I., Green composites from sustainable cellulose nanofibrils: A review. *Carbohydr. Polym.*, 87, 2, 963–979, 2012.
66. Abdul Khalil, H.P.S., Davoudpour, Y., Nazrul Islam, Md., Mustapha, A., Sudesh, K., Dungani, R., Jawaid, M., Production and modifcation of nanofbrillated cellulose using various mechanical processes: A review. *Carbohydr. Polym.*, 99, 649–665, 2014.
67. Ribeiro, R.S.A., Pohlmann, B.C., Calado, V., Bojorge, N., Pereira, N., Jr., Production of nanocellulose by enzymatic hydrolysis: Trends and challenges. *Eng. Life Sci.*, 19, 279–291, 2019.
68. Siró, I. and Plackett, D., Microfbrillated cellulose and new nanocomposite materials: A review. *Cellulose*, 17, 459–494, 2010.
69. Osong, S.H., Norgren, S., Engstrand, P., Processing of wood-based microfbrillated cellulose and nanofbrillated cellulose, and applications relating to papermaking: A review. *Cellulose*, 23, 93–123, 2015.
70. Kumar, A., Dhiman, A., Suhag, R., Sehrawat, R., Upadhyay, A., McClements, D.J., Comprehensive review on potential applications of microfluidization in food processing. *Food Sci. Biotechnol.*, 31, 1, 17–36, 2021.
71. Klemm, D., Kramer, F., Moritz, S., Lindström, T., Ankerfors, M., Gray, D., Dorris, A., Nanocelluloses: A new family of nature-based materials. *Angew. Chem., Int. Ed.*, 50, 5438–5466, 2011.
72. Yang, X., Reid, M.S., Olsén, P., Berglund, L.A., Eco-friendly cellulose nanofibrils designed by nature: Effects from preserving native state. *ACS Nano*, 14, 1, 724–735, 2020.
73. Jozala, A.F., de Lencastre-Novaes, L.C., Lopes, A.M., de Carvalho Santos-Ebinuma, V., Mazzola, P.G., Pessoa, A., Jr, Grotto, D., Gerenutti, M., Chaud, M.V., Bacterial nanocellulose production and application: A 10-year overview. *Appl. Microbiol. Biotechnol.*, 100, 2063–2072, 2016.
74. Singhania, R.R., Patel, A.K., Tsai, M.-L., Chen, C.-W., Di Dong, C., Genetic modification for enhancing bacterial cellulose production and its applications. *Bioengineered*, 12, 1, 6793–6807, 2021.
75. Chunyan, Z., Industrial-scale production and applications of bacterial cellulose. *Front. Bioeng. Biotechnol.*, 8, 605374, 2020.
76. Prakash Menon, M., Selvakumar, R., Suresh kumar, P., Ramakrishna, S., Extraction and modification of cellulose nanofibers derived from biomass for environmental application. *RSC Adv.*, 7, 42750–42773, 2017.

77. Pang, M., Huang, Y., Meng, F., Zhuang, Y., Liu, H., Du, M., Ma, Q., Wang, Q., Chen, Z., Chen, L. *et al.*, Application of bacterial cellulose in skin and bone tissue engineering. *Eur. Polym. J.*, 122, 109365, 2020.
78. Lahiri, D., Nag, M., Dutta, B., Dey, A., Sarkar, T., Pati, S., Edinur, H.A., Abdul Kari, Z., Mohd Noor, N.H., Ray, R.R., Bacterial cellulose: Production, characterization, and application as antimicrobial agent. *Int. J. Mol. Sci.*, 22, 23, 12984, 2021.
79. Yang, Q., Wang, J., Yao, Y. *et al.*, Mechanical and thermal properties enhancement and swelling behavior of bacterial cellulose/collagen/polyvinyl alcohol nanofiber hydrogel film. *Fibers Polym.*, 23, 305–314, 2022.
80. Varshney, S., Mishra, N., Gupta, M., Progress in nanocellulose and its polymer based composites: A review on processing, characterization, and applications. *Polym. Compos.*, 42, 3660–3686, 2021.
81. Ilyas, R.A., Sapuan, S., Ishak, M., Zainudin, E.S., Mahamud, A., Nanocellulose reinforced starch polymer composites: A review of preparation, properties and application, 2018.
82. Liao, C., Li, Y., Tjong, S.C., Graphene nanomaterials: Synthesis, biocompatibility, and cytotoxicity. *Int. J. Mol. Sci.*, 19, 3564, 2018.
83. Shams, S.S., Zhang, R., Zhu, J., Graphene synthesis: A review. *Mater. Sci. Pol.*, 33, 566–578, 2015.
84. Coroş, M., Pogăcean, F., Măgeruşan, L., Socaci, C., Pruneanu, S., A brief overview on synthesis and applications of graphene and graphene-based nanomaterials. *Front. Mater. Sci.*, 13, 23–32, 2019.
85. Hajian, A., Lindstrom, S.B., Pettersson, T., Hamedi, M.M., Wagberg, L., Understanding the dispersive action of nanocellulose for carbon nanomaterials. *Nano Lett.*, 17, 1439–1447, 2017.
86. Malho, J.-M., Laaksonen, P.I., Walther, A., Ikkala, O., Linder, M.B., Facile method for stiff, tough, and strong nanocomposites by direct exfoliation of multilayered graphene into native nanocellulose matrix. *Biomacromolecules*, 13, 1093–1099, 2012.
87. Zhang, X., Lu, Z., Zhao, J., Li, Q., Zhang, W., Lu, C., Exfoliation/dispersion of low- temperature expandable graphite in nanocellulose matrix by wet co-milling. *Carbohydr. Polym.*, 157, 1434–1441, 2017.
88. Yao, Q., Fan, B., Xiong, Y., Jin, C., Sun, Q., Sheng, C., 3D assembly based on 2D structure of cellulose nanofibril/graphene oxide hybrid aerogel for adsorptive removal of antibiotics in water. *Sci. Rep.*, 7, 45914, 2017.
89. Wicklein, B., Kocjan, A., Salazar-Alvarez, G., Carosio, F., Camino, G., Antonietti, M., Bergström, L., Thermally insulating and fire-retardant lightweight anisotropic foams based on nanocellulose and graphene oxide. *Nat. Nanotechnol.*, 10, 277–283, 2015.
90. Liu, P., Zhu, C., Mathew, A.P., Mechanically robust high flux graphene oxide-nanocellulose membranes for dye removal from water. *J. Hazard. Mater.*, 371, 484–493, 2019.

91. Xu, T., Jiang, Q., Ghim, D., Liu, K.K., Sun, H., Derami, H.G., Wang, Z., Tadepalli, S., Jun, Y.S., Zhang, Q., Catalytically active bacterial nanocellulose-based ultrafiltration membrane. *Small*, 14, 1704006, 2018.
92. Jun, Y.-S., Wu, X., Ghim, D., Jiang, Q., Cao, S., Singamaneni, S., Photothermal membrane water treatment for two worlds. *Acc. Chem. Res.*, 52, 1215–1225, 2019.
93. Jiang, Q., Ghim, D., Cao, S., Tadepalli, S., Liu, K.-K., Kwon, H., Luan, J., Min, Y., Jun, Y.-S., Singamaneni, S., Photothermally active reduced graphene oxide/bacterial nanocellulose composites as biofouling-resistant ultrafiltration membranes. *Environ. Sci. Technol.*, 53, 412–421, 2018.
94. Liu, X., Shen, H., Song, S., Chen, W., Zhang, Z., Accelerated biomineralization of graphene oxide–incorporated cellulose acetate nanofibrous scaffolds for mesenchymal stem cell osteogenesis. *Colloids Surf. B Biointerfaces*, 159, 251–258, 2017.
95. Jia, L., Huang, X., Tao, Q., Enhanced hydrophilic and antibacterial efficiencies by the synergetic effect TiO2 nanofiber and graphene oxide in cellulose acetate nanofibers. *Int. J. Biol. Macromol.*, 132, 1039–1043, 2019.
96. Jasmani, L. and Adnan, S., Preparation and characterization of nanocrystalline cellulose from Acacia mangium and its reinforcement potential. *Carbohydr. Polym.*, 161, 166–171, 2017.
97. Jahan, Z., Niazi, M.B.K., Gregersen, Ø.W., Mechanical, thermal and swelling properties of cellulose nanocrystals/PVA nanocomposites membranes. *J. Ind. Eng. Chem.*, 57, 113–124, 2018.
98. Rescignano, N., Fortunati, E., Montesano, S., Emiliani, C., Kenny, J.M., Martino, S., Armentano, I., PVA bio-nanocomposites: A new take-off using cellulose nanocrystals and PLGA nanoparticles. *Carbohydr. Polym.*, 99, 47–58, 2014.
99. Enayati, M.S., Behzad, T., Sajkiewicz, P., Rafienia, M., Bagheri, R., Ghasemi-Mobarakeh, L., Kolbuk, D., Pahlevanneshan, Z., Bonakdar, S.H., Development of electrospun poly (vinyl alcohol)-based bionanocomposite scaffolds for bone tissue engineering. *J. Biomed. Mater. Res. Part A*, 106, 1111–1120, 2018.
100. Mbituyimana, B., Liu, L., Ye, W., Boni, B.O.O., Zhang, K., Chen, J., Thomas, S., Vasilievich, R.V., Shi, Z., Yang, G., Bacterial cellulose-based composites for biomedical and cosmetic applications: Research progress and existing products. *Carbohydr. Polym.*, 273, 118565, 2021.
101. Shah, N., Ul-Islam, M., Khattak, W.A., Park, J.K., Overview of bacterial cellulose composites: A multipurpose advanced material. *Carbohydr. Polym.*, 98, 1585–1598, 2013.
102. Saibuatong, O.-A. and Phisalaphong, M., Novo aloe vera–bacterial cellulose composite film from biosynthesis. *Carbohydr. Polym.*, 79, 455–460, 2010.
103. Ciechanska, D., Multifunctional bacterial cellulose/chitosan composite materials for medical applications. *Fibres Text. East. Eur.*, 12, 69–72, 2004.
104. Kim, J., Cai, Z., Chen, Y., Biocompatible bacterial cellulose composites for biomedical application. *J. Nanotechnol. Eng. Med.*, 1, 011006, 2010.

105. Ul-Islam, M., Shah, N., Ha, J.H., Park, J.K., Effect of chitosan penetration on physico-chemical and mechanical properties of bacterial cellulose. *Korean J. Chem. Eng.*, 28, 1736–1743, 2011.
106. Luo, H., Xiong, G., Huang, Y., He, F., Wang, Y., Wan, Y., Preparation and characterization of a novel COL/BC composite for potential tissue engineering scaffolds. *Mater. Chem. Phys.*, 110, 193–196, 2008.
107. Zhang, S. and Luo, J., Preparation and properties of bacterial cellulose/alginate blend bio-fibers. *J. Eng. Fibers Fabr.*, 6, 155892501100600309, 2011.
108. Maneerung, T., Tokura, S., Rujiravanit, R., Impregnation of silver nanoparticles into bacterial cellulose for antimicrobial wound dressing. *Carbohydr. Polym.*, 72, 43–51, 2008.
109. Bloor, D., Donnelly, K., Hands, P., Laughlin, P., Lussey, D., A metal–polymer composite with unusual properties. *J. Phys. D Appl. Phys.*, 38, 2851, 2005.
110. Islam, M.S., Molla, M., Sarker, M., Karim, M.M., Masum, S., Yeum, J., Fabrication of pllulan/silver nanoparticle composite nanospheres using electrospray technique for antibacterial applications. *Int. J. Basic Appl. Sci.*, 11, 36–40, 2011.
111. Qian, L. and Hinestroza, J.P., Application of nanotechnology for high performance textiles. *J. Text. Appar. Technol. Manage.*, 4, 1–7, 2004.
112. Priya, S.G., Jungvid, H., Kumar, A., Skin tissue engineering for tissue repair and regeneration. *Tissue Eng. Part B Rev.*, 14, 1, 105–118, 2008.
113. Groeber, F., Holeiter, M., Hampel, M., Hinderer, S., Schenke-Layland, K., Skin tissue engineering–*in vivo* and *in vitro* applications. *Adv. Drug Delivery Rev.*, 63, 4-5, 352–366, 2011.
114. Wei, Z., Wu, C., Li, R., Yu, D., Ding, Q., Nanocellulose based hydrogel or aerogel scaffolds for tissue engineering. *Cellulose*, 28, 7497–7520, 2021.
115. Xu, X., Zhou, J., Jiang, Y., Zhang, Q., Shi, H., Liu, D., 3D printing process of oxidized nanocellulose and gelatin scaffold. *J. Biomater. Sci. Polym. Ed.*, 29, 12, 1498–1513, 2018.
116. Tortorella, S., Maturi, M., Dapporto, F. *et al.*, Surface modification of nanocellulose through carbamate link for a selective release of chemotherapeutics. *Cellulose.*, 15, 8503–8511, 2020.
117. Stanisławska, A., Staroszczyk, H., Szkodo, M., The effect of dehydration/rehydration of bacterial nanocellulose on its tensile strength and physicochemical properties. *Carbohydr. Polym.*, 236, 116023, 2020.
118. Ghafari, R., Jonoobi, M., Amirabad, L.M., Oksman, K., Taheri, A.R., Fabrication and characterization of novel bilayer scaffold from nanocellulose based aerogel for skin tissue engineering applications. *Int. J. Biol. Macromol.*, 136, 796–803, 2019.
119. Pramanik, R., Ganivada, B., Ram, F., Shanmuganathan, K., Arockiarajan, A., Influence of nanocellulose on mechanics and morphology of polyvinyl alcohol xerogels. *J. Mech. Behav. Biomed. Mater.*, 90, 275–283, 2019.
120. Luo, H., Li, J., Zhou, F., Advances in hard tissue engineering materials—Nanocellulose-based composites. *Pap. Biomater.*, 3, 4, 62–76, 2018.

121. Ratcliffe, A., Tissue engineering of vascular grafts. *Matrix Biol.*, 19, 4, 353–357, 2000.
122. Hubbell, J.A., Massia, S.P., Desai, N.P., Drumheller, P.D., Endothelial cell-selective materials for tissue engineering in the vascular graft via a new receptor. *Biotechnol. (Nature Publishing Company)*, 9, 6, 568–572, 1991.
123. Luo, H., Cha, R., Li, J., Hao, W., Zhang, Y., Zhou, F., Advances in tissue engineering of nanocellulose-based scaffolds: A review. *Carbohydr. Polym.*, 224, 115144, 2019.
124. Ghorbani, M., Nezhad-Mokhtari, P., Sohrabi, H., Roshangar, L., Electrospun chitosan/nanocrystalline cellulose-graft-poly (N- vinylcaprolactam) nanofibers as the reinforced scaffold for tissue engineering. *J. Mater. Sci.*, 55, 5, 2176–2185, 2020.
125. Pooyan, P., Kim, I.T., Jacob, K.I., Tannenbaum, R., Garmestani, H., Design of a cellulose-based nanocomposite as a potential polymeric scaffold in tissue engineering. *Polymer*, 54, 8, 2105–2114, 2013.
126. Chen, Q., de Larraya, U.P., Garmendia, N., Lasheras-Zubiate, M., Cordero-Arias, L., Virtanen, S., Boccaccini, A.R., Electrophoretic deposition of cellulose nanocrystals (CNs) and CNs/alginate nanocomposite coatings and free standing membranes. *Colloids Surf. B, Biointerfaces*, 118, 41–48, 2014.
127. He, X., Xiao, Q., Lu, C., Wang, Y., Zhang, X., Zhao, J., Zhang, W., Zhang, X., Deng, Y., Uniaxially aligned electrospun all-cellulose nanocomposite nanofibers reinforced with cellulose nanocrystals: Scaffold for tissue engineering. *Biomacromolecules*, 15, 2, 618–627, 2014.
128. Mali, P. and Sherje, A.P., Cellulose nanocrystals: Fundamentals and biomedical applications. *Carbohydr. Polym.*, 275, 118668, 2022.
129. Huang, C., Hao, N., Bhagia, S., Li, M., Meng, X., Pu, Y.J., Yong, Q., Ragauskas, A.J., Porous artificial bone scaffold synthesized from a facile *in situ* hydroxyapatite coating and crosslinking reaction of crystalline nanocellulose. *Materialia*, 4, 237–246, 2018.
130. Herdocia-Lluberes, C.S., Laboy-López, S., Morales, S., Gonzalez-Robles, T.J., González-Feliciano, J.A., Nicolau, E., Evaluation of synthesized nanohydroxya patite-nanocellulose composites as biocompatible scaffolds for applications in bone tissue engineering. *J. Nanomater.*, 2015, 1–9, 2015.
131. Huang, C., Fang, G., Zhao, Y., Bhagia, S., Meng, X., Yong, Q., Ragauskas, A.J., Bio-inspired nanocomposite by layer-by-layer coating of chitosan/hyaluronic acid multilayers on a hard nanocellulose-hydroxyapatite matrix. *Carbohydr. Polym.*, 222, 115036, 2019.
132. Sukul, M., Nguyen, T.B.L., Min, Y.-K., Lee, S.-Y., Lee, B.T., Effect of local sustainable release of BMP2-VEGF from nano-cellulose loaded in sponge biphasic calcium phosphate on bone regeneration. *Tissue Eng. Part A*, 21, 1822–1836, 2015.
133. Patel, D.K., Dutta, S.D., Shin, W.C., Ganguly, K., Lim, K.T., Fabrication and characterization of 3D printable nanocellulose-based hydrogels for tissue engineering. *RSC Adv.*, 11, 7466–7478, 2021.

134. Sarkar, C., Chowdhuri, A.R., Kumar, A., Laha, D., Garai, S., Chakraborty, J., Sahu, S.K., One pot synthesis of carbon dots decorated carboxymethyl cellulose-hydroxyapatite nanocomposite for drug delivery, tissue engineering and Fe^{3+} ion sensing. *Carbohydr. Polym.*, 181, 710–718, 2018.
135. Yarbrough, J.M., Zhang, R., Mittal, A., Vander Wall, T., Bomble, Y.J., Decker, S.R., Himmel, M.E., Ciesielski, P.N., Multifunctional cellulolytic enzymes outperform processive fungal cellulases for coproduction of nanocellulose and biofuels. *ACS Nano*, 11, 3101–3109, 2017.
136. Habibi, Y., Key advances in the chemical modification of nanocelluloses. *Chem. Soc. Rev.*, 43, 1519–1542, 2014.
137. Dias, O.A.T., Konar, S., Leão, A.L., Yang, W., Tjong, J., Sain, M., Current state of applications of nanocellulose in flexible energy and electronic devices. *Front. Chem.*, 8, 420, 2020.
138. Mohd Amin, K., Annamalai, P.K., Morrow, I.C., Martin, D., Production of cellulose nanocrystals via a scalable mechanical method. *RSC Adv.*, 5, 57133–57140, 2015.
139. Pradhan, D., Jaiswal, A.K., Jaiswal, S., Emerging technologies for the production of nanocellulose from lignocellulosic biomass. *Carbohydr. Polym.*, 285, 119258, 2022.
140. Trache, D., Tarchoun, A.F., Derradji, M., Hamidon, T.S., Masruchin, N., Brosse, N., Hussin, M.H., Nanocellulose: From fundamentals to advanced applications. *Front. Chem.*, 8, 392, 2020.
141. Chinga-Carrasco, G., Ehman, N.V., Pettersson, J., Vallejos, M.E., Brodin, M.W., Felissia, F.E., Håkansson, J., Area, M.C., Pulping and pretreatment affect the characteristics of bagasse inks for three-dimensional printing. *ACS Sustainable Chem. Eng.*, 6, 3, 4068–4075, 2018.
142. Trache, D., Tarchoun, A.F., Derradji, M., Hamidon, T.S., Masruchin, N., Brosse, N., Hussin, M.H., Nanocellulose: From fundamentals to advanced applications. *Front. Chem.*, 8, 392, 2020.
143. Chinga-Carrasco, G., Ehman, N.V., Pettersson, J., Vallejos, M.E., Brodin, M.W., Felissia, F.E., Håkansson, J., Area, M.C., Pulping and pretreatment affect the characteristics of Bagasse inks for three-dimensional printing. *ACS Sustain. Chem. Eng.*, 6, 3, 4068–4075, 2018.

8

Application of Nanocellulose for Wound Dressings

B.A. Aderibigbe

Department of Chemistry, University of Fort Hare, Alice Campus, Eastern Cape, South Africa

Abstract

Nanocellulose-based biomedical materials offer distinct properties, such as excellent biocompatibility, reduced cytotoxic effects, biodegradability, and good mechanical properties. It is affordable and readily available, making it a sustainable biomaterial for biomedical applications. One of the outstanding biomedical applications of cellulose-based materials is in the design of wound dressings. Wound healing is very complex and using an appropriate wound dressing is crucial to afford accelerated healing with reduced scar formation. There are several commercially available wound dressings; however, some of them are not biocompatible and do not promote an accelerated wound-healing process. Nanocellulose-based wound dressings exhibit excellent features that are useful for the treatment of chronic, infected, high exuding and bleeding wounds. The surfaces of nanocellulose-based biomaterials have been modified with bioactive molecules for improved cellular adhesion, proliferation, and differentiation processes that are useful in skin regeneration. Different forms of wound dressings have been prepared from nanocellulose-based biomaterials. However, the development of nanocellulose-based wound dressings is still in its infancy. There have been reports (*in vitro* and *in vivo*) on the efficacy of nanocellulose-based dressings for the treatment of bleeding, exuding, chronic, acute wounds, and skin regeneration. This chapter will give a comprehensive report on the different forms of nanocellulose-based wound dressings and future perspectives.

Keywords: Nanocellulose, wound dressings, chronic wounds, burn wounds, microbial infections, antibiotics, nanoparticles, biomaterials

Email: blessingaderibigbe@gmail.com

Mainak Mukhopadhyay and Debalina Bhattacharya (eds.) Nanocellulose: A Biopolymer for Biomedical Applications, (193–220) © 2024 Scrivener Publishing LLC

8.1 Introduction

Cellulose is an important component that makes up the skeletal part of plants. Cellulose is a readily available polymer-based biomaterial with properties crucial for biomedical applications. Its structure is composed of repeated units of D-glucose building blocks [1]. It is extracted from plants, algae, wood, and bacteria [2]. Some of its unique features are good strength, low density, excellent stiffness, chemical inertness, easy surface modification, and a low coefficient of thermal expansion [3]. Its amorphous and crystalline domains are visible in native cellulose. However, the crystallinity varies in the range of 40–70%, which is influenced by its source and extraction method [3, 4]. The density of the amorphous domain is low when compared to the crystalline domain, making it react with other molecular groups. The resistance of the crystalline domains to treatments such as mechanical, chemical, and enzymatic is significant when compared to the amorphous domains [3, 4]. Some factors influence the morphology, size, and features of cellulose, such as the origin and the processing conditions.

Nanocellulose is cellulose in the nanometer range, and it has emerged as a unique biomaterial due to some of its features, including a low coefficient of thermal expansion, good biocompatibility, low toxicity, high Young's modulus, hydrogen-bonding capacity, high tensile strength, surface area-to-volume ratio, renewability, and eco-friendliness [3, 5]. It is divided into three classes, namely: (1) nanofibrils; (2) nanocrystals; and (3) bacterial nanocellulose [6]. The nanofibrils are characterized by longer nanofibers with an entangled network and are also flexible. They are also characterized by lower crystallinity [3]. It is prepared from lignocellulosic biomass via selected treatments, such as chemical, enzymatic, and mechanical treatments [3, 7]. Cellulose nanocrystals are less flexible with a rod-like shape and a 54–88% crystallinity index range [3, 8]. Bacterial nanocellulose has ribbon-shaped nanofibers, and the particle size range in diameter is 20–100 nm [3, 9]. It is obtained from the fermentation of carbohydrates from plants and sugar via microorganisms. It has a high purity and water content of over 99% [10].

Nanocellulose has been extensively researched to create wound dressings, such as nanofibers, hydrogels, films, membranes, hydrocolloids, and bandages to treat acute, chronic, highly oozing, and bleeding wounds because of its special properties [11]. Bacterial nanocellulose's high content purity, tensile strength, and flexibility have been explored to treat wounds with reduced scarring, provide an optimal moist environment suitable for accelerated wound healing, and reduce pain and fluid loss, making nanocellulose an important biomaterial for the development of wound dressings [11, 12]. This chapter will report on the efficacy of nanocellulose-based wound dressings.

8.2 Types of Wounds

On the basis of how long they take to recover, wounds are categorized. Wounds might be acute or chronic. Acute wounds heal in a predictable amount of time; however, chronic wounds do not heal for an extended length of time (Figure 8.1) [13]. Hemostasis/inflammation, proliferation, and remodeling are key stages in the wound-healing process. The hemostasis phase follows an injury and includes blood vessel constriction to stop bleeding, platelet activation, and aggregation, which releases platelet-derived growth factor and transforming growth factor (TGF), as well as the release of inflammatory cells. With cellular migration, germs and cell debris are eliminated throughout the inflammatory phase. An inflammatory response is brought on by neutrophils secreting pro-inflammatory cytokines. Leukocytes, regulatory T lymphocytes, and other elements are responsible for controlling tissue inflammation [14]. The transition of the

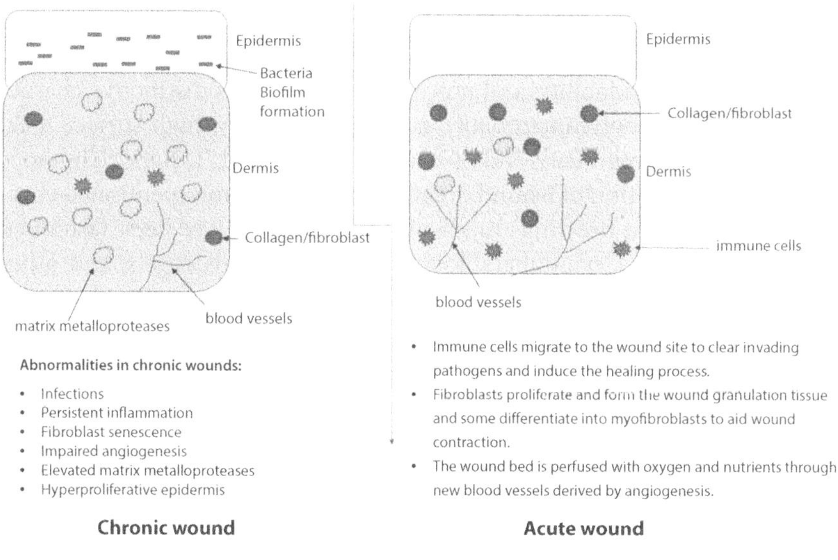

Figure 8.1 Acute wound and chronic wound healing.

wound-healing process from the inflammatory phase to the proliferative phase depends on macrophages. The lack of macrophages during the inflammatory or proliferative phase is the cause of hemorrhage or impaired tissue growth [14]. Re-epithelialization, the development of granulation tissue, and neovascularization all occur during the proliferative phase.

At the maturation and remodeling phase, the wound reaches its maximum strength. A scar replaces granulation tissue, and the formation of type I fibrillar collagen is replaced by type III reticular collagen. At this phase, the zinc-dependent endopeptidases produced by epidermal cells are crucial in tissue remodeling [15, 16]. However, chronic wounds, such as diabetic foot, venous leg, and pressure ulcers, do not adhere to the abovementioned normal healing phases. They are characterized by a heavy inflammatory infiltrate, particularly neutrophils; hyperpigmentation resulting from the recruitment of melanocytes at the wound site; partial proliferative activation; epidermal hyperproliferation; and non-healing ulcers. Persistent inflammation is reported to be the hallmark of most chronic wounds, and a large influx and retention of innate immune cells inhibit the repair processes. Some important roles of the immune cells are inhibited in chronic wounds, such as bactericidal and phagocytic activities that result in necrotic debris being formed at the wound edge [13].

8.3 Nanocellulose-Based Wound Dressings

Nanocellulose offers exceptional biological properties, including biocompatibility, biodegradability, and non-toxicity, as well as distinctive characteristics, such as polyfunctionality, ease of availability, high surface area, moldability, hydrophilicity, and mechanical strength [17]. Nanocellulose is useful for the creation of wound dressings due to the aforementioned characteristics. Wound dressings made of nanocellulose have been fashioned into sponges, films, nanofibers, foams, composites, hydrogels, and other materials (Figure 8.2).

8.3.1 Hydrogels

Hydrogels are gels made of polymers that are three-dimensional and have a porous structure that helps them absorb water [18]. The number of polar functional groups that crosslink together affects how hydrophilic a hydrogel is. They are similar to the extracellular matrix in that they have a high water content, are biocompatible, may expand and contract, and are not sticky, making them excellent for use as wound dressings [19]. The way

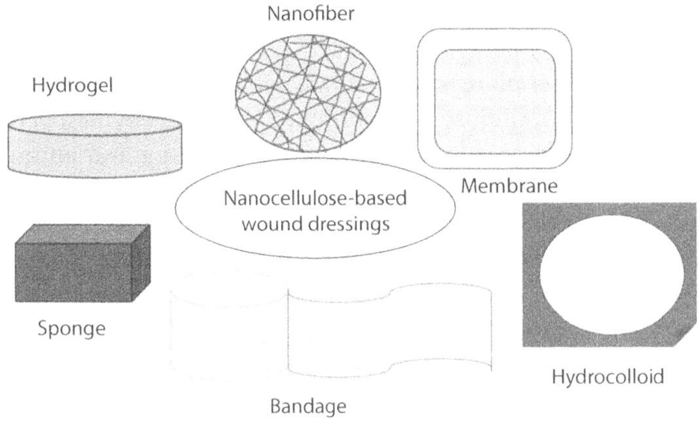

Figure 8.2 Different forms of wound dressings prepared from nanocellulose.

hydrogels help wounds heal is through their ability to exchange moisture, which creates an environment that is good for wound healing [18, 19]. Additionally, its high moisture content has a numbing effect that lessens pain [20]. Due to their lack of adhesion, they can be easily removed from the wound without causing additional damage. It is simple to keep an eye on the wound-healing process due to their transparency [18].

Several researchers have reported the creation of hydrogels from nanocellulose. Liu *et al.* [21] reported a nanocellulose dopamine-based hydrogel. Tetracycline hydrochloride was loaded into the hydrogels. The hydrogel exhibited excellent pH and near-infrared drug release capability. Its potential use in the treatment of infected wounds was demonstrated by its high antibacterial activity against *Escherichia coli* and *Staphylococcus aureus*, which are prevalent bacterial strains in chronic wounds. Additionally, the hydrogel accelerated *in vivo* wound healing. Because of the fibroblasts' increased activity, there were fewer inflammatory cells and less granulation tissue, which suggested that the collagen fibers had been replaced by structured collagen fibrils [21]. The phenyl bisphosphonate bismuth (III) complex was added by Maliha *et al.* to a nanocellulose hydrogel that had been oxidized with tetramethylpiperidine-1-oxyl. *Acinetobacter baumannii* and *Pseudomonas aeruginosa* were significantly killed by the incorporation of 9 g/g of the bismuth complex into the hydrogels, whereas vancomycin-resistant enterococci and methicillin-resistant *Staphylococcus aureus* were bacteriostatically inhibited. The hydrogel was non-toxic to mammalian fibroblast cells and murine L929 cell lines [22]. Basu *et al.* developed ion-crosslinked nanocellulose hydrogels using divalent copper or calcium ions. Calcium and copper-crosslinked hydrogels inhibited *S. epidermidis* and *P. aeruginosa* growth [23]. In order to speed up skin regeneration, Patel *et al.* reported an injectable and sticky hydrogel made from spherical nanocellulose and reinforced with carboxymethyl chitosan. Rat *in vivo* research revealed effective wound healing and antimicrobial capabilities [24]. Nanocellulose hydrogels with diosgenin-modified cross-linked gelatin interpenetration were studied by Erdagi *et al.* Neomycin and diosgenin together had a considerable inhibitory effect on the two bacteria strains utilized in the study, *E. coli* and *S. aureus*, ranging from 50 to 88%. The concentrations of nanocellulose and diosgenin had an impact on the hydrogels' swelling behaviors. According to reports, the hydrogel's pore diameters fall between 10 and 30 micrometers. Nearly 100% of the cells in the hydrogels survived for 3 days of incubation with human healthy dermal fibroblast cell lines *in vitro*. When neomycin was added to the hydrogels, the hydrogels exhibited strong antibacterial activity. According to the release studies, the first release was rapid for 15 minutes and continued for 9 to 12 h. The crosslinked hydrogels may be used as bandages for wounds [25]. In order to transport human dermal fibroblasts to the wound *in vivo*, Loh *et al.* explored bacterial nanocellulose acrylic acid hydrogels that are not biodegradable. Over 50% of human dermal fibroblasts were transported by the hydrogel in just 24 h, and skin microstructure started to form on day 7 after that. The loaded human dermal fibroblasts are essential for enhancing skin quality and reducing the chance of over-granulation, which happens when fibroblast proliferation is stimulated in an attempt to seal a wound [26].

Chen *et al.* reported graphene oxide-bacterial nanocellulose grafted poly(acrylic acid) composite hydrogels as wound dressings. They were prepared by an electron-beam irradiation technique. An increased concentration of graphene oxide in the hydrogel reduced the density of crosslinking and increased the pore sizes. The water vapor transmission rate was increased due to the increased pore sizes. The mechanical properties of the hydrogel were retained, but their bioadhesion hardness was lowered due to the high concentration of graphene oxide, revealing good durability and making them easily removable. The hydrogels were biocompatible with dermal fibroblast cell attachment [27]. Basu *et al.* also developed Ca^{2+}-crosslinked nanofibrillated cellulose hydrogels as hemostatic wound dressings. The hydrogel promoted blood coagulation and accelerated clot formation. Loading kaolin in the hydrogel further enhanced and boosted hemostasis, while the hydrogel containing collagen induced blood activation. The nanofibrillated cellulose hydrogels demonstrate great potential for the management of bleeding wounds [28]. Shahriari-Khalaji *et al.* enhanced the antimicrobial activity of bacterial nanocellulose-based and alginate-based hydrogels. The limitation associated with bacterial nanocellulose is the lack of antibacterial activity. The hydrogels were prepared by a vacuum suction method, and they were crosslinked via immersion in different cation solutions (cerium, manganese, silver, copper, cobalt, and zinc). The concentration of 0.05 M of the cross-linker was useful for preparing hydrogels that displayed significant antibacterial activity. The sequence of antibacterial activity of the hydrogels was bacterial nanocellulose-alginate-based-cobalt > bacterial nanocellulose-alginate-based-silver > bacterial nanocellulose-alginate-based-zinc > bacterial nanocellulose-alginate-based-copper > bacterial nanocellulose-alginate-based-manganese > bacterial nanocellulose-alginate-based-cerium. The blood clotting capability of the hydrogels was high in the first 15 min for the hydrogels bacterial nanocellulose-alginate-based-zinc and bacterial nanocellulose-alginate-based-copper that exhibited rapid blood clotting. The hydrogels were non-toxic on fibroblast cells with significantly accelerated healing *in vivo* for hydrogels [29]. Ren *et al.* prepared a hydrogel in which ε-polylysine-grafted nanocellulose polydopamine nanoparticles were loaded in genipin crosslinked gelatin. The hydrogel's broad-spectrum antibacterial activity, good antioxidant performance, and capability to induce cell proliferation were significant. Full-thickness *in vivo* studies on methicillin-resistant *Staphylococcus aureus*-infected skin revealed accelerated healing with reduced inflammation [30].

8.3.2 Sponges

Sponges are characterized by a macroporous structure that provides high exudate retention capacity and promotes high oxygen permeability, and high flexibility [31]. Cheng *et al.* developed sponges with Janus characters from cellulose nanofibers. The hydrophobic layer was prepared from vinyl trimethoxy silane and cellulose nanofibers, while the hydrophilic layer was prepared using a combination of chitosan, cellulose nanofibers, and crosslinking with γ-glycidoxypropyltrimethoxysilane. The hydrophilic layer of the sponge absorbed water from the blood and worked in synergy with the hemostatic chitosan-rich layer to accelerate blood clotting. The hydrophobic layer exerted pressure and hindered blood penetration. The sponges were effective in controlling bleeding in a femoral artery injury model *in vivo*, with a 50% reduction in blood loss. Extended survival time was reported in a carotid artery injury model [32]. The sponge's effective hemostasis effect was due to the hydrophilic layer's capability to accelerate blood clotting, while the hydrophobic layer exerted proper pressure and prevented further blood permeation (Figure 8.3).

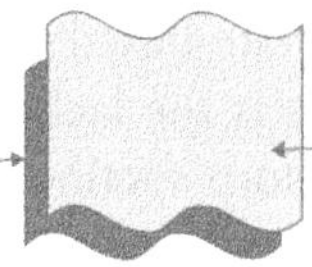

Figure 8.3 Nanocellulose-based sponge with Janus character.

Sultana *et al.* prepared freeze-dried sponges from tempo-oxidized nanocellulose and chitosan. It was loaded with Lawson for controlled delivery during wound healing. Increasing the content of chitosan decreased the pore diameter tempo-oxidized nanocellulose, enhancing the mechanical stability of the sponges. The rate of degradation of the sponges was high *in vitro*, and the drug release at a pH of 5.5 was rapid. The *in vivo* studies on a rat tail amputation model revealed enhanced hemostasis. The high swelling and rapid degradation of the freeze-dried sponges maintained rapid hemostasis properties *in vivo* in the rat tail amputation study. *In vivo* studies on a rat cutaneous full-thickness punch biopsy model revealed thick granulation tissue, collagen, and angiogenesis after 2 weeks [33]. Zheng *et al.* prepared an injectable hemostatic sponge characterized by a high blood absorption ratio, good shape memory property, high elasticity, and antibacterial activity on *E. coli*. It was prepared from cellulose nanocrystals, the skin secretion of *Andrias davidianus*, and cellulose nanofibers. *In vivo* studies in an incompressible animal model showed that the sponge displayed excellent hemostatic properties. The sponge also displayed excellent shape memory in water and excellent stability, which are attributed to the hydrogen-bonded networks and the additional ionic bonds between the carboxyl and amine groups [34].

Sponge made of bacterial cellulose, chitosan, and collagen was created by Yaun *et al.* Collagen was added to the sponges in order to improve their ability to maintain hemostasis. The unique characteristics of the wound dressings were discovered through *in vivo* tests, including strong mechanical strength, exceptional hemostatic efficiency, exceptional biodegradation, and broad-spectrum antibacterial activities. The sponges proved to be absorbable hemostats for the management of internal hemorrhage in deep, non-compressible wounds because they demonstrated a strong adherence of platelets and erythrocytes and decreased blood loss [35]. Chloramphenicol was loaded into poly(vinyl alcohol)-nanocellulose fibril-based sponges, according to Balla *et al.* The high compressive strength of the sponges was improved by the dispersion of nanocellulose fibrils in the sponges. The release profiles indicated a 3-day controlled release. By forming a tough network, cellulose nanofibril reinforcement boosted mechanical properties—compressive strength—while decreasing water sensitivity [36]. Gentamicin was covalently grafted to a nanocellulose-based sponge by Xiao *et al.* via an enamine bond. The sponge was made via multi-crosslinking cellulose nanofibers, cellulose acetoacetate, and 3-aminopropyl(triethoxy)silane. When tested against the bacterial strains *E. coli* and *S. aureus*, the sponge demonstrated good antibacterial properties. The sponge's density of 0.0174 g/cm^3 indicated that it was remarkably light [37]. A microfibrillated cellulose-reinforced sponge made of carboxymethyl chitosan and oxidized starch was described by Long *et al.* The sponge's quick water absorption and blood-triggered shape restitution properties in more than 3 s were made possible by its 1.2 weight percent microfibrillated cellulose content. Recombinant humanized collagen type III was combined with silver nanoparticles to modify the sponge and add antibacterial properties. The sponge's antibacterial efficacy against *S. aureus* and *E. coli* demonstrated that it might be used to heal chronic wounds. The presence of silver nanoparticles also reduced inflammation, enhanced angiogenesis, and increased cell proliferation, according to *in vivo* experiments [38].

8.3.3 Nanofibers

Because of their high nanoporosity, aspect ratio, and surface area, nanofibers have received extensive research attention as wound dressings. The optimum quantity of moisture is provided for fibroblasts to function and cells to multiply, which aids in the wound's ability to heal [39]. These features also aid and stimulate drug administration, inhibit bacterial invasion, and permit gas permeability and a good water vapor transfer rate. The most popular method for making them is electrospinning because it is affordable, simple, and repeatable. However, the properties of electrospun nanofibers are influenced by the conductivity, surface tension, viscoelasticity, surface tension, and volatility of the polymer solution. The properties of the electrospun nanofibers are significantly influenced by a number of additional factors, including the applied voltage (kV), flow rate, distance between the needle tip and the collector, humidity, and temperature [39, 40].

Silva and his colleagues made biodegradable nanofibrous patches out of nanofibrillated cellulose and lysozyme, a protein found in hen egg whites. The nanofibrous patch had good antimicrobial and antioxidant activity. These are biological properties that can help treat wounds that are infected. The patches were biocompatible with L929 fibroblast cells and induced an almost complete wound [41]. Shahriari-Khalaji *et al.* reported grafting aminoalkylsilane to a bacterial nanocellulose membrane, which was bridged with Pullulan-ZnO-NP hybrid electrospun nanofibers. The obtained patches displayed good blood clotting capability and excellent antibacterial activity. ZnO continued to be released from the membrane. When tested *in vitro* with L929 fibroblast cells, it proved to be biocompatible in nature. A faster healing process with re-epithelialization, the growth of tiny blood vessels, and collagen synthesis was demonstrated in *in vivo* investigations using a rat model. The cost of the nanofiber dressing was reasonable [42]. Chitosan, polyacrylamide, and cellulose nanocrystals were used by Ribeiro *et al.* to create nanofibers (ethylene oxide). The nanofibers included an extract made from the acacia plant. The thermal characteristics, physical integrity, water vapor transmission rate, and diameter of the nanofibers were all improved by the cellulose nanocrystals in a special way. After being loaded into the nanofibers, the plant extract's antibacterial action remained intact. The antifungal activity did, however, increase. The nanofibers had a steady release of the extract for 24 h and were biocompatible, making them ideal for lowering the frequency of dressing replacement [43]. By electrospinning chitosan and bacterial nanocellulose at the same time using separate syringes and coaxial electrospinning, Ardila *et al.* created chitosan-bacterial nanocellulose. This process created core-shell structures for both materials. The employed coaxial electrospinning method produced high-quality nanofibers. The nanofibers' strong antibacterial activity against *E. coli* (99.9%) and average diameter of 85 nm were two characteristics that set them apart. It is claimed that the chitosan amino group protonation at the surface of the nanofibers is responsible for the nanofibers' purported antibacterial mode of action. The nanofibrous network additionally encourages strong contact with the bacterial cell wall [44].

Wang *et al.* loaded an antioxidant, procyanidins obtained from sea buckthorn, into *Apocynum venetum* cellulose nanofiber. The bilayer nanofiber was prepared by emulsion electrospinning. The nanofibers were uniform and smooth, with a core-sheath structure. The core was composed of extract-loaded-polyvinyl pyrrolidone, and the shell was prepared from polylactic acid. The encapsulation efficiency of nanofiber membranes was over 83%. The *in vitro* drug release profile of the nanofiber was characterized by an initial burst release. A sustained drug release occurred for 30 h after a burst release. The nanofibers also displayed excellent antioxidant activity. The nanofibers exhibited a high antioxidant effect, with a DPPH clearance of over 88.62%. The effect of the cellulose in the nanofiber network included a prolonged release of the antioxidant, which promoted cell proliferation and wound healing [45]. Wang also reported the structure, morphology, and antibacterial activities of *Apocynum venetum* poly(lactic acid) and nanocellulose nanofibers. Different forms of cellulose nanofibers from *Apocynum venetum* were prepared from the phloem and straw. The straw produced finer and longer-length networks, while the phloem produced short, rod-shaped network structures. Combining both sources of cellulose for the preparation of the nanofibers improved the wettability and permeability of PLA. The antibacterial activity of the nanofibers prepared from the straw was enhanced by over 90% when compared to the nanofibers prepared from the phloem against *E. coli* [46]. Sugha *et al.* developed spray-coated cellulose nanofibers with a size of 70 nm. The surface was coated with montmorillonite to reduce the pore size and enhance the stiffness. Incorporating AgNPs into the nanofibers enhanced their antioxidant properties. The nanofibers also exhibited a high-water absorption rate, an important feature for treating exuding wounds [47].

8.3.4 Hydrocolloids

Hydrocolloid dressings are classified as interactive wound dressings. They inhibit bacterial invasion, are permeable to water vapor, absorb wound exudates, and also exhibit properties of debridement. They are suitable for moderately exuding wounds, such as minor burn wounds, pressure sores, and pediatric wound care. When in contact with wound exudate, they created gels and also offered a wet environment beneficial to tissue granulation process. They do not hurt when you take them out of the wounds. Their disadvantage, however, is that they must be used as secondary dressings and are not appropriate for wounds with excessive exudate [48]. On hydrocolloids based on nanocellulose, there are not many research publications.

Lungu *et al.* reported hydrocolloids (alginate and pectin) combined with wood-obtained nanocellulose fibrils. Surface-functionalized cellulose nanofibrils with carboxyl groups were used for a non-conventional physical gelation approach at pH <3 using citric acid. The wound dressings displayed higher stability and induced a high rate of proliferation [49]. Kong *et al.* explored the mechanisms underlying adhesion loss in hydrocolloid wound dressings. Adhesion loss is a serious hydrocolloid restriction. Over the course of 10 h, a hydrocolloid dressing containing sodium carboxymethyl cellulose displayed a number of instances of CMC swelling, dissolving, and oozing. Due to the dressing's released CMC, the surface energy increased from 14.5 to 80.7 mN/m with an adhesion loss. A hydrocolloid dressing that adheres to the skin absorbs wound exudate during the healing process because of its hydrophilic fillers. The hydrophilic fillers break down and flow to the dressing's surface during a protracted healing phase as a result of more exudate being absorbed. Thick hydrophilic polymer solution coatings improve surface energy. Reduced adhesion force and loss of adherence occur as a result of the hydrocolloid dressing's higher surface energy than the skin's [50]. Collado-Boira *et al.* combined carboxymethylcellulose cellulose fibers with hydrocolloid dressing to treat irritating contact dermatitis, a common side effect of ostomies brought on by the peristomal skin coming into contact with the acidic stoma effluent. Clinical evaluation of the hydrocolloid wound dressing was studied on three individuals with severe irritating contact dermatitis-related peristomal skin lesions, ranging in age from 70 to 81. Using a bandage composed of hydrocolloid and carboxymethylcellulose cellulose fibers allowed the lesions to heal after 4 weeks [51].

8.3.5 Films/Membranes

Film wound dressings let gas through, and because they are clear, it is easy to see how the wound is healing. They are not permeable to bacteria and are conformable and adhesive [52]. However, upon removal, they can cause further trauma and are not effective for treating highly exuding wounds. They are suitable for wounds such as lacerations, minor burns, and abrasions. They are also appropriate as waterproofs for primary dressings, such as foam. They are also suitable for radiation dermatitis [52]. Some researchers have reported that nanocellulose-based films are suitable for accelerating the healing of infected and exuding wounds.

Abdollahi *et al.* employed a solution-casting method to develop gelatin/Persian gum/bacterial nanocellulose films. The film was loaded with *Teucrium polium* extract and frankincense essential oil. The dispersion of the bacterial nanocellulose in the film matrix was uniform and confirmed by a scanning electron microscopy analysis. The presence of Persian gum and bacterial nanocellulose in the films reduced the rate of biodegradation, thereby enhancing stability. The incorporation of essential oils in the films decreased their swelling capability. The films displayed good anti-inflammatory and antibacterial activity, resulting from the loading of the plant extract and essential oil. The films also exhibited non-hemolytic activity [53]. Taheri *et al.* dispersed varied amounts of tannic acid and bacterial nanocellulose in chitosan/gelatin films for reinforcement. The dispersion of nanocellulose in the film matrix was uniform and revealed by SEM images. A high rate of wound contraction was visible *in vivo* in Wistar rats. The presence of nanocellulose in the films induced complete epithelialization with thick dermal formation and promoted fibroblast proliferation [54]. Fu *et al.* employed bacterial nanocellulose biosynthesized by *Gluconacetobacter xylinus* for the development of films. The films displayed low cytotoxicity, revealing good biocompatibility. The films also induced good proliferation and adhesion in NIH/3T3 cells *in vitro*. Rapid healing and a minimal inflammatory response were observed in *in vivo* experiments on large-area full-thickness skin defects on the back of C57BL/6 mice during animal surgery [55].

ZnO-loaded halloysite nanotubes were integrated by Jaberifard *et al.* into dialdehyde-corn starch-gelatin-bacterial nanocellulose sheets. Thermal stability was improved by increasing the amount of bacterial nanocellulose in the films from 0.03 wt% to 0.12 wt%. The films dramatically increased NIH-3T3 fibroblast cells' ability to proliferate and develop, which is due to ZnO-loaded halloysite nanotubes and bacterial nanocellulose. The films had remarkable antibacterial activity against *S. aureus* and *E. coli* [56]. Nanocrystalline cellulose and calcium peroxide particles were added by Akhavan-Kharazian *et al.* to gelatin/chitosan films. The nanocrystalline cellulose and calcium peroxide particles enhanced the mechanical characteristics of the films. Nanocrystalline cellulose and calcium peroxide particles were present in the films, which caused a decrease in swelling and WVTR. The films were non-toxic and had effective antibacterial activity against *E. coli* [57]. Riccio *et al.* loaded either silver sulfadiazine or betamethasone into chitosan-betamethasone nanocellulose-based films prepared by solvent casting. The films were flexible with high exudate absorption ability and displayed sustained drug release by diffusion. The films were permeable, reduced swelling capability, and allowed gas exchanges useful to prevent microbial growth. The wound dressings were found to be effective for both infectious and dry wounds [58]. Nezhad-Mokhtari *et al.* used bacterial nanocellulose obtained from Acetobacter for the reinforcement of gelatin/dialdehyde-modified guar gum films loaded with honey. Increasing the honey content of the films resulted in improved adhesion. The films promoted a significant proliferation of NIH-3T3 fibroblast cells *in vitro* and good degradation behavior. The films loaded with 15 wt% honey inhibited *S. aureus* and *E. coli* bacteria [59]. Lima *et al.* incorporated Pinhão seed of *Araucaria angustifolia* nanocellulose into polyvinyl alcohol film. The *in vitro* scratch assay of the film showed complete closure in 72 h. The addition of nanocellulose to the films enhanced the modulus elasticity and non-toxic nature [60].

Additionally, the use of nanocellulose-prepared membranes as wound dressings has been documented. The creation and characterization of low-cost membranes made from wood cellulose nanofibrils was described by Claro *et al.* The membranes were acceptable for re-epithelialization over a 4-day period because they had strong adherence to the moist wound surfaces. There was no allergic or inflammatory reaction to the membranes. However, due to the membrane's limited porosity, exudate accumulated between the membrane and the wound [61]. Different types of bacteria may biosynthesize bacterial nanocellulose, which is a suitable biomaterial for designs of wound dressings, according to Yuan *et al.* When compared to the lower surfaces, the higher surfaces of the membranes were denser and more compact. The development of fibroblasts on the membrane's lower surfaces showed that they were more biocompatible than the top surface. A study conducted *in vivo* revealed that the membranes with thick upper surfaces promoted faster healing [62]. The effectiveness of physically immobilizing a commercial laccase on bacterial nanocellulose membranes was examined by Sampaio *et al.* They became more susceptible to Gram-positive bacteria compared to Gram-negative bacteria as a result of the immobilization of laccase on the membrane, which had an antibacterial effect. It was not cytotoxic to the membranes [63]. Bacterial nanocellulose/chitosan/poly(vinyl alcohol) membranes made by an immersion approach were described by Ju *et al.* Large pores could be seen in the membranes. Chitosan and poly(vinyl alcohol) decreased the tensile strength and Young's modulus of wet composite membranes. In contrast to bacterial nanocellulose, which had no discernible antibacterial effects, the membranes demonstrated good hydrophilicity and considerable antibacterial activity against *E. coli* and *S. aureus* [64]. Dexpanthenol was found to be loaded into a multi-layered membrane made of oxidized bacterial cellulose by Fonseca *et al.* Alginate and chitosan were coated onto the layers of the membrane using a spin-assisted layer-by-layer construction method. Thermal stability of the membranes was good. The membranes have a 240–250% moisture absorption and a Young's modulus of 4 GPa. These membranes had a sustained drug release profile, were effective against *S. aureus*, and did not harm HaCaT cells. After 24 h, the *in vitro* wound healing assay revealed good cell migration capacity and full gap closure [65].

Aloe vera extract's impact on bacterial nanocellulose membranes was studied by Piaia *et al.* When cultured on membranes, the extract's presence increased the metabolic activity and proliferation of human fibroblasts and keratinocytes, demonstrating their ability to stimulate the healing of skin tissue [66]. *In vivo* testing of the pinus nanocellulose membrane's ability to cure second-degree burns in male Wistar rats was conducted by Coelho *et al.* On day 20, the membrane encouraged neovascularization and the development of a superficial neoepithelium layer on the wound surface, which sped up healing [67].

8.3.6 3D-Printed Wound Dressings

3D bioprinting technology has attracted huge attention in medicine. It is a technique that involves the fabrication of a complex matrix called bioink.

Some parameters influence the high functional bioink integrity, including the physicochemical properties—crosslinking degree, viscosity, gelation time, and shear-thinning—and printing parameters—duration of printing, diameter, feed rate, nozzle temperature, etc. [68]. 3D printing technology is appropriate for subtle micro-component design to regulate the delivery of different bioactive agents. During printing procedures, the wound dressing scaffolds can be loaded with nanoparticles and bioactive agents [69]. The technique is reliable and affordable. It is useful for tailoring the stiffness and geometry of scaffolds to the shape of the skin lesion. It offers excellent control of the scaffold's geometry, porosity, and mechanical stability [70]. 3D printing has been explored for the design of nanocellulose-based wound dressings by some researchers.

Rees *et al.* investigated how nanocellulose, a bioink, may be used to modify film surfaces. Using kraft pulp fibers from *Pinus radiata* that had been bleached, nanocellulose was created. The pretreatment of the nanocellulose materials included TEMPO-mediated oxidation (C-TEMPO) and a mixture of carboxymethylation and periodate oxidation (also termed C-Periodate). The substrate and bioink employed were C-TEMPO and C-Periodate, respectively. Due to its better consistency and suitable rheology, C-Periodate nanocellulose was effective as a bioink. The 3D structure collapsed as a result of the C-TEMPO nanocellulose's poor consistency and viscosity. The macropores were blocked by the crosslinker $CaCl_2$. The 3D structures with porosity created by the nanocelluloses are helpful for the release of antibacterial agents. The nanocelluloses prevented the growth of bacteria [71]. According to Xu *et al.*, double crosslinking using *in situ* Ca2+ and 1,4-butanediol diglycidyl ether produced 3D-printed nanocellulose hydrogel scaffolds. The 1% nanocellulose-printed hydrogel scaffolds had mechanical strengths ranging from 3 to 8 kPa. The hydrogel scaffolds enabled the multiplication of fibroblast cells, demonstrating their potential as scaffolds for skin regeneration and wound healing [72]. In 3D-printed scaffolds, Resndahl *et al.* examined cellulose nanofibrils produced by 2,2,6,6-tetramethylpyperidine-1-oxyl-mediated oxidation. The scaffolds promoted the proliferation of fibroblasts and increased the expression of marker genes important for wound healing [73]. According to Xu *et al.*, gelatin methacrylate and 2,2,6,6 tetramethylpiperidine-1-oxyl radical-oxidized cellulose nanofibrils (1 w/v%) were combined to create 3D scaffolds. The scaffolds demonstrated strong mechanical strength between 2.5 and 5 kPa, as well as good structural stability. When the scaffolds were grown with 3T3 fibroblasts, they were biocompatible. Fibroblast growth was induced by the scaffolds [74].

8.3.7 Composites

Nanocellulose has also been designed in the form of composites for wound dressings. Dehkordi *et al.* employed nanocrystalline cellulose as a reinforcement agent in hyaluronic acid-based composites loaded with chitosan nanoparticles to improve mechanical properties. The *in vivo* studies showed that wounds covered with the composite reached almost 100% wound closure with significant re-epithelialization. The composite significantly lowered inflammation with the formation of granulation tissue [75]. Yuan *et al.* used novel electrophoresis-based technology to prepare bacterial nanocellulose-based composites functionalized with collagen and lactoferrin. The water-vapor transmission rate revealed their capability to provide a moderately moist environment for the wound bed to accelerate epithelial cell migration and the wound-healing process. Loading lactoferrin in the composites enhanced the bacterial activity of the composites, with a bacterial viability of over 77%. *In vivo* studies in rat models revealed the formation of a new epithelium in 9 days [76]. Yang *et al.* employed bacterial cellulose produced from *Acetobacter xylinum* in combination with methylglyoxal to prepare a composite via a dip-coating method. The composite was loaded with Manuka honey. The composites exhibited high-temperature stability and mechanical properties. The composite showed a high zone of inhibition diameter of 17.1 mm for *Staphylococcus aureus*, 15.5 mm for *Escherichia coli*, and 12.3 mm for *Pseudomonas aeruginosa* [77]. Vilela *et al.* prepared a nanocomposite from crosslinking poly([2-(methacryloyloxy)ethyl]trimethylammonium chloride] and bacterial nanocellulose. The nanocomposite displayed interesting features, such as high water uptake capacity, a Young's modulus of 2.4 GPa, and UV–A and UV–B blocking properties, and was non-cytotoxic on human keratinocytes. The fungal inactivation capability of the nanocomposite against *C. albicans* revealed their potential use as wound dressings [78]. Sun *et al.* prepared nanocellulose-polyethylene glycol-based nanocomposites. The mechanical properties of the nanocomposites after wetting revealed that a higher degree of fibrillation lowered the water absorption. The nanocomposites were ductile, flexible, and conformed to the skin, properties useful for wound dressings [79]. Poonguzhali *et al.* used a solution casting method to prepare chitosan-poly(vinyl pyrrolidone)-nanocellulose composites. Increasing the concentration of nanocellulose in the composites decreased their thermal stability. The composites revealed the capability to provide moisture to the wound bed. The nanocomposites also exhibited enhanced swelling capability, good antibacterial activity, and blood compatibility. Incorporating 5% wt of nanocellulose in the nanocomposites increased their swelling capability significantly. The tensile strength of the nanocomposites was in the range of 35.6±5.8–39.6±6.9 MPa [80].

8.3.8 Other Scaffolds

Other scaffolds have been designed using nanocellulose. Aerogels have been prepared from nanocellulose for the treatment of wounds. Aerogels are highly porous materials with high surface areas resulting from their solid structures that are not regular and can be tailored using selected preparation conditions. It can be prepared in different forms depending on the shape of the mold used during gelation. Rapid matrix erosion when placed in liquids and rapid drug release limit the use of hydrophilic aerogels. To overcome the aforementioned limitation, the functionalization of the surface with hydrophobic groups is important, resulting in a hydrophobic aerogel matrix that limits the rapid penetration of liquid into the pores and prevents the burst release of drugs [81].

Saini *et al.* made a microfibrillated cellulose solution by using *in situ* fibrillation of cellulose in cinnamon extract. This was used to make a viscous gel that was freeze-dried to make a flexible composite aerogel. The optimal concentration of carboxylated cellulose nanocrystals used for the preparation of aerogel with improved mechanical properties (compressive strength of 37 kPa and Young's modulus of 110 kPa) was 0.3 weight %. The addition of cinnamon extract enhanced the antibacterial property of the aerogel. The prolonged and excellent antimicrobial activity of the aerogel was significant against *E. coli* and *S. aureus*. The aerogel prepared was bio-based, durable, flexible, and showed the capability to overcome long-term side effects associated with scaffolds used for wound dressings [82]. Ghafari *et al.* used a one-step freeze-drying technique to prepare a bilayer aerogel-based scaffold, cellulose nanofiber-poly(vinyl) alcohol. The scaffolds were characterized by interconnected pores. The use of a lower concentration of the polymers produced a highly porous scaffold that exhibited reduced mechanical strength and increased water uptake. The bilayered scaffold mimicked the epidermis and dermis with porosities of 89% and 95% and pore sizes of 19.72 and 90.71 μm, respectively. The porosity and pore size of the two layers are ideal for accelerated wound healing. The high pore size resulted in reduced mechanical properties. Using a high concentration of polymers resulted in increased crystallinity [83].

Other wound dressing scaffolds were prepared from nanocellulose. Holzer *et al.* reported bacterial nanocellulose scaffolds that produce a cooling effect on burn wounds. The scaffold's high water content provided a cooling effect on burn wounds with the potential to reduce intradermal damage to the skin. The scaffold reduced the temperature significantly in the simulated burn skin model to a maximum temperature difference of 6.4°C, with the potential to reduce necrosis and dermal–epidermal separation. Hypothermia is a major problem encountered when cooling burn wounds and can result in high mortality. However, cooling a burn wound is important for reducing skin damage and scar formation. The cooling effect of the bacterial nanocellulose-based scaffold was achieved through evaporation because the dressing was stored at room temperature [84].

Nanocellulose-based cryogels have been reported. By using low-temperature polymerization reactions, cryogels are gel networks. They have a three-dimensional structure that can support the proliferation of cells. Features such as pore size, density, and thickness affect how they perform. Some of the unique features of cryogels are their high water content, non-toxicity, blood compatibility, resistance to degradation, etc. [85]. Ghafari *et al.* developed a bilayer cryogel structure simulating the epidermis and dermis layers. A single step of freeze-drying using cellulose nanofibers and poly(vinyl) alcohol produced it. The pore size of the scaffold was 19.72 μm for the simulated epidermis and 90.71 μm for the dermal layers. The pore size is important for cell adhesion, proliferation, and the transportation of nutrients. The biodegradability and degradation rate of the scaffold were influenced by the pore size. Increasing porosity increased the biodegradability of the scaffolds. The ability of cells to penetrate the bilayer scaffold is important for cell migration and proliferation [86].

8.4 Commercialized Nanocellulose-Based Wound Dressings

There are some commercialized nanocellulose-based wound dressings. Epicite hydro® is prepared from bacterial nanocellulose. It has a nanofiber network structure with good flexibility and mechanical stability. It has a water content of over 95% and is biocompatible. Its high water content is useful for prolonged dermal hydration and provides a moist environment for the wound bed, thereby hindering excessive fluid loss from the wound. It is non-adhesive, has a soft texture, and is semi-transparent [87]. It has been evaluated in children with burn wounds. There were no observed infections in the wounds, and the reepithelialization was rapid with the absence of hypertrophic scars [87, 88]. Dermafill™ is prepared from bacterial cellulose derived from *Acetobacter xylinum* and has also been evaluated for the management of non-healing lower extremity ulcers, where it accelerated the rate of wound closure [89].

Bionext® is a membrane nanofiber prepared from pure cellulose obtained from *Acetobacter xylinum*. Its capability to regulate water moisture, protect wounds from the invasion of microorganisms, and promote cell metabolism makes it an effective wound dressing for the treatment of burns, diabetic wounds, ulcers, etc. Its selective permeability promotes the uptake of gases. It can stay intact for over 15 days, thereby overcoming the need for frequent changing [90]. PRISMA™ Matrix is a freeze-dried composite containing collagen (55%), oxidized, regenerated cellulose (44%), and silver (1%). It provides a moist environment for the wound bed and promotes the formation of granulation tissue and epithelialization [91]. Xcell® is used to treat chronic venous lower-leg ulcers. It was reported to absorb moisture and is effective for treating hard-to-heal chronic wounds [92]. Other commercialized nanocellulose-based wound dressings are Bioprocess®, ConvaTec, MTA protective tissue, and Epiprotect®. These wound dressings mimic extracellular matrix and improve epithelialization and skin regeneration [93].

8.5 Future Perspective and Conclusion

Nanocellulose has been widely investigated for the design of wound dressings in different forms, such as nanofibers, hydrogels, films, membranes, hydrocolloids, etc., to treat acute, chronic, high-exuding, and bleeding wounds. Furthermore, bacterial nanocellulose's high content purity, tensile strength, and flexibility have been exploited to design wound dressings that reduced scarring, provided an optimal moist healing environment, accelerated the wound-healing process, and reduced fluid loss and pain [94]. The above-mentioned features make nanocellulose an important biomaterial for the development of wound dressings. Despite the widely reported studies on nanocellulose-based wound dressings, very few commercialized nanocellulose-based products have been reported. The reports of 3D-printed nanocellulose scaffolds were few. Their use in 3D bioprinting needs further investigation and evaluation. Nanocellulose has been widely reported as a biocompatible biomaterial with low toxic and immunogenic effects [94]. However, some studies have shown that nanocellulose can induce an inflammatory reaction [95, 96]. The pro-inflammatory response of nanocellulose can be overcome by the introduction of chemical groups to its surface. However, more immunological investigations of nanocellulose are crucial to fully understanding their effects on innate and adaptive immunity. There is a need for more *in vivo* studies using appropriate animal models to fully understand the complex immune cell interactions with nanocellulose and the safe use of nanocellulose in the design of wound dressings.

Acknowledgments

The financial support of Govan Mbeki Research and Development Centre, University of Fort Hare, South African Medical Research Council, and National Research Foundation, South Africa towards this research is hereby acknowledged.

References

1. Santos, D., Silva, U.F., Duarte, F.A., Bizzi, C.A., Flores, E.M., Mello, P.A., Ultrasound-assisted acid hydrolysis of cellulose to chemical building blocks: Application to furfural synthesis. *Ultrason. Sonochem.*, 40, 81–88, 2018.
2. Leong, S.L., Tiong, S.I.X., Siva, S.P., Ahamed, F., Chan, C.H., Lee, C.L., Chew, I.M.L., Ho, Y.K., Morphological control of cellulose nanocrystals via sulfuric acid hydrolysis based on sustainability considerations: An overview of the governing factors and potential challenges. *J. Environ. Chem. Eng.*, 10, 108145, 2022.
3. Trache, D., Tarchoun, A.F., Derradji, M., Hamidon, T.S., Masruchin, N., Brosse, N., Hussin, M.H., Nanocellulose: From fundamentals to advanced applications. *Front. Chem.*, 8, 392, 2020.
4. Tarchoun, A.F., Trache, D., Klapötke, T.M., Derradji, M., Bessa, W., Ecofriendly isolation and characterization of microcrystalline cellulose from giant reed using various acidic media. *Cellulose*, 26, 7635–7651, 2019.
5. Salimi, S., Sotudeh-Gharebagh, R., Zarghami, R., Chan, S.Y., Yuen, K.H., Production of nanocellulose and its applications in drug delivery: A critical review. *ACS Sustain. Chem. Eng.*, 7, 15800–15827, 2019.
6. Lee, S.H., Kim, H.J., Kim, J.C., Nanocellulose applications for drug delivery: A review. *J. For. Environ. Sci.*, 35, 141–149, 2019.
7. Copenhaver, K., Li, K., Wang, L., Lamm, M., Zhao, X., Korey, M., Neivandt, D., Dixon, B., Sultana, S., Kelly, P., Gramlich, W.M., Pretreatment of lignocellulosic feedstocks for cellulose nanofibril production. *Cellulose*, 29, 1–42, 2022.
8. Naz, S., Ali, J.S., Zia, M., Nanocellulose isolation characterization and applications: A journey from non-remedial to biomedical claims. *Biodes. Manuf.*, 2, 187–212, 2019.
9. Trache, D., Nanocellulose as a promising sustainable material for biomedical applications. *AIMS Mater. Sci.*, 5, 201–5, 2018.

10. Gatenholm, P. and Klemm, D., Bacterial nanocellulose as a renewable material for biomedical applications. *MRS Bull.*, 35, 208–213, 2010.
11. Sindhu, K.A., Prasanth, R., Thakur, V.K., Medical applications of cellulose and its derivatives: Present and future, in: *Nanocellulose Polymer Nanocomposites*, pp. 437–477, Scrivener Publishers, Canada, 2014.
12. Shi, C., Wang, C., Liu, H., Li, Q., Li, R., Zhang, Y., Liu, Y., Shao, Y., Wang, J., Selection of appropriate wound dressing for various wounds. *Front. Bioeng. Biotechnol.*, 8, 182, 2020.
13. Martin, P. and Nunan, R., Cellular and molecular mechanisms of repair in acute and chronic wound healing. *Br. J. Dermatol.*, 173, 370–378, 2015.
14. Sorg, H., Tilkorn, D.J., Hager, S., Hauser, J., Mirastschijski, U., Skin wound healing: An update on the current knowledge and concepts. *Eur. Surg. Res.*, 58, 81–94, 2017.
15. Wallace, H.A., Basehore, B.M., Zito, P.M., Wound healing phases, in: *StatPearls*, StatPearls Publishing, Treasure Island (FL), 2022, PMID: 29262065.
16. Almadani, Y.H., Vorstenbosch, J., Davison, P.G., Murphy, A.M., Wound healing: A comprehensive review. *Semin. Plast. Surg.*, 35, 141–144, 2021.
17. Ullah, M.W., Rojas, O.J., McCarthy, R.R., Yang, G., Nanocellulose: A multipurpose advanced functional material. *Front. Bioeng. Biotechnol.*, 9, 1–4, 2021.
18. Gupta, A., Kowalczuk, M., Heaselgrave, W., Britland, S.T., Martin, C., Radecka, I., The production and application of hydrogels for wound management: A review. *Eur. Polym. J.*, 111, 134–151, 2019.
19. Caló, E. and Khutoryanskiy, V.V., Biomedical applications of hydrogels: A review of patents and commercial products. *Eur. Polym. J.*, 65, 252–267, 2015.
20. Edwards, J., Hydrogels and their potential uses in burn wound management. *Br. J. Nurs.*, 19, S12–16, 2010.
21. Liu, Y., Sui, Y., Liu, C., Liu, C., Wu, M., Li, B., Li, Y., A physically crosslinked polydopamine/nanocellulose hydrogel as potential versatile vehicles for drug delivery and wound healing. *Carbohydr. Polym.*, 188, 27–36, 2018.
22. Maliha, M., Brammananth, R., Dyson, J., Coppel, R.L., Werrett, M., Andrews, P.C., Batchelor, W., Biocompatibility and selective antibacterial activity of a bismuth phosphinato-nanocellulose hydrogel. *Cellulose*, 28, 4701–4718, 2021.
23. Basu, A., Heitz, K., Strømme, M., Welch, K., Ferraz, N., Ion-crosslinked wood-derived nanocellulose hydrogels with tunable antibacterial properties: Candidate materials for advanced wound care applications. *Carbohydr. Polym.*, 181, 345–350, 2018.
24. Patel, D.K., Ganguly, K., Hexiu, J., Dutta, S.D., Patil, T.V., Lim, K.T., Functionalized chitosan/spherical nanocellulose-based hydrogel with superior antibacterial efficiency for wound healing. *Carbohydr. Polym.*, 284, 119202, 2022.

25. Erdagi, S.I., Ngwabebhoh, F.A., Yildiz, U., Genipin crosslinked gelatin-diosgenin-nanocellulose hydrogels for potential wound dressing and healing applications. *Int. J. Biol. Macromol.*, 149, 651–663, 2020.
26. Loh, E.Y., Fauzi, M.B., Ng, M.H., Ng, P.Y., Ng, S.F., Amin, M.C., Insight into delivery of dermal fibroblast by non-biodegradable bacterial nanocellulose composite hydrogel on wound healing. *Int. J. Biol. Macromol.*, 159, 497–509, 2020.
27. Chen, X.Y., Low, H.R., Loi, X.Y., Merel, L., Mohd Cairul Iqbal, M.A., Fabrication and evaluation of bacterial nanocellulose/poly (acrylic acid)/ graphene oxide composite hydrogel: Characterizations and biocompatibility studies for wound dressing. *J. Biomed. Mater. Res. B Appl. Biomater.*, 107, 2140–2151, 2019.
28. Basu, A., Hong, J., Ferraz, N., Hemocompatibility of Ca2+-crosslinked nano-cellulose hydrogels: Toward efficient management of hemostasis. *Macromol. Biosci.*, 17, 1700236, 2017.
29. Shahriari-Khalaji, M., Hong, S., Hu, G., Ji, Y., Hong, F.F., Bacterial nanocellulose-enhanced alginate double-network hydrogels cross-linked with six metal cations for antibacterial wound dressing. *Polymers*, 12, 2683, 2020.
30. Ren, L., He, G., Zhou, Y., Dai, J., Miao, W., Ouyang, C., Liu, J., Chen, G., Hydrogel based on nanocellulose/polydopamine/gelatin are used for the treatment of MRSA infected wound with broad-spectrum antibacterial, anti-oxidant property and tissue suitability. *Biomater. Sci.*, 10, 3174–3187, 2022.
31. Afjoul, H., Shamloo, A., Kamali, A., Freeze-gelled alginate/gelatin scaffolds for wound healing applications: An *in vitro, in vivo* study. *Mater. Sci. Eng. C*, 113, 110957, 2020.
32. Cheng, H., Xiao, D., Tang, Y., Wang, B., Feng, X., Lu, M., Vancso, G.J., Sui, X., Sponges with janus character from nanocellulose: Preparation and applications in the treatment of hemorrhagic wounds. *Adv. Healthc. Mater.*, 9, 1901796, 2020.
33. Sultana, T., Hossain, M., Rahaman, S., Kim, Y.S., Gwon, J.G., Lee, B.T., Multi-functional nanocellulose-chitosan dressing loaded with antibacterial Lawsone for rapid hemostasis and cutaneous wound healing. *Carbohydr. Polym.*, 272, 118482, 2021.
34. Zheng, L., Wang, Q., Zhang, Y.S., Zhang, H., Tang, Y., Zhang, Y., Zhang, W., Zhang, X., A hemostatic sponge derived from skin secretion of Andrias davidianus and nanocellulose. *Chem. Eng. J.*, 416, 129136, 2021.
35. Yuan, H., Chen, L., Hong, F.F., A biodegradable antibacterial nanocomposite based on oxidized bacterial nanocellulose for rapid hemostasis and wound healing. *ACS Appl. Mater. Interfaces*, 12, 3382–3392, 2019.
36. Balla, E.D., Bikiaris, N.D., Nanaki, S.G., Papoulia, C., Chrissafis, K., Klonos, P.A., Kyritsis, A., Kostoglou, M., Zamboulis, A., Papageorgiou, G.Z., Chloramphenicol loaded sponges based on PVA/nanocellulose nanocomposites for topical wound delivery. *J. Compos. Sci.*, 5, 208, 2021.

37. Xiao, Y., Rong, L., Wang, B., Mao, Z., Xu, H., Zhong, Y., Zhang, L., Sui, X., A light-weight and high-efficacy antibacterial nanocellulose-based sponge via covalent immobilization of gentamicin. *Carbohydr. Polym.*, 200, 595–601, 2018.
38. Long, L.Y., Hu, C., Liu, W., Wu, C., Lu, L., Yang, L., Wang, Y.B., Microfibrillated cellulose-enhanced carboxymethyl chitosan/oxidized starch sponge for chronic diabetic wound repair. *Biomater. Adv.*, 135, 112669, 2022.
39. Ambekar, R.S. and Kandasubramanian, B., Advancements in nanofibers for wound dressing: A review. *Eur. Polym. J.*, 11, 304–309, 2019.
40. Sylvester, M.A., Amini, F., Tan, C.K., Electrospun nanofibers in wound healing. *Mater. Today: Proc.*, 29, 1–6, 2020.
41. Silva, N.H., Garrido-Pascual, P., Moreirinha, C., Almeida, A., Palomares, T., Alonso-Varona, A., Vilela, C., Freire, C.S., Multifunctional nanofibrous patches composed of nanocellulose and lysozyme nanofibers for cutaneous wound healing. *Int. J. Biol. Macromol.*, 165, 1198–210, 2020.
42. Shahriari-Khalaji, M., Hu, G., Chen, L., Cao, Z., Andreeva, T., Xiong, X., Krastev, R., Hong, F.F., Functionalization of aminoalkylsilane-grafted bacterial nanocellulose with ZnO-NPs-doped pullulan electrospun nanofibers for multifunctional wound dressing. *ACS Biomater. Sci. Eng.*, *7*, 3933–3946, 2021.
43. Ribeiro, A.S., Costa, S.M., Ferreira, D.P., Calhelha, R.C., Barros, L., Stojković, D., Soković, M., Ferreira, I.C., Fangueiro, R., Chitosan/nanocellulose electrospun fibers with enhanced antibacterial and antifungal activity for wound dressing applications. *React. Funct. Polym.*, *159*, 104808, 2021.
44. Ardila, N., Medina, N., Arkoun, M., Heuzey, M.C., Ajji, A., Panchal, C.J., Chitosan–bacterial nanocellulose nanofibrous structures for potential wound dressing applications. *Cellulose*, 23, 3089–3104, 2016.
45. Wang, L., Wang, C., Wang, L., Zhang, Q., Wang, Y., Xia, X., Emulsion electrospun polylactic acid/apocynum venetum nanocellulose nanofiber membranes with controlled sea buckthorn extract release as a drug delivery system. *Text. Res. J.*, 91, 1046–1055, 2021.
46. Wang, L., Wang, C., Zhang, Q., Liu, J., Xia, X., Comparison of morphological, structural and antibacterial properties of different Apocynum venetum poly (lactic acid)/nanocellulose nanofiber films. *Text. Res. J.*, 90, 593–605, 2020.
47. Subha, V., Ranu, A., Shankar, A., Kirubanandan, S., Satheeshkumar, E., Suresh, S., Pugazhendhi, A., Ilangovan, R., Functionalization of spray coated cellulose nanofiber sheet with montmorillonite (MMT) and silver nanoparticles (AgNPs) to biomedical nanocomposite as wound regeneration scaffold. *Prog. Org. Coat.*, 166, 106782, 2022.
48. Dhivya, S., Padma, V.V., Santhini, E., Wound dressings–A review. *BioMedicine*, *5*, 1–5, 2015.
49. Lungu, A., Cernencu, A.I., Dinescu, S., Balahura, R., Mereuta, P., Costache, M., Syverud, K., Stancu, I.C., Iovu, H., Nanocellulose-enriched

hydrocolloid-based hydrogels designed using a Ca2+ free strategy based on citric acid. *Mater. Des.*, 197, 109200, 2021.
50. Kong, D., Zhang, Q., You, J., Cheng, Y., Hong, C., Chen, Z., Jiang, T., Hao, T., Adhesion loss mechanism based on carboxymethyl cellulose-filled hydrocolloid dressings in physiological wounds environment. *Carbohydr. Polym., 235*, 115953, 2020.
51. Collado-Boira, E., Boldo-Roda, P., Bernat-Adell, M.D., Morar, K.G., Ayora, A.F., Medina, P.S., Effectiveness of semiocclusive sodium carboxymethyl cellulose fibers and hydrocolloid dressings for irritant peristomal dermatitis: A case series. *Adv. Skin Wound Care*, 34, 493–497, 2021.
52. Weller, C.D., Team, V., Sussman, G., First-line interactive wound dressing update: A comprehensive review of the evidence. *Front. Pharmacol.*, 11, 1–13, 2020.
53. Abdollahi, S. and Raoufi, Z., Gelatin/Persian gum/bacterial nanocellulose composite films containing Frankincense essential oil and Teucrium polium extract as a novel and bactericidal wound dressing. *J. Drug Deliv. Sci. Technol.*, 72, 103423, 2022.
54. Taheri, P., Jahanmardi, R., Koosha, M., Abdi, S., Physical, mechanical and wound healing properties of chitosan/gelatin blend films containing tannic acid and/or bacterial nanocellulose. *Int. J. Biol. Macromol.*, 154, 421–432, 2020.
55. Fu, L., Zhou, P., Zhang, S., Yang, G., Evaluation of bacterial nanocellulose-based uniform wound dressing for large area skin transplantation. *Mater. Sci. Eng. C*, 33, 2995–3000, 2013.
56. Jaberifard, F., Ghorbani, M., Arsalani, N., Mostafavi, H., A novel insoluble film based on crosslinked-starch with gelatin containing ZnO-loaded halloysite nanotube and bacterial nanocellulose for wound healing applications. *Appl. Clay Sci.*, 230, 106667, 2022.
57. Akhavan-Kharazian, N. and Izadi-Vasafi, H., Preparation and characterization of chitosan/gelatin/nanocrystalline cellulose/calcium peroxide films for potential wound dressing applications. *Int. J. Biol. Macromol.*, 133, 881–891, 2019.
58. Riccio, B.V.F., Klosowski, A.B., Prestes, E., de Sousa, T.B., de Assunção Morais, L.C., Lemes, B.M., Beltrame, F.L., Campos, P.M., Ferrari, P.C., Chitosan/nanocellulose-based bionanocomposite films for controlled betamethasone and silver sulfadiazine delivery. *J. Appl. Polym. Sci.*, 138, 50468, 2021.
59. Nezhad-Mokhtari, P., Asadi, N., Rahmani Del Bakhshayesh, A., Milani, M., Gama, M., Ghorbani, M., Akbarzadeh, A., Honey-loaded reinforced film based on bacterial nanocellulose/gelatin/guar gum as an effective antibacterial wound dressing. *J. Biomed. Nanotechnol.*, 18, 2010–2021, 2022.
60. de Lima, T.A.D.M., de Lima, G.G., Chee, B.S., Henn, J.G., Cortese, Y.J., Matos, M., Helm, C.V., Magalhães, W.L., Nugent, M.J., Characterization of gels and films produced from Pinhão seed coat nanocellulose as a potential use for wound healing dressings and screening of its compounds towards antitumour effects. *Polymers*, 14, 2776, 2022.

61. Claro, F.C., Jordão, C., de Viveiros, B.M., Isaka, L.J.E., Villanova Junior, J.A., Magalhães, W.L.E., Low cost membrane of wood nanocellulose obtained by mechanical defibrillation for potential applications as wound dressing. *Cellulose*, 27, 10765–10779, 2020.
62. Yuan, H., Chen, L., Hong, F.F., Evaluation of wet nanocellulose membranes produced by different bacterial strains for healing full-thickness skin defects. *Carbohydr. Polym.*, 285, 119218, 2022.
63. Sampaio, L.M., Padrão, J., Faria, J., Silva, J.P., Silva, C.J., Dourado, F., Zille, A., Laccase immobilization on bacterial nanocellulose membranes: Antimicrobial, kinetic and stability-properties. *Carbohydr. Polym.*, 145, 1–12, 2016.
64. Ju, S.Y., Yu, H.L., Ji, L., Wang, K., Jiang, J.X., Preparation and characterization of bacterial nanocellulose-based composite membranes. *Sci. Adv. Mater.*, 12, 802–809, 2020.
65. Fonseca, D.F., Carvalho, J.P., Bastos, V., Oliveira, H., Moreirinha, C., Almeida, A., Silvestre, A.J., Vilela, C., Freire, C.S., Antibacterial multi-layered nanocellulose-based patches loaded with dexpanthenol for wound healing applications. *Nanomaterials*, 10, 2469, 2020.
66. Piaia, L., Pittella, C.Q.P., Souza, S.S.D., Berti, F.V., Porto, L.M., Incorporation of Aloe vera extract in bacterial nanocellulose membranes. *Polímeros*, 32, 1–8, 2022.
67. Coelho, G.A., Magalhães, M.A.B., Matioski, A., Ribas-Filho, J.M., Magalhães, W.L.E., Claro, F.C., Ramos, R.K., Camargo, T.M.S.D., Malafaia, O., Pine nanocellulose and bacterial nanocellulose dressings are similar in the treatment of second-degree burn? Experimental study in rats. *ABCD. Arq. Bras. Cir. Dig. (São Paulo)*, 33, 1–7, 2020.
68. Smandri, A., Nordin, A., Hwei, N.M., Chin, K.Y., Abd Aziz, I., Fauzi, M.B., Natural 3D-printed bioinks for skin regeneration and wound healing: A systematic review. *Polymers*, 12, 1–19, 2020.
69. Tsegay, F., Elsherif, M., Butt, H., Smart 3D printed hydrogel skin wound bandages: A review. *Polymers*, 14, 1–36, 2022.
70. Miguel, S.P., Cabral, C.S., Moreira, A.F., Correia, I.J., Production and characterization of a novel asymmetric 3D printed construct aimed for skin tissue regeneration. *Colloids Surf. B*, 181, 994–1003, 2019.
71. Rees, A., Powell, L.C., Chinga-Carrasco, G., Gethin, D.T., Syverud, K., Hill, K.E., Thomas, D.W., 3D bioprinting of carboxymethylated-periodate oxidized nanocellulose constructs for wound dressing applications. *BioMed. Res. Int.*, 2015, 1–7, 2015.
72. Xu, C., Molino, B.Z., Wang, X., Cheng, F., Xu, W., Molino, P., Bacher, M., Su, D., Rosenau, T., Willför, S., Wallace, G., 3D printing of nanocellulose hydrogel scaffolds with tunable mechanical strength towards wound healing application. *J. Mater. Chem. B*, 6, 7066–7075, 2018.
73. Rosendahl, J., Zarna, C., Håkansson, J., Chinga-Carrasco, G., Gene-expression analysis of human fibroblasts affected by 3D-printed carboxylated nanocellulose constructs. *Bioengineering*, 10, 1–16, 2023.

74. Xu, W., Molino, B.Z., Cheng, F., Molino, P.J., Yue, Z., Su, D., Wang, X., Willfor, S., Xu, C., Wallace, G.G., On low-concentration inks formulated by nanocellulose assisted with gelatin methacrylate (GelMA) for 3D printing toward wound healing application. *ACS Appl. Mater. Interfaces*, 11, 8838–8848, 2019.
75. Dehkordi, N.K., Minaiyan, M., Talebi, A., Akbari, V., Taheri, A., Nanocrystalline cellulose–hyaluronic acid composite enriched with GM-CSF loaded chitosan nanoparticles for enhanced wound healing. *Biomed. Mater.*, 14, 035003, 2019.
76. Yuan, H., Chen, L., Hong, F.F., Homogeneous and efficient production of a bacterial nanocellulose-lactoferrin-collagen composite under an electric field as a matrix to promote wound healing. *Biomater. Sci.*, 9, 930–41, 2021.
77. Yang, M., Ward, J., Choy, K.L., Nature-inspired bacterial cellulose/methylglyoxal (BC/MGO) nanocomposite for broad-spectrum antimicrobial wound dressing. *Macromol. Biosci.*, 20, 1–7, 2020.
78. Vilela, C., Oliveira, H., Almeida, A., Silvestre, A.J., Freire, C.S., Nanocellulose-based antifungal nanocomposites against the polymorphic fungus Candida albicans. *Carbohydr. Polym.*, 217, 207–216, 2019.
79. Sun, F., Nordli, H.R., Pukstad, B., Gamstedt, E.K., Chinga-Carrasco, G., Mechanical characteristics of nanocellulose-PEG bionanocomposite wound dressings in wet conditions. *J. Mech. Behav. Biomed. Mater.*, 69, 377–384, 2017.
80. Poonguzhali, R., Basha, S.K., Kumari, V.S., Synthesis and characterization of chitosan-PVP-nanocellulose composites for *in-vitro* wound dressing application. *Int. J. Biol. Macromol.*, 105, 111–120, 2017.
81. Ulker, Z. and Erkey, C., An emerging platform for drug delivery: Aerogel based systems. *J. Control. Release*, 177, 51–63, 2014.
82. Saini, A., Yadav, C., Sethi, S.K., Xue, B.L., Xia, Y., Li, K., Manik, G., Li, X., Microdesigned nanocellulose-based flexible antibacterial aerogel architectures impregnated with bioactive Cinnamomum cassia. *ACS Appl. Mater. Interfaces*, 13, 4, 4874–85, 2021 Jan 19.
83. Ghafari, R., Fabrication and characterization of novel bilayer scaffold from nanocellulose based aerogel for skin tissue engineering applications. *Int. J. Biol. Macromol.*, 136, 796–803, 2019.
84. Holzer, J.C., Tiffner, K., Kainz, S., Reisenegger, P., de Mattos, I.B., Funk, M., Lemarchand, T., Laaff, H., Bal, A., Birngruber, T., Kotzbeck, P., A novel human *ex-vivo* burn model and the local cooling effect of a bacterial nanocellulose-based wound dressing. *Burns*, 46, 8, 1924–1932, 2020.
85. Bakhshpour, M., Idil, N., Perçin, I., Denizli, A., Biomedical applications of polymeric cryogels. *Appl. Sci.*, 9, 1–22, 2019.
86. Ghafari, R., Jonoobi, M., Naijian, F., Ashori, A., Mekonnen, T.H., Taheri, A.R., Fabrication and characterization of bilayer scaffolds-nanocellulosic cryogels-for skin tissue engineering by co-culturing of fibroblasts and keratinocytes. *Int. J. Biol. Macromol.*, 223, 100–7, 2022.

87. Cattelaens, J., Turco, L., Berclaz, L.M., Huelsse, B., Hitzl, W., Vollkommer, T., Bodenschatz, K.J., The impact of a nanocellulose-based wound dressing in the management of thermal injuries in children: Results of a retrospective evaluation. *Life*, 10, 1–11, 2020.
88. Luca-Pozner, V., Nischwitz, S.P., Conti, E., Lipa, G., Ghezal, S., Luze, H., Funk, M., Remy, H., Qassemyar, Q., The use of a novel burn dressing out of bacterial nanocellulose compared to the French standard of care in paediatric 2nd degree burns–a retrospective analysis. *Burns*, 48, 1472–80, 2022.
89. Portal, O., Clark, W.A., Levinson, D.J., Microbial cellulose wound dressing in the treatment of nonhealing lower extremity ulcers. *Wounds*, 21, 1–3, 2009.
90. bionext, https://www.bennetthealth.net/bionext/. Accessed 13 February 2023.
91. PROMOGRAN PRISMA™ Wound Balancing, https://www.acelity.com/healthcare-professionals/global-product-catalog/catalog/promogran-prisma-wound-balancing-matrix. Accessed 13 February 2023.
92. Frankel, V.H., Serafica, G.C., Damien, C.J., Development and testing of a novel biosynthesized XCell for treating chronic wounds. *Surg. Technol. Int.*, 12, 27–33, 2004.
93. He, W., Wu, J., Xu, J., Mosselhy, D.A., Zheng, Y., Yang, S., Bacterial cellulose: Functional modification and wound healing applications. *Adv. Wound Care*, 10, 623–40, 2021.
94. de Amorim, J.D., da Silva Junior, C.J., de Medeiros, A.D., do Nascimento, H.A., Sarubbo, M., de Medeiros, T.P., Costa, A.F., Sarubbo, L.A., Bacterial cellulose as a versatile biomaterial for wound dressing application. *Molecules*, 27, 1–25, 2022.
95. Čolić, M., Tomić, S., Bekić, M., Immunological aspects of nanocellulose. *Immunol. Lett.*, 222, 80–9, 2020.
96. Bacakova, L., Pajorova, J., Bacakova, M., Skogberg, A., Kallio, P., Kolarova, K., Svorcik, V., Versatile application of nanocellulose: From industry to skin tissue engineering and wound healing. *Nanomaterials*, 9, 1–39, 2019.

9

Use of Nanocellulose as Drug Carriers for Drug Delivery Applications

Rubai Ahmed[1‡], Swarnali Das[1‡], Jhimli Banerjee[1,2], Sovan Samanta[1], Sandeep Kumar Dash[1*] and Arindam Pramanik[3,4†]

[1]Department of Physiology, University of Gour Banga, Malda, West Bengal, India
[2]Department of Food and Nutrition, Gour Mahavidyalaya, Malda, West Bengal, India
[3]Amity Institute of Biotechnology, Amity University, Noida, India
[4]School of Medicine, University of Leeds, Leeds, United Kingdom

Abstract

Cellulose is a natural biopolymer that forms a source of sustainable and renewable raw materials for industrial usage. In recent times, cellulose at the nanoscale has revolutionized various industrial sectors, including the food sector, the technical sector, and, most importantly, the biomedical sector and pharmaceutical industry. Nanocellulose is broadly categorized into three types: bacterial nanocellulose, cellular nanofibrils, and cellulose nanocrystals. These nanocelluloses have unique characteristics that are diverse from traditional plant-based cellulosic materials, which include distinct morphology and geometric dimension, high surface area, alignment and orientation, crystallinity, rheological properties, biodegradability, non-toxicity, and biocompatibility. Various biomedical applications of nanocellulose, such as wound healing, skin and bone tissue repair, soft tissue replacement, blood vessel replacement, tissue scaffolds for cell culture, and antimicrobial biomaterials, show that nanocellulose has achieved advancement in the application of drug delivery. This includes the modification and functionalization of nanocellulose for increasing drug adsorption using physical treatment, chemical treatment, and post-chemical modification. Whereas post-chemical modification includes sulfonation and Fischer esterification among the common approaches, there are

**Corresponding author*: deep.vu@gmail.com
†*Corresponding author*: a.pramanik@leeds.ac.uk; arindampramanik87@gmail.com
‡*Authors have equally contributed*

Mainak Mukhopadhyay and Debalina Bhattacharya (eds.) Nanocellulose: A Biopolymer for Biomedical Applications, (221–266) © 2024 Scrivener Publishing LLC

many additional techniques that help to improve the drug delivery application of nanocellulose as a carrier. Various methods (1D, 2D, and 3D formulations) for drug loading or encapsulation in nanocellulose are also demonstrated in this chapter. The drugs utilized for delivery using nanocellulose include a broad class, including vitamins, cardiovascular drugs, anticancer drugs, antifungal drugs, anti-inflammatory drugs, and many more. In addition, this study describes the most recent clinical applications of nanocellulose in the biological sector. In order to demonstrate the potential of nanosized cellulose in biomedical applications, examples of recent clinical studies are provided. Thus, it is evident that nanocellulose has gained a significant status in the field of drug delivery. This chapter emphasizes on discussing the state of the art of nanocellulose for drug delivery applications.

Keywords: Bacterial nanocellulose, cellular nanofibrils, cellulose nanocrystals, drug delivery, nanocellulose

Abbreviations

BBB	blood–brain barrier
BC	bacterial cellulose
BNC	bacterial nanocellulose
BSA	bovine serum albumin
CGC	CNF grafted collagen
CMC	carboxymethylcellulose
CNC	cellulose nanocrystals
CNF	cellulose nanofibrils
DEX	dexamethasone
DMSO	dimethyl sulfoxide
DTX	docetaxel
ECMs	extracellular matrices
EMF	external magnetic field
GEM	gemcitabine
IA	itaconic acid
KT	ketorolac tromethamine
LbL	layer-by-layer
LDL	low-density lipoprotein
MCC	micro-crystalline cellulose
MFC	microfibrillated cellulose
miRNA	micro-RNA
MLS	methacrylated lignosulfonate
MNC	magnetic nanoparticles
MSNs	mesoporous silica nanoparticles
MTX	methotrexate

MX	Meloxicam
NCC	nano-crystalline cellulose
NIPAM	N-isopropyl acrylamide
PDA	polydopamine
PDNA	plasmid DNA
PEI	polyethylenimine
PTX	paclitaxel
PVA	polyvinyl alcohol
PVP	polyvinylpyrrolidone
QC	quaternized cellulose
RH	rice husks
siRNA	small interfering RNA
TC	tetracycline
TH	tetracycline hydrochloride

9.1 Introduction

Cellulose, a common organic substance which is a structural element of plant tissue and cells, can act as a biodegradable natural polymer. This is a naturally occurring long-chain polymer that indirectly plays a significant part in the human food cycle [1, 2]. This polymer has multipurpose application with its semi-synthetic derivatives in many industrial fields such as paper and wood, clothes and fibers, veterinary foods [3], cosmetics, and, very importantly, in pharmaceutical industries as an excipient. Cellulose plays a major role in a variety of pharmaceutical sectors, including osmotic drug delivery systems, delayed- and extended-release coated dosage forms, controlled- and extended-release matrices, mucoadhesives, bioadhesives, compression tablets, stabilizers and thickening agents as liquid dosage forms, semisolid preparations (as gelling agents), and various tenders [1]. In the pharmaceutical industry, drug delivery systems have been shown to improve therapeutic efficacy, decrease toxicity, increase patient compliance, and even enable whole new medical therapies [4]. Its typical goal is to enhance the therapeutic pharmacological effects and get rid of issues such as limited solubility, drug aggregation, poor biodistribution, low bioavailability, and lack of selectivity, or to lessen the negative effects of the drugs [5–7]. Numerous drug delivery systems have been recognized and are being examined to achieve a successful therapeutic impact in humans. These technologies improve drug encapsulation and distribute drugs effectively over intricate physiological barriers. For targeted drug delivery systems, a variety of carrier molecules are used, including cyclodextrins, microspheres, gels, prodrugs, liposomes, proliposomes, etc. Nanoparticles made of eco-friendly polymers demonstrate their pledge to meet the demanding standards. But in some drug delivery methods, such as stability against aerosolization-induced stresses, the capacity to transfer into an aerosol, the targeting of specific sites or cell populations, biocompatibility, a predetermined manner of drug release, and degradation that occurs within a reasonable amount of time, nanoparticles did not show their accuracy [4]. In addition to improving the accuracy and effectiveness of medications, future technologies will also have an impact on healthcare by making them more accessible, less expensive, and more effective. One of the most auspicious "green" constituents in recent years is nanocellulose, which can be found as cellulose nanofibrils (CNFs), cellulose nanocrystals (CNCs), or bacterial cellulose [8, 9]. Nanocellulose may be obtained from an extensive range of resources, such as algae, plants, and bacteria. They are divided into three primary groups based on the abstraction method and dimensions (length and diameter): CNFs, CNCs, and bacterial nanocellulose (BNC). While cellulose nanofibrils have both crystalline and amorphous parts in their complex, cellulose nanocrystals have a highly crystalline, needle-like shape. Bacterial nanocellulose has been established as the purest form of nanocellulose. Nanocellulose has been employed as a biomaterial for drug delivery owing to its biocompatibility, non-immunogenicity, high specificity, minimal risk of toxicity, flexibility for chemical modification, and strong mechanical qualities. Developed technology and built multidimensional structures of nanocellulose, such as 1D methods (microparticles and nanofibers), 2D methods (films), and 3D methods (hydrogels and aerogels), have various benefits [9, 10]. Nanocellulose's large, specific surface area enables high-specific drug loading. Also, with simple chemical surface changes, a variety of medicines can bind to nanocellulose. When the nanocellulose surface is sufficiently chemically or physically modified, controlled release can be obtained for a variety of medications. In this book chapter, we summarize the potential applications, types, toxicity, and targeted drug delivery of nanocelluloses in a systematic manner [11, 12].

9.2 Strategies for Drug Delivery

9.2.1 Active Targeting Drug Delivery

An active targeted drug delivery system works by delivering a specific quantity of a therapeutic, diagnostic, or both to a specific location of the body's unique organ that is afflicted with infection [13, 14]. Notably, active targeting is necessary to deliver medications, genes, and different theranostic approaches to the required site while avoiding normal tissues, which enhances therapeutic effectiveness and diminishes side effects. Compared to passive or free drug targeting in nanotechnology, the quantity of drug provided to the target cell can be greatly increased. According to research on cancer, following buildup in the tumor location, active targeting can even boost the drug's efficiency. This is done by covering the surface of nanocarriers with ligands that can bind to receptors that are highly or overexpressed by tumor cells. By increasing the affinity of nanocarriers for the surface of cancer cells, this strategy will boost drug penetration [15]. Researchers have concentrated NPs for surface modification using natural ligands in addition to optimizing their physicochemical features. Biological ligands can accelerate modified NPs' uptake by binding on the surface of target cells with specific receptors, a process called "active targeting" by NPs. According to *in vitro* studies, research has demonstrated different biological ligands, including proteins, polysaccharides, polymeric micelles [16], peptides, aptamers, and tiny molecules for NP-mediated active targeting drug delivery [17].

Even though this approach has potential, it has flaws that could make it less effective against other treatments. Particularly, numerous active targeting nanoparticles have shorter blood circulation times than non-targeted nanoparticles due to immunogenicity or non-specific binding, significant sensitivity to the breakdown of lysosomes, and lower tumor penetration [18].

9.2.2 Passive Targeting Drug Delivery

Physical targeting, also referred to as passive targeting, is the process of creating a drug carrier complex that resists removal by bodily processes like phagocytosis, metabolism, excretion, and opsonization. This allows the complex to stay in the bloodstream and transmit drugs to target receptors via factors like pH, temperature, molecular size, or shape [19]. According to Bazak *et al.*, in cancer research, due to the enhanced permeability and retention rate of macromolecules, which consist of nanoparticles, they can accumulate selectively in cancerous tissues via passive targeting [20]. Passive targeting takes the form of drug buildup that occurs naturally in places with leaky vasculature. Drug delivery systems should use nanoparticles that are large enough to prevent blood capillaries from quickly absorbing them. Yet, they should be tiny enough to avoid being engulfed by macrophages that are fixed to the reticuloendothelial system, including those found in the liver and spleen [19, 21].

Except for active and passive drug delivery, there are different kinds of drug delivery applied in the pharmaceutical field, like activation-modulated drug delivery, rate-programmed drug delivery systems, feed-back-regulated drug delivery, and others [22].

9.3 Application of Various Biomaterials for Drug Delivery

Biomaterials are playing a significant role in the progression of the medicinal and biological fields through their excellent potency in drug delivery [23]. Due to their high bioavailability, controlled release, superior encapsulation, and less toxic features, biodegradable nanoparticles have been used often in drug delivery systems [24]. Biocompatible and biodegradable drug carriers with controlled-release technologies can help prevent or reduce the adverse effects that could lead to drug delivery to cells with higher effectiveness during the process of health rehabilitation. The biomaterials can be metallic, ceramic, or polymeric in nature, which shows versatile efficacy towards cancer immunotherapy, autoimmune diseases, genome editing, and other pathogenic conditions [23, 25, 26].

9.3.1 Synthetic Biomaterials

Since silica has advantageous biocompatibility, chemical characteristics, and thermal stability, mesoporous silica nanoparticles (MSNs) in particular are frequently used as delivery reagents [27, 28]. For chemo-based cancer therapy, silica is more advantageous due to its particular mesoporous structure, which makes it easier to load drugs effectively and allows for their subsequent regulated release at the target region [29]. They are utilized equally frequently in the biomedical fields of diagnostics, target drug administration, biosensing, cellular absorption, etc. [27]. Due to their unique characteristics, such as their high surface-to-volume ratios, enhanced conductivity and strength, biocompatibility, ease of functionalization, and optical properties, carbon nanotubes are being researched as potential novel carriers for targeted drug and gene delivery [30, 31]. Carbon nanotubes can play a vital role in targeted drug delivery in cancer cells [30, 32]. Several cell types efficiently absorb carbon nanotubes through a variety of processes. Carbon nanotubes can play a major role as carriers of different anti-cancer drugs such as doxorubicin, docetaxel (DTX), paclitaxel (PTX), methotrexate (MTX), gemcitabine (GEM), and other anti-cancer, anti-inflammatory, and osteogenic dexamethasone (DEX) steroids. Moreover, the simple surface functionalization of carbon nanotubes has led to their usage to carry various genes such as micro-RNA (miRNA), small interfering RNA (siRNA), and plasmid DNA (PDNA) as delivery vectors of genes for several disease conditions, including cancer. Despite all of these reassurances, scientists are nonetheless extremely concerned about the environmental effects of carbon nanotube nanotoxicology, especially since these materials are not biodegradable. Further investigation is required to advance the field of carbon nanotubes in medicine [30]. There are different synthetic polymers used for controlled drug delivery, such as temperature-sensitive synthetic polymers [PEO, PPO, and Poly(N-isopropylacrylamide)] and pH-sensitive synthetic polymers (poly(acrylic acid), poly(methacrylic acid), and Poly(vinylacetaldiethylaminoacetate)) [26].

9.3.2 Natural Biomaterials

Polymeric nanoparticles can act as a significant device to develop targeted drug delivery and drug bioavailability for different diseases [33, 34]. For instance, several types of drug delivery systems are made using natural polymeric materials such as fibrin, gelatin, hyaluronic acid, methylcellulose, dextran, gellan, pullulan, calcium phosphate, chitosan, calcium phosphate, xanthan gum, and alginate [26, 35]. Studies show that a variety of factors, including drug diffusion through the polymeric wall or nanosphere matrix of the nanocapsules, erosion of the polymeric matrix, or drug desorption from the particle surface, affect the release of drugs from polymeric NPs [34, 36]. When it comes to nanocapsules, the medicine dissolved in the oily nucleus should theoretically be released from this vesicular structure after diffusing through the polymeric wall, displaying zero-order kinetics [37]. In recent advanced cancer treatments, polymeric nanoparticles (NPs) have shown their efficacy for the regulated administration of cancer drugs. Polymeric nanocarriers, also known as polymeric nanomedicine, have the potential to conjugate or encapsulate with anticancer medications and exhibit prolonged release of therapeutics with less cytotoxicity and enhanced tumor retention, but up to this point, drug delivery systems require more study [38, 39]. As well as polymeric nanoparticles, lipidic nanoparticles are also rapidly used in targeted drug delivery [27, 30, 40]. Liposomes were developed as programs for intracellular drug delivery and targeting of several active principles because of their amphiphilic properties, low toxicity, biodegradability, and biocompatibility. Examples include liposomes distributed in polymer formats (microparticles, gels, and hydrogels) containing amalgamations of antitumor, anti-inflammatory, anesthetic, antibiotic, or antimycotic drugs [41]. Lipid nanoparticles also showed huge potential in the delivery of nucleic acid drugs, as demonstrated by the approval of mRNA COVID-19 vaccines and two RNA therapies. Lipid nanoparticles in drug delivery systems have numerous desirable properties, together with improved biocompatibility, simplicity in preparation, scale-up viability, nontoxicity, and targeted drug delivery. Fully realizing the potential of lipid nanoparticles for future clinical and pharmacological applications, however, requires extensive study on structure–function connections, large-scale manufacturing, and targeted distribution [42]. Many different diseases, including infections, cancer, and brain disorders, have been studied in relation to drug-loaded lipid nanoparticles [40]. Apo E, a low-density lipoprotein (LDL), has been demonstrated to enhance nanoparticle penetration through the BBB, enabling targeted brain drug administration, according to Hartl *et al.* [43]. Several Apo E-coated LNPs were developed by researchers for brain cancer treatment. In a prior research study, camptothecin-loaded LNPs were developed using a polysorbate 80 coating, and Apo E would then selectively bind to the LNPs, enabling targeted distribution across the BBB [44]. A drug named donepezil (an anticholinergic drug) is delivered by the same strategy to treat Alzheimer's disease [42]. Collagen, gelatin, and silk are examples of non-structural proteins that are attracting attention for their capacity to alter the functional characteristics of biomaterials. Protein-based substances obtained from natural sources are having an ever-growing impact on biomedicine [45].

Despite those biomaterials, different types of other biocompatible molecules (magnetic nanoparticles, dendrimers, solid lipid nanoparticles, and quantum dots) are used in advanced drug delivery. Due to nanocellulose's distinct physical characteristics, concentrated surface area, minimal cytotoxicity risk, and superior biological capabilities, it has recently gained a lot of interest in drug delivery techniques [46]. Nanocellulose-based systems play an important role as transporters in drug (cancer drugs) and antimicrobial dispensing by allowing for sustained and controlled release of drugs and antimicrobial agents, respectively. Having excellent biocompatibility, nanocellulose has demonstrated its potency for various applications and administration methods [46, 47]. Advanced molecular research has revealed that different forms of nanocellulose can be used as drug excipients for drug delivery, tissue scaffolds for cellular growth, enzyme and protein recognition, and immobilization. Along with that, at the macroscopic level, they are substitutes for blood vessels and soft tissues, compounds that help skin and bone tissue recover, and antimicrobial materials [48, 49]. Except for bone tissue engineering, nanocellulose also acts as an efficient biosensor and helps in wound healing [49]. The functional modification of nanocellulose will also affect its potential biological application [48].

9.4 Different Methods Involved in Nanocellulose Synthesis

9.4.1 Mechanical Synthesis

In order to prepare lignocellulosic biomass by the mechanical process that yields nanocellulose, plant material must first undergo chemical pretreatment with enzymes and acids, as well as physical treatment like decrystallization, derivatization, or grinding [50]. An example of a commonly used grinding device for cellulose fiber fibrillation is a machine having two grinding stones with the ability to rotate counterclockwise. The cellulose slurry is transferred between two grinding stones, one that is static and the other that rotates at roughly 1500 rpm, repeatedly until the requisite nanoscale dimensions are attained and further size reduction is no longer possible. The fundamental idea is that the shearing strength produced by the grinding pebbles causes the cell membrane to break down [50–52]. Wet grinding gets much more attention because of its ability to prevent agglomeration and promote efficient dispersion of the produced nanocellulose particles [52]. Cellulosic fibers that have been frozen with liquid nitrogen and then crushed to release the fibrillated fibers are referred to as "cryo-crushing". A standard cryo-crusher has two grinding stones, a stator, and a rotor that can spin at 1500 rpm. The source of the native cellulose affects the generated fibrillated cellulose fibers' diameters, which are typically in the nanoscale (5–100 nm) range. This technique's primary flaw is its high energy requirement [52, 53]. In order to synthesize nanofibrils from cellulosic fibers, a sonicator uses ultrasound power. Mechanical synthesis using a sonicator can be completed using both direct and indirect procedures. Indirect sonication involves transmitting ultrasonic energy from the probe through a water medium into a single sample vessel or several sample tubes, while direct sonication involves transmitting amplified vibration energy directly into the sample with higher intensities. This procedure is mostly used for the synthesis of nanofibrillated cellulose [52]. Refining is another mechanical procedure used for nanocellulose synthesis. The pulp is refined in a machine between a stainless-steel roll with bars and a revolving disc with a smooth bed, where the pulp is dispersed uniformly over the disc wall. Although both elements rotate in the same direction but at different speeds [52, 54], by rupturing the intra- and intermolecular hydrogen bonds, high-pressure homogenization in combination with ultrasonication similarly causes the relaxation of fibers [50, 55]. There are other physical or mechanical procedures, such as refining and high-pressure homogenization, and ball milling is used for the synthesis of nanocellulose [50] (Figure 9.1).

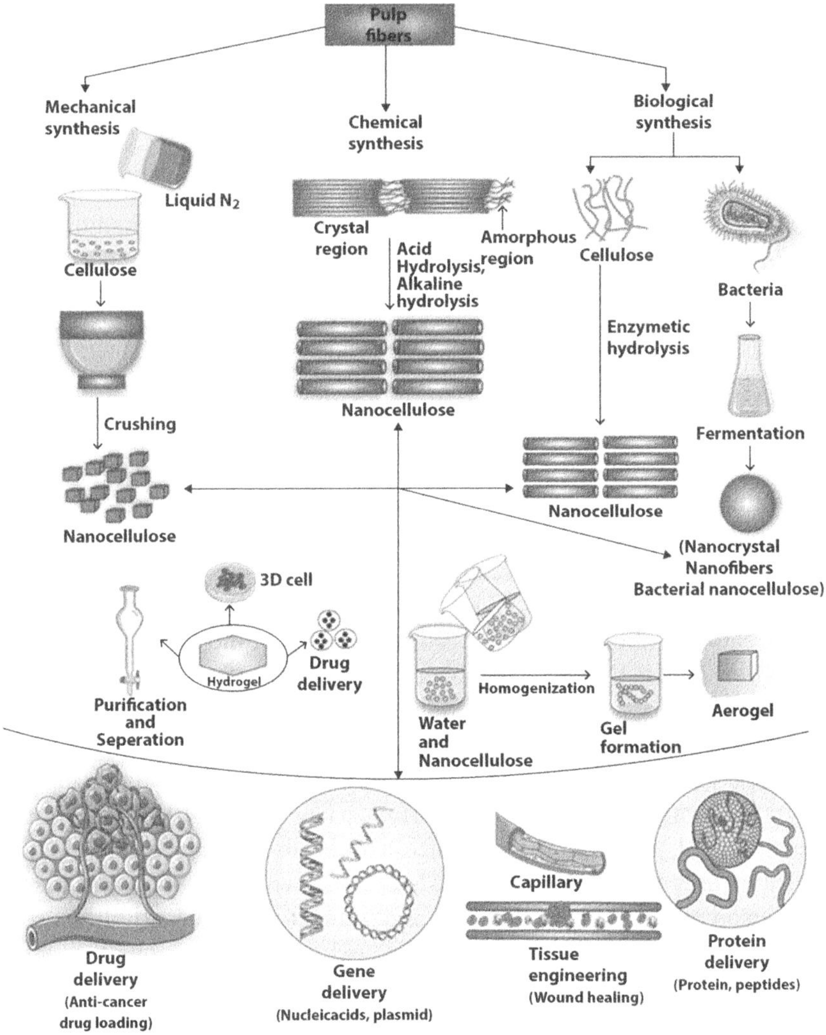

Figure 9.1 Figure depicted multiple nanocellulose manufacturing techniques using diverse methodologies, including mechanical, chemical, and biological from pulp fiber (forest residue, agricultural residue, industrial residue, and algae), and diverse roles in the field of drug delivery, tissue engineering, wound healing, and protein delivery via aerogel and hydrogel.

9.4.2 Chemical Synthesis

For the chemical synthesis of nanocellulose, different processes are used, such as alkaline pre-treatment, oxidative pre-treatment, enzymatic pre-treatment, acid hydrolysis [50], TEMPO-mediated oxidation, the carboxymethylation process, the acetylation process, etc. [52] (Figure 9.1). Pre-treatment with chemicals like oxidative, alkali, enzymatic, and lastly, acids is necessary for the isolation of crystalline cellulose regions through acid hydrolysis to remove the lignin, hemicellulose, wax, fat, and pectins from the cellulose structure [50]. Due to its abundant supply, good mechanical characteristics, low toxicity, bioresorbability, and rich surface chemistry, nanocellulose has emerged as a sustainable bionanomaterial for the production of smart materials. According to Jamwal *et al.*, cellulose from fallen and dried pine needles was hydrolyzed using a mixed acidic process to create spherical nanocellulose [56]. This study found that, for removing cationic dyes from wastewater, hydrogel is a quick and effective adsorbent. The key step in chemical treatment procedures is acid hydrolysis, which mostly produces nanocrystalline cellulose as a by-product. It is typically preceded by pre-treatment procedures such as raw cellulosic ingredient purification, alkali treatment to increase surface area and make the fibers more prone to hydrolysis while also disrupting lignin structures, and treatment with a bleaching agent such as $NaClO_2$, O_3, H_2O_2, O_2, Cl_2, or ClO_2 to remove lignin and other impurities in order to obtain cellulose [52]. After acid hydrolysis and bleaching or mechanical disintegration, the biomass is treated chemically with an alkaline solution to eliminate hemicellulose from its structure, but this has different disadvantages, like crystallinity reduction, side reactions, impurity isolation, loss of physicochemical properties, the need for additional treatments, etc. [57]. HBr, H_3PO_4, H_2SO_4, and HCl are common mineral acids used to produce crystalline nanocellulose. The removal of amorphous regions from cellulosic material is made easier by acid hydrolysis, which produces highly crystalline nanocellulose with 100–200 nm in length and 5–15 nm in width [10]. By adjusting the acids and their concentrations, as well as how ionic media react with cellulose to produce nanocellulose, it is possible to alter the crystallinity and shape of crystalline nanocellulose [10]. In cellulose isolations, the well-known ionic liquids 1-butyl-3-methylimidazolium hydrogen sulfate ($bmimH_2SO_4$), 1-allyl-3-methylimidazolium chloride (AmimCl), 1-butyl-3-methylimidazolium acetate, tetrabutylammonium acetate, 2-hydroxyethyl ammonium hydrogen sulfate, and 1-butyl-3-methylimidazolium chloride (BmimCl) are used [58]. Other benefits of the transition metal salt treatment over mineral acid treatment include less easy reaction-induced separation, less reactor corrosion, and reduced safety problems. The amorphous areas of cellulose are removed using metal salts as a catalyst [10].

9.4.3 Biological Synthesis

Some bacterial genera from the *Rhizobium*, *Achromobacter*, *Acetobacter*, *Agrobacterium*, *Azotobacter*, *Escherichia*, *Salmonella*, and *Sarcina* families generate bacterial nanocellulose by using glucose and other carbon sources [50, 59, 60]. The purest form of cellulose is BNC, which is synthesized by a variety of microbial strains. It is formed inside microbes' cells as 1,4-glucan chains, which emerge as protofibrils through terminal complexes found in the microbe's cell wall. These protofibrils then solidify to create ribbon-shaped microfibrils, which, in turn, form pellicles made up of bundles [50]. The current state of BNC research is primarily concerned with its low-cost production [61], the exploration and genetic engineering of different microbial strains for productivity improvements, and the enhancement of its existing structural features as well as the addition of new features by creating its composites with various functional materials [50]. To enhance BNC synthesis via microbial cells, different strategies are used, such as genetic engineering technologies, submerged and static fermentation, strain improvement, co-culturing of microbes, spin filter bioreactors, rotary biofilm contactors, and silicon sheath reactors [61]. Researchers are examining the impact of various additives on the production and structural characteristics of bacterial nanocellulose, including carboxymethyl cellulose, oligosaccharides, lignosulfonate, agar, sodium alginate, xanthan, alcohols, dextran, acetan, and yeast extracts [50]. These studies have shown that it is probable to control molar mass circulation, structural features, kinetics, yield, and production of other metabolic products in addition to bacterial nanocellulose production by choosing suitable cultivation conditions, additives, substrates, and active bacterial strains [50, 61]. Researchers are also looking at several waste products for low-cost production of nanofibers, including fruit juices, agricultural wastes, sugarcane molasses, industrial wastes, and food wastes [50]. Except for mechanical, chemical, and biological synthesis procedures, there are other synthesis procedures named top-down approach, bottom-up approach, cell-free synthesis, etc. (Figure 9.1).

9.5 Types of Nanocellulose

9.5.1 Cellulose Nanofibers (CNFs)

Various organic spin-offs of corn, banana, rice, wood and wood pulp, sugar cane, and wheat are employed for the synthesis of CNF [49]. CNF is a short and flexible fibril in comparison to BNC and CNCs, with 1–10 nm in length and 20–100 nm in width that differs based on cellulose sources and the process of defibrillation. Although CNF has far less crystallinity than CNC and BNC, it has both crystalline and amorphous domains. CNF accounts for the thixotropic and shear-thinning qualities; however, it is frequently unstable in solution [49, 62]. The most popular technique for separating CNF is mechanical treatment, although there are other methods, like chemical and enzymatic treatment [48, 49]. The preferred methods for CNF synthesis are fiber delamination by ultrasound, homogenization by high pressure, and ball-mining. During the production of CNF, a lot of energy is needed to repeatedly homogenize the fiber-water solution [49, 63]. Another common chemical pre-treatment method for the synthesis of CNF involves cellulose oxidation, using the 2,2,6,6-tetramethylpiperidine-1-oxyl radical (TEMPO) [49].

9.5.2 Cellulose Nanocrystals (CNCs)

Acid hydrolysis of cellulose substances distributed in water is a distinctive method for generating CNCs [49]. Wood pulp-derived nanocrystals are thinner and shorter than microfibrillated cellulose-derived nanocrystals (MFCs) [61]. Releasing the crystalline portions of the semicrystalline cellulosic fibers, mineral acid hydrolysis produces CNCs. The first step in this type of chemical process is the release of polysaccharides destined for the surface of fibrils; then, the more easily reachable amorphous territories are slashed and demolished to release rod-like crystalline cellulose sections [61, 63]. The dimension of the freed crystals is highly influenced by the crystallinity of the cellulose sources, which is variable. However, they tend to clump together in very hydrophobic solutions. CNCs exhibit some solubility in aqueous-based mixes and in organic solvents with high dielectric constants, such as ethylene glycol and dimethyl sulfoxide (DMSO) [61]. The crystalline parts of cellulose are often left alone as intense sulfuric acid is employed to dissolve the amorphous portions of the material. The sulfate groups stay attached at the fiber surface as contaminants even though this process creates a stiff rod-like CNCs with over 90% purity [49]. CNCs typically range in size from 200 to 500 nm in length and 3–35 nm in diameter. It is generally known that CNCs can be isolated through acid hydrolysis using different mineral acids such as hydrochloric acid, sulfuric acid, and phosphoric acid, as well as techniques like enzymatic processes. During hydrolysis, the amorphous portion of cellulose fibers cleaves, producing a rod-like nanostructure and a hard, highly crystalline nanostructure. Throughout the acid hydrolysis procedure, the inclusion of acid advances the surface charge of CNCs, enabling colloidal dispersal to form quickly [49]. Elastic moduli have been estimated to range from 105 to 168 GPa, while the crystalline proportion of cellulose can range from 50 to 90% depending on the source. The longest variety of CNC, with dimensions of 5–50 nm in diameter and 100–300 nm in length, is gained by tunicates [49]. In particular concentrations, CNC has a shear-weakening or thinning characteristic and behaves like liquid or molten crystals, producing irregular nematic phases [64]. In additive manufacturing, hydrogen bond networks that allow for stress transmission occur when CNC is applied to polymer materials [49, 65].

9.5.3 Microfibrillated Cellulose (MFC)

MFC may be obtained from sugar beet, wood, potato tubers, hemp, etc. Intense mechanical processing is required for MFC manufacturing by cellulose fiber fibrillation into nanoscale components. However, chemical treatments might be used before mechanical fibrillation, depending on the source matter and level of processing. These chemical procedures attempt to create pulped cellulose that has been purified, like bleached cellulose pulp, so that it can be processed further. There are examples of cellulose microfibril isolation processes that combine enzymatic pre-treatment with mechanical treatments to reduce energy consumption. MFC is created by mechanically putting suspensions of wood-based cellulose fiber through high-pressure homogenizers (MFC). Fibers are delaminated by mechanical treatment, which also releases microfibrils that are around 20 nm in diameter. The main barrier to commercial success has been the manufacturing of MFC's extremely high energy expenditure, which is over 25,000 kWh per ton as a result of the necessity for several runs through homogenizers [61]. Siqueira *et al.* (2010) claim that TEMPO-facilitated oxidation at room temperature with 500 rpm stirring produced customized MFC. The best conditions were found at pH 10, which produced cellulose nanofibers that were 3–4 nm wide and a few microns long [61, 66].

9.5.4 Bacterial Nanocellulose (BNC)

In addition to having plant origins, some bacteria from the species *Agrobacterium*, *Rhizobium*, *Acetobacter*, *Alcaligenes*, *Pseudomonas*, or *Sarcina* can release cellulose fibers as extracellular structures. *Acetobacter xylinum* (*Gluconacetobacter xylinus*), a Gram-negative species of acetic acid-producing bacteria, is the most effective generator of bacterial cellulose [63]. It can be possible to regulate the molar mass and supramolecular formation by choosing the substrates, various additives, growth conditions, and then the bacterial strain [61]. Consequently, both the process of production and key characteristics of cellulose can be controlled, such as yield, kinetics, and other metabolic products. Aerobic bacteria, like those of the genus *Gluconacetobacter* that produce acetic acid, produce materials also known as microbial cellulose, bacterial cellulose, or biocellulose as a basic element of their biofilms. BNC is created as a nanomaterial and polymer using biotechnological assemblage procedures from minimal-molecular-weight carbon sources, including D-glucose, as opposed to MFC and CNCs materials, which are extracted from cellulose sources. In terms of molecular composition, bacterial cellulose is considerably different from cellulose of plant origin. The crystallinity of BNC is also higher, approximately 60%–90%. Bacterial cellulose is distinguished by its extraordinarily high water content of 90% or more and by its remarkable purity (with no associated compounds like pectin, hemicelluloses, and lignin) [61]. Acetobacter produces BNC, a homopolymer of 1,4-linked glucose, employing straightforward culturing techniques to enable affordable and ecologically approachable small- and large-scale production [67, 68]. Ultra-unpolluted or pure cellulose can be produced with a variety of physio-chemical properties reliant on the growth medium and methods of fermentation. After biosynthesis, glucan polymers from bacterial cellulose are formed in the method to form a nanostructured network that is semi-transparent, mechanically extremely robust, elastic, and soft, and has a high capacity to store water and exchange gasses [68]. BNC has been used in several medical applications, including wound dressing, tissue engineering, drug delivery systems, and implants for bones, cartilages, teeth, corneas, blood vessels, heart valves, tympanic membranes, and urinary tracts. It is generally regarded as safe and one of the most biocompatible materials [68, 69].

9.6 Different Methods Used for Drug Loading/ Encapsulation of Drug in Nanocellulose

Drug carriers often need to maintain exact drug release performance and keep their pharmaceuticals safe. As a result, it ought to be developed to enhance regulation of the medication release rate, boost usage, and lessen the harmful impacts of pharmaceuticals [70, 71]. Large-surface-area, highly affine nanocellulose-based microspheres have become a major focus of research on pharmacological sustained-release systems. Furthermore, nanocellulose can be altered using a variety of methods due to the significant number of surface hydroxyl groups. Surface modification is a useful technique for controlling the structure–function relationship of nanocellulose [72]. It can also be utilized to modify the ratio of drug loading for synthetically produced drug-loaded bacterial cellulose microparticles and nonionized hydrophobic drugs, which are used in a range of biological applications [9].

9.6.1 1D Formulation Strategies

Advanced technology designed the multidimensional structure of nanocellulose, including 1D (nanofibers and microparticles), for increased drug delivery efficacy. The emulsion technique is frequently utilized for preparing drug nanoparticles (1D) for its promising drug delivery system with distinctive characteristics, like the calm encapsulation of several compounds and colloidal stability [9]. In aqueous solutions, it is frequently emulsified and polymerized using high-performance surfactants and emulsion stabilizers [73]. Drugs are additionally preserved and protected against gastrointestinal hydrolysis and enzymatic breakdown by being encapsulated in an emulsion. Interfacial colloidal nanoparticle complexation was used to create nanocomposite filaments with a variety of features (antibacterial capabilities, drug loading performance, conductivity, etc.); these filaments were made from colloidal nanocelluloses that had opposing charges to one another [9]. In a research study, CNC was employed as a chemical in calcium cross-linked alginate microspheres, where its presence stabilized the cross-linked alginate matrix and increased theophylline drug encapsulation potency while obtaining regulated drug release behavior [74]. According to a previous study, curcumin-containing Pickering emulsion was stabilized with Fe_3O_4@cellulose nanocrystals for magnetically triggered drug release. MTT assessment of this study demonstrated that in the presence of a magnetic field, the curcumin-loaded nanomaterials could successfully stop the proliferation of human colon cancer cells [9, 75].

9.6.2 2D Formulation Strategies

Being endowed with a number of qualities including strong biocompatibility, degradability, and good permeability, and outstanding mechanical properties, nanocellulose can act as excellent film-forming material [9, 76].

Nanocellulose is an effective film-forming substance that can be utilized to make a variety of composite materials, such as films for sustained drug release, films for bone repair, hemostatic films, etc. [9]. As drug carriers in 2D formulation, nanocellulose often takes the form of combined films or composite films. The nanocellulose-based composite films have the potency to load a variety of antibiotics, anti-inflammatory drugs, antivirals, and several other drugs [9, 77]. According to a study, LbL (layer-by-layer)-assembled thin films for drug administration have been created by utilizing the complementary hydrogen and electrostatic bonding interface between negatively charged CNCs and positively charged chitosan. The loaded drug was released over time in a sustained way in both the acidic pH of 6.4 and the physiological pH of 7.4 in the PBS buffer. The pH of tumor cells' extracellular area is lower, so more doxorubicin is released when the pH is acidic, which is beneficial for cancer treatment. Although CNCs is mostly hydrophilic in nature, it has a surface that is more hydrophobic, allowing it to interact with molecules that are hydrophobic. In this study, the results demonstrated that curcumin, which is a water-insoluble drug, can interact well with nanocrystalline cellulose, and the conjugates of curcumin with nanocrystalline cellulose provide a stable dispersion in water. So, to create thin films, these conjugates can be combined in an LbL way [9, 78]. In a study, researchers added honey to CNCs composite sheets. Researchers found that the drug release ratio steadily increased and followed first-order kinetics, remaining constant for roughly 48 h *in vitro*. According to this study, nanocellulose flakes have substantial research value in several drug delivery fields [9, 79]. Saidi *et al.* demonstrated in a study that diclofenac is delivered under control using bacterial cellulose/poly(N-methacryloyl glycine) (PMGly) composite films. Methacryloyl glycine monomer was *in situ* polymerized to create the composite film within the network of bacterial cellulose. By merely submerging wet composite films in solution of the drug, diclofenac was loaded into bacterial cellulose/PMGly films. The composite films indicated a reduced release rate of drug (9%) under a pH of 2.1 over the course of 24 h, compared to a significantly larger release (85%) under a pH of 7.4. Because of its regulated and pH-sensitive drug delivery qualities, the composite film has the capability to be employed for both transdermal and oral drug delivery [80]. In a research study, Poonguzhali *et al.* created alginate/CNC composite films for *in vitro* drug release. According to the findings, CNC could enhance the composite films' mechanical and swelling properties, which exhibited persistent drug release patterns [81].

9.6.3 3D Formulation Strategies

As extracellular matrices (ECMs), hydrogels with nanocellulose-based composites have highly porous architectures, excellent mechanical power, high water retention, a high level of selective surface area, and efficient biocompatibility. In research, the networks of the nanocellulose-based composite aerogels and hydrogels were utilized for 3D drug delivery. Physical crosslinking and chemical crosslinking are commonly used methods for creating nanocellulose-based composite hydrogels where function groups like carboxyl groups and hydroxyl groups play a major role [9, 82–84]. The functional groups and morphological features of various hydrogel types can have an impact on the drug dispersion pathways throughout the release and adsorption processes; therefore, the use of sensitive hydrogels as biomaterials in focused or continuous drug delivery systems is possible [9, 85]. When the exterior environment varies, such as pH, temperature, electric field, and light, the smart reactive hydrogel will shrivel or develop as needed due to the presentation of complexation, hydrogen bonds, non-covalent interactions, ions, and electrostatic interactions [9, 86]. In order to create the composite hydrogel, Treesuppharat *et al.* copolymerized bacterial cellulose and gelatin. The produced hydrogels possessed the necessary mechanical qualities as well as thermal stability due to the homogeneous size and shape of the bacterial cellulose chains [87]. Moreover, a bacterial cellulose hydrogel is used to load bovine serum albumin for drug delivery [9]. Water absorption data showed that, even after 48 h, bacterial cellulose/acrylic acid composite hydrogels swelled most when the pH was 7. So, it indicates that composite hydrogels contain both thermoresponsive and pH-responsive drug release qualities [9]. Physical crosslinking was used by Liu *et al.* to create a porous polydopamine (MPDA)@graphene oxide (GO)/CNFs composite hydrogel for controlled drug release [9, 88]. In contrast to regular hydrogels, injectable hydrogels have also received much research attention as a crucial component of drug delivery systems, where they are used in drug administration and tissue engineering [9, 89]. Doxorubicin, tetracycline, and bovine serum albumin (BSA) were released *in vitro* experiments using the CNC hydrogels as drug carriers in normal saline and simulated gastric juice [9]. Drugs under long-lasting electromotive force can be maintained by magnetic hydrogel, which also increases drug loading effectiveness, lowers drug dosage, and lessens harm to healthy organs [9]. The permeable substance with low density and high porosity known as nanocellulose-based aerogels is frequently used to describe how various drugs can be stored [9]. The creation of hydrogels via a process of chemical or physical crosslinking precedes the development of nanocellulose-based aerogels. The hydrogels are subsequently transformed into aerogels by a variety of drying processes, including freeze-drying, evaporation, and supercritical drying [90]. Physical and chemical cross-linking processes must be controlled before drying in order to regulate the development of the three-dimensional system and the characteristics of the permeable material. In a study, polyethyleneimine was used to crosslink TEMPO-mediated bacterial cellulose composite aerogel [9, 91]. In a different experiment, aspirin was simply adsorbed into the composite aerogel. Aspirin's cumulative release ratio from composite aerogel was 80.6% in the simulated intestinal fluid condition at a pH of 7.5 during the investigation of drug release behaviors. Sultan *et al.* demonstrated a study where nanocellulose was used to construct 3D-printed hydrogel scaffolds for bone and cartilage regeneration. This research proposes a toxic-free, biobased, low-temperature tissue engineering method that is scalable, cost-effective, and uses biobased formulations. It can also be applied to other biomaterials and other applications [92].

9.7 Importance of Nanocellulose as Drug Carrier

9.7.1 Biodegradability and Biocompatibility

Cellulose is a renowned natural polymer with a biodegradable nature. In addition to nontoxicity and biocompatibility, materials used in biomedical applications must also be biodegradable. Researchers discovered that cellulose can chemically decompose at high temperatures [9, 93]. Moreover, the cellulose nanoscales employed in a variety of biomedical applications still retain their biodegradable properties. Yet, scientific investigations cannot clear the doubt that nanocellulose is biodegradable in human and animal tissues because cells lack the ability to produce cellulose. The leisurely degradation of unmodified cellulose in the human body may be explained by the cellulose chains' nonenzymatic, spontaneous biodegradability [9].

9.7.2 Toxicological Concern

Particularly for applications involving drug delivery, nanocellulose toxicity is a significant consideration. Nevertheless, defining its toxicity has generated contradictions among scientists and society. Many factors, including form, size, surface functionalization, morphology, crystallinity, extraction techniques, colloidal stability, and cellulose sources, have been identified as potentially influencing nanocellulose toxicity [9, 12, 94].

Several polymer-based nanoparticles may have negative impacts on cells that take time to manifest. According to research, bacterial cellulose is found to be safe and does not exhibit any cytotoxicity in mouse subcutaneous tissue (*in vivo* study), which is helpful in the creation of tissue-engineered grafts. A review found multiple papers discussing CNCs *in vitro* and, more significantly, *in vivo* research. They discovered that CNCs have very low toxicity across a number of research models. However, they did mention that, infrequently, it can result in very harmful consequences [95]. In an *in vivo* investigation, Alexandrescu *et al.* investigated the cytotoxicity of this form of nanocellulose (CNFs). This study demonstrates both direct and indirect interactions with 3T3 cells, and cytotoxicity testing revealed that the samples did not pose a significant risk to the cells [96]. Additionally, earlier *in vivo* research found that, in comparison to their CNCs sample, CNFs produced low toxicity to embryonic zebrafish [97]. Research studies (*in vitro* and *in vivo*) were conducted on the toxicity of bacterial nanocellulose fibers. Their results showed that the nanofibers were inefficient at changing biochemical parameters in the bodies of mice without having any detrimental effects on the organ and were noncytotoxic to endothelial cells of the human umbilical vein [12]. Moreover, Seabra *et al.* (2018) evaluated a number of articles where the results showed that cellulose nanocrystals exhibit relatively low toxicity in a number of study models, while they also highlighted that, in specific circumstances, it may result in significant harmful effects [95]. A survey of the literature reveals studies on the cytotoxicity of crystalline microcellulose in the cutaneous, oral, and pulmonary regions. Even though this review found no toxic effect when crystalline microcellulose was administered orally and topically, contradictory results for pulmonary administration were found. So, more research is required to determine the toxicity of crystalline nanocellulose in the body when ingested or applied topically, as well as the identification of inflammatory or oxidative stress responses at the cellular level [98]. CNCs are an excellent choice for nanocarriers in biotechnology and nanomedicine. The majority of investigations verified that they do not show any noticeable cyto- or genotoxicity. Hosseinidoust *et al.* (2015) supported this finding but also highlighted that, in the same way other nanoparticles can stress cells, cellulose nanocrystals can also show toxic effects due to their high charge density [99]. A study was performed by De Loid *et al.* for the toxicological evaluations of depleted nanocellulose. No considerable differences in serum indicators or histopathology were seen between the control group and the rats fed nanocellulose suspension in this study. The investigation indicated that ingested nanocellulose was essentially safe until consumed in tiny amounts, but the materials' long-term impacts and their effects *in vivo* are not yet clear [100]. It is unable to provide a realistic picture of the affinity or compatibility of the chosen nanocellulose-based materials because the majority of toxicity investigations are performed through cell tissue culture [101]. The proinflammatory activities of nanocellulose can be considerably decreased by adding various chemical groups to its surface. Thus, more immunological research on materials made of nanocellulose is required in order to comprehend how they affect innate and adaptive immunity. Nanocellulose-based compounds are still preferred over other types of materials nonetheless, as a result of their low cytotoxicity [9, 12].

9.8 Drug Delivery via Different Nanocelluloses

9.8.1 Nanocellulose Aerogels

Aerogel is a unique creation, synthesized in conjunction with the adaptable approach of sol-gel chemistry with a variety of sources of macromolecules, inorganic materials, and composites, with some incredible properties such as extremely low density, heat resistance, high porosity, a large surface area, a poor dielectric constant, and a reduced refractive index. Aerogel-based biomaterials resemble the properties of the biological ECMs, offer structural support, and have biochemical impacts on surrounding cells in tissues [102, 103]. To form nanocellulose-based aerogels, both CNCs and CNFs, sublimating the liquid portion of traditional gels, and freeze-drying or critical point drying are commonly used to produce lightweight, highly porous materials with an average 10-nanometer pore size [104]. The characteristics of cellulose, including its strong interaction with water, could make it swell easily, and its capacity to protect and release substances under the right circumstances could aid in the ability of nanocellulose aerogels to load drugs efficiently and to postpone the release of drugs by interacting with drug molecules [105]. A similar prior study also reported that due to their physical structure, polysaccharide-based aerogel-drug combinations with concentrations of 0.07 to 0.46 g/cm^3 may provide improved drug loading capacity and bioavailability as well as the potential for controlled release [106]. Similarly, Chen *et al.* (2018) reported that bacterial cellulose with PEI (polyethylenimine) grafts was an ideal sample with high adsorption capacities, sustained, pH-dependent drug release, and low cytotoxicity. Aerogels were generated by freezing the

PEI-bacterial cellulose further, and they were successfully employed as a vehicle to carry proteins and medicines under varied circumstances [91].

9.8.2 pH-Responsive Hydrogels

Hydrogels are made of polymeric materials that can hold a lot of water and also have some special properties, like being soft, flexible, and safe for living things. The chemical cross-linking, which is developed to generate hydrogels with exceptional mechanical strength, should be removed due to the reported toxicity of the cross-linkers and physically cross-linked procedures used as an alternative. Now, hydrogels are made into contact lenses, hygiene products, scaffolds for tissue engineering, and wound healing. Besides these, hydrogels have generated a lot of interest as drug delivery platforms due to their biodegradable nature and the fact that drug release is regulated and responsive to stimuli, particularly pH [107–110].

These hydrophilic, renewable, biocompatible, and biodegradable celluloses with a high aspect ratio and surface area are in the form of fibrils (CNF), crystals (CNC), or bacteria (BNC) and can be chemically changed along with functional groups or by anchoring biomolecules to make these cellulose-based gels. Nanocellulose hydrogels have been shown to mimic the ECM's characteristics while exhibiting low cytotoxicity in 3D cell culture [111].

According to a prior study, the swelling–loading method was used to add nepafenac (NP) to a three-dimensional cross-linked hydrogel system that contains carboxymethyl chitosan (CMC) and poloxamer made of poly(ethylene oxide), poly(propylene oxide), and poly(ethylene oxide) (PEO–PPO–PEO) for the *in vitro* drug release assay. In comparison to the other situation, the hydrogels were discovered to have large pores at 35°C and pH 7.4, which caused more swelling and the maximal release behavior of the model drug from the hydrogel [112]. Xu *et al.* demonstrated in their study that the macromolecular crosslinker used in the hydrogels was poly(L-lactide)-co-polyethyleneglycol-co-poly(L-lactide) dimethacrylates, which were then copolymerized with acrylic acid and N-isopropylacrylamide. The gels completely degraded within 2 weeks at pH 7.4, which was achieved by adjusting the crosslinking density and pH [113]. In a similar study also observed by Jin *et al.* (2018), N-isopropyl acrylamide (NIPAM), itaconic acid (IA), and methacrylated lignosulfonate (MLS) were copolymerized to create hydrogel, and the multifunctional MLS acted as a new macro-cross-linker with pH sensitivity between 3.0 and 9.1 [114].

9.8.3 Injectable Hydrogels—Implants

Because of the improved properties of injectable hydrogels, such as favorable carrier properties in three dimensions, low invasiveness, biocompatibility, adaptability in shape for administration, and injectability under mild conditions with a specialized procedure that necessitates improved gelation kinetics, their importance in tissue engineering and drug delivery fields is increasing [115, 116]. Hydrogels showed minimal spreading and depletion, low protein levels, low cell adhesion, and no discernible inflammatory response. After injecting them, they were eliminated via redistribution to the cecum, coupled with intestinal clearance and disintegration, but have potential for medicinal applications due to their substantial amount of water and mechanical similarity to natural tissues. Hydrogels can be made injectable, which lowers the cost of surgery and the time it takes to recuperate and use them as potential components of the drug delivery system [89, 117]. According to several studies, the injectable hydrogel is also used in wound healing, cancer treatment, bone tissue engineering, and drug delivery [116, 118, 119].

9.8.4 Magnetic Nanocellulose

The combination of magnetic nanoparticles (MNC) and CNCs, which were extracted from rice husks (RHs) using the co-precipitation process, had the intention of increasing mechanical strength and controlling drug [120, 121]. Lin *et al.* (2019) reported that through an external magnetic field (EMF) and altering the amount of Fe_3O_4 nanoparticles in the hydrogel, it is possible to regulate the effectiveness of drug release dose and rate [122]. Prior study indicates that hydrogels with higher magnetic sensitivity demonstrated decreased drug release [121]. Methotrexate (MTX) was embedded in a magnetic nanocarrier that was created by coating biocompatible tris (2-aminoethyl) amine functionalized nanocrystalline cellulose (AMFC) on Fe_3O_4 nanoparticles at pH 7.4. This MTX-loaded nanocarrier had a superior release at pH 5.4 and less cytotoxicity [123].

9.9 Nanocellulose-Based Drug Delivery in Pathological Disorder

The distinctive properties of nanocelluloses, which have recently undergone rapid development in the science field due to their unique characteristics including low risk of toxicity, biocompatibility, biodegradability, and recyclability, along with customizable surface features, are also beneficial to enhance the drug delivery system as drug excipients and also as drug delivery matrices [105, 124]. The use of nanocellulose as a carrier to modulate the rate of drug release has proven to be effective for different drug delivery routes, particularly oral, transdermal, and local against a variety of diseases [46].

9.9.1 Inflammation

An orally administered crystalline drug, Meloxicam (MX), has low solubility and anti-inflammatory, analgesic, and antipyretic effects encapsulated in NCC and MCC carriers. Compared to ground MX powder, cogrinding with MCC did not improve MX solubility, and compared to powdered MX, cogrinding with NCC increased solubility by up to 20%, making NCC an appropriate carrier for water-insoluble drugs [125]. Due to their hydrophilic nature, the free OH groups in BNC have strong hydrogen bonds, making them the best candidates for interacting with drugs and their regulated delivery [126]. A potential candidate for transdermal drug delivery systems is the CNF/chitosan and the CFNs coated with the nonsteroidal anti-inflammatory drug (NSAID) KT to lessen the side effects and increase their proper effectiveness. The mechanical properties and crystallinity of the transdermal film formulation were enhanced with an increase in CNF concentration, whereas the KT release rate from the film was reduced [127]. Another typical non-steroidal anti-inflammatory drug, Diclofenac sodium salt, encapsulated within bacterial cellulose membranes, was evaluated as a novel nanostructured transdermal delivery system in an *in vitro* study using Franz cells. The ease of fabrication and assembly of the drug-loaded membranes clearly demonstrates the tremendous prospects [128]. As per the report of Guo *et al.* (2017), droplet precipitation was used to successfully create cellulose/alginate (C/Alg) beads for metformin hydrochloride medication release (MH), and in simulated intestinal fluid, cellulose-containing micro- and nanostructures increased cellular uptake and delayed the quick release of alginate, which indicate that cellulose might be helpful to modify the inflating property [129] (Figure 9.2).

9.9.2 Cancer

According to a prior study, the KU-7 bladder cancer cells treated with the surface modification of NCC through Cetyltrimethylammonium

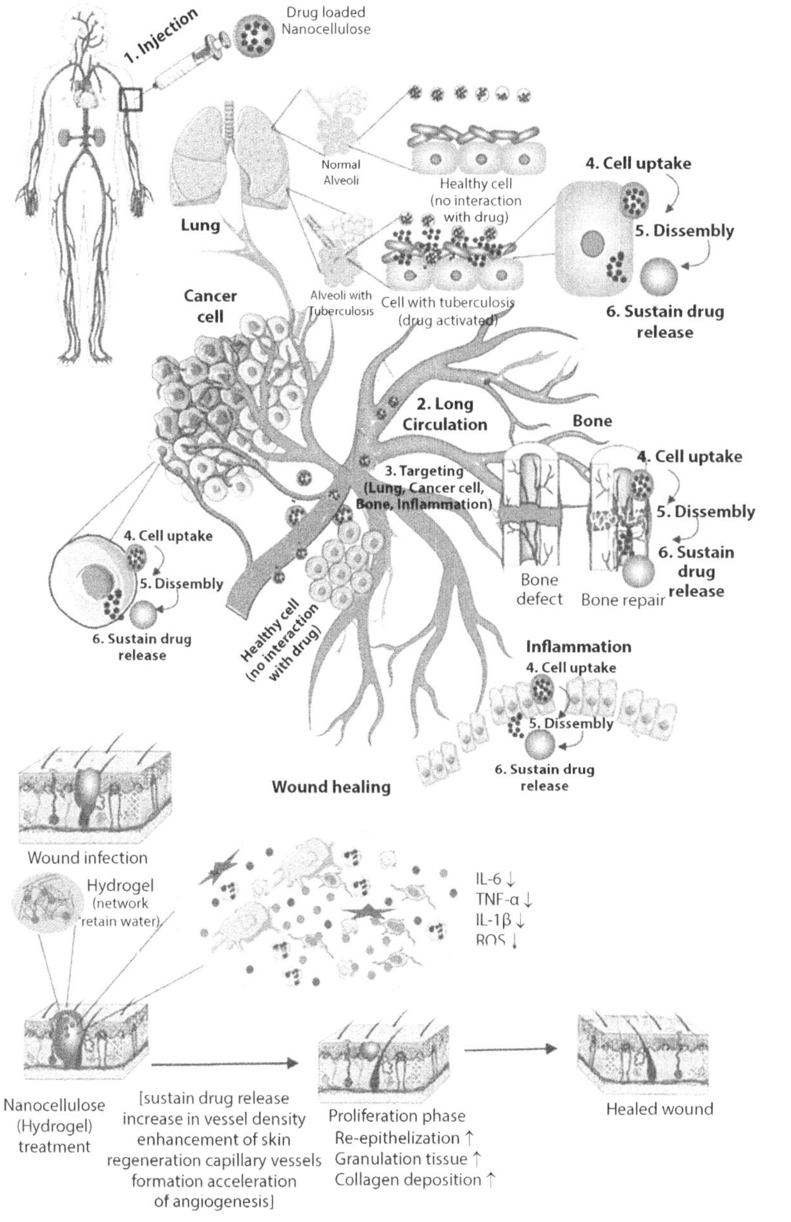

Figure 9.2 Schematic representation of nanocellulose entrapped drug delivery to the targeted organ. (1) Drug-loaded nanocellulose injection, (2) long circulation, (3) targeting the specific organ (lung, bone, cancer cell, and inflammatory organ), (4) cellular uptake, (5) disassembling the drug from nanocellulose, and (6) sustained drug release to the specific targeted organ. Nanocellulose (hydrogel) is also involved in wound healing by sustained drug release, increased vessel density, regeneration of skin, and acceleration of angiogenesis.

bromide (CTAB) control were exposed to the nonionized hydrophobic anticancer medicines docetaxel, paclitaxel, and etoposide to visualize the release amount of each gradually over several days [130]. As per the report of Mohanta *et al.*, due to the high elastic modulus and tensile strength of NCC when layer by layer assembled, it can act as a carrier along with negatively charged chitosan of an anticancer drug, doxorubicin hydrochloride, and hydrophobic curcumin and release both of them in a sustained manner [78]. A similar kind of result on a colorectal and prostate cancer cell line study also suggested that CNCs have the capacity to inhibit colon and prostate cancer via encapsulated docetaxel and etoposide [131]. Weng *et al.* reported the drug release capacity of CMC (carboxymethylcellulose) against cancer through both an *in vivo* and an *in vitro* model. As per their report, the anticancer drug doxorubicin encapsulated in CMC showed the efficacy of local release and killing the cancer cell, although in an *in vivo* study, they were suitable for distribution via microcatheters, effective at occluding arteries, and biodegradable inside arteries [132]. According to You *et al.* (2016), in an *in vivo* mouse model with liver cancer, quaternized cellulose (QC) and rigid rod-like cationic CNCs regulated by the cross-linking agent nanocomposite hydrogel encapsulated with doxorubicin, injected dermally alongside the tumor, had a lot of potential for usage in subcutaneous and prolonged anticancer drug administration to improve treatment efficacy and patient compliance [133]. Using anticancer medicines, Fe_3O_4-Ag_2O quantum dots/CNF nanocomposites were developed to improve the drug delivery mechanism, which could help treat skin cancer illnesses [134]. According to the prior study, the efficiency of folic acid conjugated with CNCs for the targeted delivery of chemotherapeutic drugs to folate receptor-positive cancer cells in human (DBTRG-05MG, H_4) and rat (C6) brain tumor cells was higher than that of nontargeted CNCs [135]. In a prior study, a hydrogel-based NCC encapsulated the ibuprofen to treat cancer tissue and resulted in the sustained release of the drug due to their high physical involvement [120]. Similar kinds of results were reported in *in vivo* and *in vitro* studies that proposed that when CMC was encapsulated with docetaxel (DTX) and treated against cancer cells, the *in vitro* study resulted in enhanced cytotoxicity and the *in vivo* study resulted in tumor growth suppression when compared to free DTX [136, 137] (Figure 9.2).

9.9.3 Bone-Related Disorder

Several studies show the possibility of using various nanocellulose forms as drug delivery methods to treat conditions affecting the bones. For example, due to their hydrophilic nature, the free OH groups in BNC have strong hydrogen bonds, making them ideal prospects for interacting with drugs and regulated delivery. Via the use of fluorescein-5' isothiocyanate labeling, they clarified the site-specific uptake that led to a minimum amount of imprecise cellular receiving, demonstrating that nanocellulose is a viable candidate for drug delivery. The berberine hydrochloride and berberine sulfate coats of BNC enhance the drug release duration, cell-specific antigen and receptor, along with cell uptake specificity [126, 138–141]. CNCs were covalently conjugated with 3-aminopropylphosphoric acid (ApA) and sodium alendronate (Aln), including the bisphosphonate groups, and found to have significant potential in drug delivery in osteoblast cells (hFOB 1.19, ATCC CRL-11372TM) in a cell culture study to treat some bone disorders [142]. In another study, BC was encapsulated with hydroxyapatite (HA), and the high presence of calcium and phosphate can cause osteoblast growth and bone formation [143].

9.9.4 Pulmonary Disorder

To increase mechanical strength and control drug release of theophylline, polysaccharide nanocrystal CNCs were added to alginate-based nanocomposite microspheres, including rod-like cellulose nanocrystals, chitin whiskers, and platelet-like starch nanocrystals, in the treatment of respiratory illness in the chosen model [74, 140]. Rifamycin is used to treat tuberculosis, although it can have significant negative side effects due to issues such as poor water solubility and a minimal biological half-life. Rifampicin is administered orally to provide controlled release and shield it from the harsh gastric environment in order to overcome these issues. As a potential drug delivery method for rifampicin, alginate (Alg), alginate-cellulose (Alg-Cel), and alginate-C-16 TMABr-modified cellulose (Alg-MCel) beads were used to improve the efficacy of this drug [144].

9.9.5 Wound Healing

A study by Liu *et al.* suggests, through an *in vivo* study, that the CNF and polydopamine (PDA) are cross-linked with calcium ions. In order to investigate hydrogel as a prospective drug delivery carrier, tetracycline hydrochloride (TH) was utilized as the model drug and the hydrogel's synergistic effect in facilitating wound healing was determined [145]. A similar study of the effect of local administration of poloxamer-loaded BNC in a shell-less egg model of a hen concluded that a ready-to-use system for octenidine delivery that combines BNC with Poloxamer could be used to treat dermal wounds for an extended period of time [146]. In order to treat wound healing, the biocompatible water-soluble polymer poly(vinyl pyrrolidone) and chitosan, both with weak mechanical properties and a loss of structural integrity, were combined. Previous studies by Poonguzhali *et al.* further showed that the rate of healing was further raised to 97% in adult Albino rats treated with Chitosan/PVP-3% stearic acid (CP-S) in an *in vivo* analysis. The migration and proliferation of keratinocytes and fibroblasts on the wound site may be the cause of how much faster the wound heals when nanocellulose is present [81]. Abu *et al.* (2020) also identified a nanocellulose film coated with honey for its kinetic and antimicrobial activity by using polyvinylpyrrolidone (PVP) as a binder, and in this *in vitro* study, they offered the nanocellulose film to be used for the treatment of chronic wounds [79]. A similar kind of study had also been found in which cellulose nanocrystal film was used as an antimicrobial drug with curcumin to enhance the lasting efficacy of the film in a diabetic rat model, and it was discovered that epithelial tissue covered the lesion, implying that it could be employed as an antimicrobial drug delivery system in diabetic wound healing [147] (Figure 9.2).

9.9.6 Optical-Related Disease

An optical drug delivery system is also an approach to overcome the challenge of efficient drug delivery, and as per the study of Orasugh *et al.* (2019), CNF-grafted collagen (CGC) affected the effectiveness of poloxame rifampicin 407-based *in situ* gels in altering ketorolac tromethamine (KT) release kinetics with a longer precorneal residence period and sustained bioavailability of KT [148]. As a drug delivery vehicle and a corneal regeneration implant for ophthalmic use, Tummala *et al.* amplified polyvinyl alcohol (PVA) and is combined with hydrogel-based nanocellulose, and the resulting material should have high water content and a specific structure that combines soft and rigid elements as building blocks. Furthermore, chitosan-poly(acrylic acid) hydrogel lenses dissolved due to the lysozyme hydrolysis of chitosan chains [149].

9.10 Commercialized and Research-Associated Nanocellulose Products Reported in Clinical Trials

Nanoparticles could be used to send anti-acne drugs directly to the follicles that are affected by acne. In order to accomplish this objective, nanoparticles that are capable of delivering the anti-acne compound known as α-mangostin have been produced using ethyl cellulose–methyl cellulose (EC-MC) [150]. Application of EC-CM nanoparticles twice daily for 2 weeks on 20 healthy human volunteers revealed no detectable skin inflammatory responses. After drug quantification in the roots of hairs taken from the skin of six human volunteers who received topical treatment, it was demonstrated that α-mangostin was successfully delivered [150]. The manufacture of synthetic tears is another fascinating use for CMC. Artificial tears, namely, those containing sodium hyaluronate 0.1% and CMC 0.5%, have been the subject of research into their efficacy in reducing postoperative ocular discomfort and bolstering tear-film integrity in eyes undergoing cataract surgery [151]. In yet another intriguing clinical experiment, the effectiveness of hydroxypropylmethylcellulose (HPMC) in the treatment of allergic rhinitis was investigated [152]. It has been postulated that inhaled allergens and other compounds can compromise the health of the nasal mucosal lining, making it easier for the body to come into contact with potentially dangerous substances emanating from the outside environment. Eventually, this can lead to allergic rhinitis, a condition characterized by chronic nasal inflammation and mucosal congestion. In order to reduce inflammation and mucosal congestion, decongestants such as oxymetazoline nasal spray, 0.05%, are frequently employed. Due to HPMC's well-known mucoprotective effects based on its viscoelastic qualities, it has been used in conjunction with oxymetazoline in some cases. Forty patients with allergic rhinitis were randomly assigned to either a group that received local oxymetazoline treatment followed by HPMC or a placebo group that received lactose powder instead of HPMC. Results of this experiment suggest that at day 15, the baseline peak inspiratory nasal flow (PNIF) in the HPMC group was higher than in the other groups, demonstrating a more significant improvement in nasal congestion [152].

Table 9.1 Recent updates on clinical trials conducted using nanocellulose-based strategies for the different biological applications.

Sl. no.	NCT no. ClinicalTrials.gov	Targeted disease and condition	Study description	Phase	Study start date	First posted	Last posted
1.	NCT03980600	Skin transplantation	Nanofibrillar cellulose wound dressing and to investigate the final product FibDex in the treatment of split thickness skin graft donor sites	Completed	February 16, 2015	June 10, 2019	June 10, 2019
2.	NCT00591916	Burn	Determine the short-term and long-term outcomes of microbial nanocellulose and fine mesh gauze impregnated with hyaluronan and thrombin	Phase III	January 2006	January 11, 2008	December 11, 2012
3.	NCT05629091	Burn	Clinical investigation of the nanofibrillar cellulose wound dressing FibDex® for the treatment of superficial dermal burns in pediatric and adult patients	Not applicable	November 30, 2022	November 29, 2022	November 29, 2022
4.	NCT05666999	Skin transplantation	Post-market clinical follow-up study for a CE marked wound care product (FibDex®) to compare the clinical performance and safety in its established intended use of split thickness skin graft donor site wounds	Not applicable	January 2, 2023	December 28, 2022	December 28, 2022
5.	NCT05079763	Acute radiation dermatitis	Clinically assess the potential of bacterial cellulose-monolaurin hydrogel, compared to placebo cream, to prevent high-grade ARD among Filipinos with breast cancer up to 4 weeks after last radiotherapy session	Phase II	September 1, 2021	October 15, 2021	September 16, 2022

(Continued)

Table 9.1 Recent updates on clinical trials conducted using nanocellulose-based strategies for the different biological applications. (*Continued*)

Sl. no.	NCT no. ClinicalTrials.gov	Targeted disease and condition	Study description	Phase	Study start date	First posted	Last posted
6.	NCT00987727	Dry eye syndromes Keratoconjunctivitis sicca	Evaluate the safety and efficacy of OPTIVE* MD for ocular surface integrity in symptomatic dry eye condition in absence of obvious eyelid inflammation	Phase IV	November 2009	September 21, 2011	September 21, 2011
7.	NCT03903592	Venous leg ulcer	A marine polysaccharide dressing to a carboxymethylcellulose dressing to determine which dressing better manages these wounds with regard to wound size and periwound skin condition	Not applicable	September 30, 2019	April 4, 2019	November 3, 2021
8.	NCT02758327	Dry eye syndrome	To demonstrate a change in osmolarity over time when moderate to severe dry eye subjects are treated with TheraTears lubricating drops	Phase IV	October 2015	May 2, 2016	August 18, 2017
9.	NCT02773251	Prostate cancer Postoperative adhesion	To assess the anti-adhesive effect of treatment with hyaluronic acid-carboxymethylcellulose following laparoscopic pelvic surgery (radical prostatectomy)	Phase IV	November 2011	May 16, 2016	May 17, 2016
10.	NCT02365675	Pemphigus Pemphigoid	To compare the efficacy of four dressings (covers), namely, gauze with petrolatum, cellulose acetate with petrolatum, pure carboxymethylcellulose with silver, and nanocrystalline silver to improve the new growth of skin and reduce pain and itch in persons suffering from pemphigus and pemphigoid	Not Applicable	January 2015	February 19, 2015	February 19, 2015

The combination of FibDex® by UPM-Kymmene Corporation and nanofibrillar cellulose (NFC), which may result in rapid wound healing at skin graft donor sites, is comparable to or even preferable to the copolymer dressing in comparison [153]. Polydopamine (PDA) filled with the anticancer drug was a pyrene-modified cellulose nanocrystal (Py-CNC) and agarose (AG) composite hydrogel with multifunctional features. The ability to deal with the complicated microenvironment of the tumor was improved by paclitaxel [154]. Avicel® (FMC BioPolymer) is commonly used for tableting medications because it has a non-toxic effect and can bind readily with MCC since it can bind with many additives without the use of any glue [155]. In addition, there have been and continue to be a great deal of clinical investigations conducted on the application of nanocellulose in the scientific discipline of biology. There is some representation of them in Table 9.1.

9.11 Future Perspective

This study suggested the importance of nanocellulose in drug delivery systems and mentioned here the use of several techniques to modify the different nanocelluloses from different sources and the route of administration of drug-loaded nanocellulose. In the process of adapting the properties of nanocellulose, such as to carry and release drug at a targeted area in a sustained manner, increase hydrophobic drug solubility, improve the efficacy of this drug, sustain bioavailability, etc., most advanced technology should be used. Some research is required to determine the level of toxicity of nanocellulose against humans and animals, which is why chemical modification should be included in the toxicity assessment. Because there are insufficient clinical trials of FDA-approved drugs for commercialization, further clinical trial research is required. The loading of drugs to control drug distribution into nanocellulose-based hydrogels is most difficult; therefore, special attention should be paid to its preparation, such as inventing an innocuous solvent, a green synthesis process, and a non-toxic crosslinking agent for further design. In biomedical applications, nanocelluloses produced naturally are safe and nontoxic, and their functionalization and surface modification aid in improving their interaction and particular application with the human body at target sites. Numerous investigations are needed to better understand their mechanism of action and examine the potential pharmacological side effects and cytotoxicity of this intriguing natural biopolymer. Several studies reported that nanocellulose works as a multifunctional nanoparticle and may also act as a theranostic treatment for cancer.

9.12 Conclusion

The revolution of cellulose nanomaterials gained an achievement in the science domain of drug delivery design and materials. Nanocellulose heralds another age of nanomaterials that has become one of the most significant nanosized biological materials of the present generation, considering its predominant mechanical, physical, chemical, and environmentally friendly properties. These unique qualities of nanocellulose expand its range of potential uses in a variety of industries, such as biomedicine, films and bio-based packaging, water treatment, hydrogels, energy, and composites. With its unique development in the field of science, nanocellulose combines with drug delivery and aids several pathological disorders in a very effective manner. Numerous studies have been undertaken to assess the potential of nanocellulose as multifunctional nanoparticles owing to some unique properties such as reduced toxicity, biocompatibility, biodegradability, recyclability, and customizable surface features, which made it an excellent choice. It was developed in a variety of forms, including microparticles, films, and gels, and administered through a variety of routes, such as topical, transdermal, oral, etc. Nanocellulose-based drug delivery demonstrated the sustained and pH-dependent release of a drug in a specific target area and increased the drug's potential in a controlled manner. Although there is debate concerning nanocellulose's toxicity, this material exhibits significant potential for the development of a promising generation of controlled drug delivery systems.

Acknowledgments

University of Gour Banga is acknowledged for providing the opportunity to work in the Department of Physiology. Parts of the Figures 9.1 and 9.2 were drawn by using pictures from Servier Medical Art. Servier Medical Art by Servier is licensed under a Creative Commons Attribution 3.0 Unported License (https://creativecommons.org/licenses/by/3.0/).

References

1. Shokri, J. and Adibkia, K., Application of cellulose and cellulose derivatives in pharmaceutical industries, in: *Cellulose-Medical, Pharmaceutical and Electronic Applications*, IntechOpen, London, UK, 2013.

2. Aziz, T., Farid, A., Haq, F., Kiran, M., Ullah, A., Zhang, K., Li, C., Ghazanfar, S., Sun, H., Ullah, R., A review on the modification of cellulose and its applications. *Polymers*, 14, 3206, 2022.
3. Seddiqi, H., Oliaei, E., Honarkar, H., Jin, J., Geonzon, L.C., Bacabac, R.G., Klein-Nulend, J., Cellulose and its derivatives: Towards biomedical applications. *Cellulose*, 28, 1893–1931, 2021.
4. Tiwari, G., Tiwari, R., Sriwastawa, B., Bhati, L., Pandey, S., Pandey, P., Bannerjee, S.K., Drug delivery systems: An updated review. *Int. J. Pharm. Investing.*, 2, 2, 2012.
5. Li, C., Wang, J., Wang, Y., Gao, H., Wei, G., Huang, Y., Yu, H., Gan, Y., Wang, Y., Mei, L., Recent progress in drug delivery. *Acta Pharm. Sin. B*, 9, 1145–1162, 2019.
6. Pramanik, A., Xu, Z., Ingram, N., Coletta, P.L., Millner, P.A., Tyler, A.I.I., Hughes, T.A., Hyaluronic-acid-tagged cubosomes deliver cytotoxics specifically to CD44-positive cancer cells. *Mol. Pharmaceutics*, 19, 4601–4611, 2022.
7. Pramanik, A., Xu, Z., Shamsuddin, S.H., Khaled, Y.S., Ingram, N., Maisey, T., Tomlinson, D., Coletta, P.L., Jayne, D., Hughes, T.A., Tyler, A.I.I., Millner, P.A., Affimer tagged cubosomes: Targeting of carcinoembryonic antigen expressing colorectal cancer cells using *in vitro* and *in vivo* models. *ACS Appl. Mater. Interfaces*, 14, 11078–11091, 2022.
8. Trache, D., Tarchoun, A.F., Derradji, M., Hamidon, T.S., Masruchin, N., Brosse, N., Hussin, M.H., Nanocellulose: From fundamentals to advanced applications. *Front. Chem.*, 8, 392, 2020.
9. Huo, Y., Liu, Y., Xia, M., Du, H., Lin, Z., Li, B., Liu, H., Nanocellulose-based composite materials used in drug delivery systems. *Polymers*, 14, 2648, 2022.
10. Patil, T.V., Patel, D.K., Dutta, S.D., Ganguly, K., Santra, T.S., Lim, K.-T., Nanocellulose, a versatile platform: From the delivery of active molecules to tissue engineering applications. *Bioact. Mater.*, 9, 566–589, 2022.
11. Lee, S.-H., Kim, H.-J., Kim, J.-C., Nanocellulose applications for drug delivery: A review. *J. For. Environ. Sci.*, 35, 141–149, 2019.
12. Salimi, S., Sotudeh-Gharebagh, R., Zarghami, R., Chan, S.Y., Yuen, K.H., Production of nanocellulose and its applications in drug delivery: A critical review. *ACS Sustain. Chem. Eng.*, 7, 15800–15827, 2019.
13. Anarjan, F.S., Active targeting drug delivery nanocarriers: Ligands. *Nano-Struct. Nano-Objects*, 19, 100370, 2019.
14. Pramanik, A., Laha, D., Chattopadhyay, S., Dash, S.K., Roy, S., Pramanik, P., Karmakar, P., Targeted delivery of "copper carbonate" nanoparticles to cancer cells *in vivo*. *Toxicol. Res.*, 4, 1604–1612, 2015.
15. Attia, M.F., Anton, N., Wallyn, J., Omran, Z., Vandamme, T.F., An overview of active and passive targeting strategies to improve the nanocarriers efficiency to tumour sites. *J. Pharm. Pharmacol.*, 71, 1185–1198, 2019.

16. Hua, Q., Qiang, Z., Chu, M., Shi, D., Ren, J., Polymeric drug delivery system with actively targeted cell penetration and nuclear targeting for cancer therapy. *ACS Appl. Bio Mater.*, 2, 1724–1731, 2019.
17. Yoo, J., Park, C., Yi, G., Lee, D., Koo, H., Active targeting strategies using biological ligands for nanoparticle drug delivery systems. *Cancers*, 11, 640, 2019.
18. Chen, W.C., Zhang, A.X., Li, S.-D., Limitations and niches of the active targeting approach for nanoparticle drug delivery. *Eur. J. Nanomed.*, 4, 89–93, 2012.
19. Kumar Khanna, V., Targeted delivery of nanomedicines. *Int. Sch. Res. Notices*, 2012, 571394, 2012.
20. Bazak, R., Houri, M., El Achy, S., Hussein, W., Refaat, T., Passive targeting of nanoparticles to cancer: A comprehensive review of the literature. *Mol. Clin. Oncol.*, 2, 904–908, 2014.
21. Narum, S.M., Le, T., Le, D.P., Lee, J.C., Donahue, N.D., Yang, W., Wilhelm, S., Passive targeting in nanomedicine: Fundamental concepts, body interactions, and clinical potential, in: *Nanoparticles for Biomedical Applications*, pp. 37–53, Elsevier, Netherlands, 2020.
22. Bennet, D. and Kim, S., Polymer nanoparticles for smart drug delivery, in: *Application of Nanotechnology in Drug Delivery*, p. 8, 2014.
23. Fenton, O.S., Olafson, K.N., Pillai, P.S., Mitchell, M.J., Langer, R., Advances in biomaterials for drug delivery. *Adv. Mater.*, 30, 1705328, 2018.
24. Kumari, A., Yadav, S.K., Yadav, S.C., Biodegradable polymeric nanoparticles based drug delivery systems. *Colloids Surf. B: Biointerfaces*, 75, 1–18, 2010.
25. Adeosun, S.O., Ilomuanya, M.O., Gbenebor, O.P., Dada, M.O., Odili, C.C., Biomaterials for drug delivery: Sources, classification, synthesis, processing, and applications, in: *Advanced Functional Materials*, pp. 141–167, 2020.
26. Kim, J.K., Kim, H.J., Chung, J.-Y., Lee, J.-H., Young, S.-B., Kim, Y.-H., Natural and synthetic biomaterials for controlled drug delivery. *Arch. Pharmacal Res.*, 37, 60–68, 2014.
27. Bharti, C., Nagaich, U., Pal, A.K., Gulati, N., Mesoporous silica nanoparticles in target drug delivery system: A review. *Int. J. Pharm. Investig.*, 5, 124, 2015.
28. Murugan, B., Sagadevan, S., Lett, A., Fatimah, I., Fatema, K.N., Oh, W.-C., Mohammad, F., Johan, M.R., Role of mesoporous silica nanoparticles for the drug delivery applications. *Mater. Res. Express*, 7, 102002, 2020.
29. Gao, Y., Gao, D., Shen, J., Wang, Q., A review of mesoporous silica nanoparticle delivery systems in chemo-based combination cancer therapies. *Front. Chem.*, 8, 598722, 2020.
30. Zare, H., Ahmadi, S., Ghasemi, A., Ghanbari, M., Rabiee, N., Bagherzadeh, M., Karimi, M., Webster, T.J., Hamblin, M.R., Mostafavi, E., Carbon nanotubes: Smart drug/gene delivery carriers. *Int. J. Nanomed.*, 16, 1681, 2021.
31. Debnath, S.K. and Srivastava, R., Drug delivery with carbon-based nanomaterials as versatile nanocarriers: Progress and prospects. *Front. Nanotechnol.*, 3, 644564, 2021.

32. Zhang, W., Zhang, Z., Zhang, Y., The application of carbon nanotubes in target drug delivery systems for cancer therapies. *Nanoscale Res. Lett.*, 6, 1–22, 2011.
33. Begines, B., Ortiz, T., Pérez-Aranda, M., Martínez, G., Merinero, M., Argüelles-Arias, F., Alcudia, A., Polymeric nanoparticles for drug delivery: Recent developments and future prospects. *Nanomaterials*, 10, 1403, 2020.
34. Zielińska, A., Carreiró, F., Oliveira, A.M., Neves, A., Pires, B., Venkatesh, D.N., Durazzo, A., Lucarini, M., Eder, P., Silva, A.M., Polymeric nanoparticles: Production, characterization, toxicology and ecotoxicology. *Molecules*, 25, 3731, 2020.
35. Hakim, L.K., Yazdanian, M., Alam, M., Abbasi, K., Tebyaniyan, H., Tahmasebi, E., Khayatan, D., Seifalian, A., Ranjbar, R., Yazdanian, A., Biocompatible and biomaterials application in drug delivery system in oral cavity. *Evid.-Based Complementary Altern. Med.*, 2021, 9011226, 2021.
36. Shen, J. and Burgess, D.J., *In vitro* dissolution testing strategies for nanoparticulate drug delivery systems: Recent developments and challenges. *Drug Deliv. Transl. Res.*, 3, 409–415, 2013.
37. Lee, J.H. and Yeo, Y., Controlled drug release from pharmaceutical nanocarriers. *Chem. Eng. Sci.*, 125, 75–84, 2015.
38. Xiao, X., Teng, F., Shi, C., Chen, J., Wu, S., Wang, B., Meng, X., Imeh, A.E., Li, W., Polymeric nanoparticles—Promising carriers for cancer therapy. *Front. Bioeng. Biotechnol.* 10, 1024143, 2022.
39. Rizi, H.A.Y., Shin, D.H., Rizi, S.Y., Polymeric nanoparticles in cancer chemotherapy: A narrative review. *Iran. J. Public Health*, 51, 226, 2022.
40. Scioli Montoto, S., Muraca, G., Ruiz, M.E., Solid lipid nanoparticles for drug delivery: Pharmacological and biopharmaceutical aspects. *Front. Mol. Biosci.*, 7, 587997, 2020.
41. Mogoşanu, G.D., Grumezescu, A.M., Bejenaru, L.E., Bejenaru, C., Natural and synthetic polymers for drug delivery and targeting, in: *Nanobiomaterials in Drug Delivery*, pp. 229–284, Elsevier, Netherlands, 2016.
42. Xu, L., Wang, X., Liu, Y., Yang, G., Falconer, R.J., Zhao, C.-X., Lipid nanoparticles for drug delivery. *Adv. NanoBiomed Res.*, 2, 2100109, 2022.
43. Hartl, N., Adams, F., Merkel, O.M., From adsorption to covalent bonding: Apolipoprotein E functionalization of polymeric nanoparticles for drug delivery across the blood–brain barrier. *Adv. Ther.*, 4, 2000092, 2021.
44. Martins, S.M., Sarmento, B., Nunes, C., Lúcio, M., Reis, S., Ferreira, D.C., Brain targeting effect of camptothecin-loaded solid lipid nanoparticles in rat after intravenous administration. *Eur. J. Pharm. Biopharm.*, 85, 488–502, 2013.
45. Troy, E., Tilbury, M.A., Power, A.M., Wall, J.G., Nature-based biomaterials and their application in biomedicine. *Polymers*, 13, 3321, 2021.
46. Kupnik, K., Primožič, M., Kokol, V., Leitgeb, M., Nanocellulose in drug delivery and antimicrobially active materials. *Polymers*, 12, 2825, 2020.

47. Sadoughi, F., Hallajzadeh, J., Asemi, Z., Mansournia, M.A., Yousefi, B., Nanocellulose-based delivery systems and cervical cancer: Review of the literature. *Curr. Pharm. Des.*, 27, 4707–4715, 2021.
48. Das, S., Ghosh, B., Sarkar, K., Nanocellulose as sustainable biomaterials for drug delivery. *Sens. Int.*, 3, 100135, 2022.
49. Randhawa, A., Dutta, S.D., Ganguly, K., Patil, T.V., Patel, D.K., Lim, K.-T., A review of properties of nanocellulose, its synthesis, and potential in biomedical applications. *Appl. Sci.*, 12, 7090, 2022.
50. Yang, G., Ullah, M.W., Zhijun, S., *Nanocellulose: Synthesis, structure, properties and applications*, World Scientific, Singapore, 2021.
51. Kargarzadeh, H., Ioelovich, M., Ahmad, I., Thomas, S., Dufresne, A., Methods for extraction of nanocellulose from various sources, in: *Handbook of Nanocellulose and Cellulose Nanocomposites*, vol. 1, pp. 1–49, 2017.
52. Etuk, V.E., Oboh, I.O., Etuk, B.R., Johnson, E.O., Egemba, K., Nanocellulose: Types, sythesis and applications, *The European Conference on Sustainability, Energy & the Environment Official Conference Proceedings*, 2018.
53. Chirayil, C.J., Mathew, L., Thomas, S., Review of recent research in nano cellulose preparation from different lignocellulosic fibers. *Rev. Adv. Mater. Sci.*, 37, 20–28, 2014.
54. Kerekes, R.J., Characterizing refining action in PFI mills. *Tappi J.*, 4, 9–14, 2005.
55. Wang, H., Zhang, X., Jiang, Z., Li, W., Yu, Y., A comparison study on the preparation of nanocellulose fibrils from fibers and parenchymal cells in bamboo (Phyllostachys pubescens). *Ind. Crops Prod.*, 71, 80–88, 2015.
56. Jamwal, P., Chauhan, G.S., Kumar, P., Kumari, B., Kumar, K., Chauhan, S., A study in the synthesis of new Pinus wallichiana derived spherical nanocellulose hydrogel and its evaluation as malachite green adsorbent. *Sustain. Chem. Pharm.*, 32, 100950, 2023.
57. Khine, Y.Y. and Stenzel, M.H., Surface modified cellulose nanomaterials: A source of non-spherical nanoparticles for drug delivery. *Mater. Horiz.*, 7, 1727–1758, 2020.
58. Gonçalves, A.P., Oliveira, E., Mattedi, S., Jose, N.M., Separation of cellulose nanowhiskers from microcrystalline cellulose with an aqueous protic ionic liquid based on ammonium and hydrogensulphate. *Sep. Purif. Technol.*, 196, 200–207, 2018.
59. Ahrem, H., Pretzel, D., Endres, M., Conrad, D., Courseau, J., Müller, H., Jaeger, R., Kaps, C., Klemm, D.O., Kinne, R.W., Laser-structured bacterial nanocellulose hydrogels support ingrowth and differentiation of chondrocytes and show potential as cartilage implants. *Acta Biomater.*, 10, 1341–1353, 2014.
60. Wang, J., Huang, S., Li, Y., Wei, Y., Xi, X., Cai, K., Microstructure, mechanical and bio-corrosion properties of Mn-doped Mg–Zn–Ca bulk metallic glass composites. *Mater. Sci. Eng.: C*, 33, 3832–3838, 2013.

61. Islam, M.U., Ullah, M.W., Khan, S., Shah, N., Park, J.K., Strategies for cost-effective and enhanced production of bacterial cellulose. *Int. J. Biol. Macromol.*, 102, 1166–1173, 2017.
62. Tanpichai, S., Quero, F., Nogi, M., Yano, H., Young, R.J., Lindström, T., Sampson, W.W., Eichhorn, S.J., Effective Young's modulus of bacterial and microfibrillated cellulose fibrils in fibrous networks. *Biomacromolecules*, 13, 1340–1349, 2012.
63. Klemm, D., Kramer, F., Moritz, S., Lindström, T., Ankerfors, M., Gray, D., Dorris, A., Nanocelluloses: A new family of nature-based materials. *Angew. Chem. Int. Ed.*, 50, 5438–5466, 2011.
64. Lagerwall, J.P.F., Schütz, C., Salajkova, M., Noh, J., Hyun Park, J., Scalia, G., Bergström, L., Cellulose nanocrystal-based materials: From liquid crystal self-assembly and glass formation to multifunctional thin films. *NPG Asia Mater.*, 6, e80–e80, 2014.
65. Frost, B.A., Sutliff, B.P., Thayer, P., Bortner, M.J., Foster, E.J., Gradient poly (ethylene glycol) diacrylate and cellulose nanocrystals tissue engineering composite scaffolds via extrusion bioprinting. *Front. Bioeng. Biotechnol.*, 7, 280, 2019.
66. Siqueira, G., Bras, J., Dufresne, A., Cellulosic bionanocomposites: A review of preparation, properties and applications. *Polymers*, 2, 728–765, 2010.
67. Abol-Fotouh, D., Hassan, M.A., Shokry, H., Roig, A., Azab, M.S., Kashyout, A.E.-H.B., Bacterial nanocellulose from agro-industrial wastes: Low-cost and enhanced production by Komagataeibacter saccharivorans MD1. *Sci. Rep.*, 10, 3491, 2020.
68. Barja, F., Bacterial nanocellulose production and biomedical applications. *J. Biomed. Res.*, 35, 310, 2021.
69. Sharma, C. and Bhardwaj, N.K., Bacterial nanocellulose: Present status, biomedical applications and future perspectives. *Mater. Sci. Eng.: C*, 104, 109963, 2019.
70. Mohammad, F., Arfin, T., Al-Lohedan, H.A., Sustained drug release and electrochemical performance of ethyl cellulose-magnesium hydrogen phosphate composite. *Mater. Sci. Eng.: C*, 71, 735–743, 2017.
71. Xie, J., Wang, C., Ning, Q., Gao, Q., Gao, C., Gou, Z., Ye, J., A new strategy to sustained release of ocular drugs by one-step drug-loaded microcapsule manufacturing in hydrogel punctal plugs. *Graefe's Arch. Clin. Exp. Ophthalmol.*, 255, 2173–2184, 2017.
72. Mondal, S., Preparation, properties and applications of nanocellulosic materials. *Carbohydr. Polym.*, 163, 301–316, 2017.
73. Ulbrich, K., Hola, K., Subr, V., Bakandritsos, A., Tucek, J., Zboril, R., Targeted drug delivery with polymers and magnetic nanoparticles: Covalent and non-covalent approaches, release control, and clinical studies. *Chem. Rev.*, 116, 5338–5431, 2016.

74. Lin, N., Huang, J., Chang, P.R., Feng, L., Yu, J., Effect of polysaccharide nanocrystals on structure, properties, and drug release kinetics of alginate-based microspheres. *Colloids Surf. B: Biointerfaces*, 85, 270–279, 2011.
75. Low, L.E., Tan, L.T.-H., Goh, B.-H., Tey, B.T., Ong, B.H., Tang, S.Y., Magnetic cellulose nanocrystal stabilized Pickering emulsions for enhanced bioactive release and human colon cancer therapy. *Int. J. Biol. Macromol.*, 127, 76–84, 2019.
76. Tan, H.-F., Ooi, B.S., Leo, C.P., Future perspectives of nanocellulose-based membrane for water treatment. *J. Water Process Eng.*, 37, 101502, 2020.
77. Lin, N. and Dufresne, A., Nanocellulose in biomedicine: Current status and future prospect. *Eur. Polym. J.*, 59, 302–325, 2014.
78. Mohanta, V., Madras, G., Patil, S., Layer-by-layer assembled thin films and microcapsules of nanocrystalline cellulose for hydrophobic drug delivery. *ACS Appl. Mater. Interfaces*, 6, 20093–20101, 2014.
79. Abu, T.M., Zahan, K.A., Rajaie, M.A., Leong, C.R., Ab Rashid, S., Hamin, N.S.M.N., Tan, W.N., Tong, W.Y., Nanocellulose as drug delivery system for honey as antimicrobial wound dressing. *Mater. Today: Proc.*, 31, 14–17, 2020.
80. Saïdi, L., Vilela, C., Oliveira, H., Silvestre, A.J.D., Freire, C.S.R., Poly (N-methacryloyl glycine)/nanocellulose composites as pH-sensitive systems for controlled release of diclofenac. *Carbohydr. Polym.*, 169, 357–365, 2017.
81. Poonguzhali, R., Khaleel Basha, S., Sugantha Kumari, V., Synthesis of alginate/nanocellulose bionanocomposite for *in vitro* delivery of ampicillin. *Polym. Bull.*, 75, 4165–4173, 2018.
82. Zhang, X., Elsayed, I., Navarathna, C., Schueneman, G.T., Hassan, E.I.B., Biohybrid hydrogel and aerogel from self-assembled nanocellulose and nanochitin as a high-efficiency adsorbent for water purification. *ACS Appl. Mater. Interfaces*, 11, 46714–46725, 2019.
83. Jiang, F. and Hsieh, Y.-L., Cellulose nanofibril aerogels: Synergistic improvement of hydrophobicity, strength, and thermal stability via cross-linking with diisocyanate. *ACS Appl. Mater. Interfaces*, 9, 2825–2834, 2017.
84. Nicu, R., Ciolacu, F., Ciolacu, D.E., Advanced functional materials based on nanocellulose for pharmaceutical/medical applications. *Pharmaceutics*, 13, 1125, 2021.
85. Li, Z., Chen, Z., Chen, H., Chen, K., Tao, W., Ouyang, X.-k., Mei, L., Zeng, X., Polyphenol-based hydrogels: Pyramid evolution from crosslinked structures to biomedical applications and the reverse design. *Bioact. Mater.*, 17, 49–70, 2022.
86. Hu, S., Zhi, Y., Shan, S., Ni, Y., Research progress of smart response composite hydrogels based on nanocellulose. *Carbohydr. Polym.*, 275, 118741, 2022.
87. Treesuppharat, W., Rojanapanthu, P., Siangsanoh, C., Manuspiya, H., Ummartyotin, S., Synthesis and characterization of bacterial cellulose and gelatin-based hydrogel composites for drug-delivery systems. *Biotechnol. Rep.*, 15, 84–91, 2017.

88. Liu, Y., Fan, Q., Huo, Y., Liu, C., Li, B., Li, Y., Construction of a mesoporous polydopamine@ GO/cellulose nanofibril composite hydrogel with an encapsulation structure for controllable drug release and toxicity shielding. *ACS Appl. Mater. Interfaces*, 12, 57410–57420, 2020.
89. Bertsch, P., Schneider, L., Bovone, G., Tibbitt, M.W., Fischer, P., Gstöhl, S., Injectable biocompatible hydrogels from cellulose nanocrystals for locally targeted sustained drug release. *ACS Appl. Mater. Interfaces*, 11, 38578–38585, 2019.
90. Barrios, E., Fox, D., Li Sip, Y.Y., Catarata, R., Calderon, J.E., Azim, N., Afrin, S., Zhang, Z., Zhai, L., Nanomaterials in advanced, high-performance aerogel composites: A review. *Polymers*, 11, 726, 2019.
91. Chen, X., Xu, X., Li, W., Sun, B., Yan, J., Chen, C., Liu, J., Qian, J., Sun, D., Effective drug carrier based on polyethylenimine-functionalized bacterial cellulose with controllable release properties. *ACS Appl. Bio Mater.*, 1, 42–50, 2018.
92. Ghilan, A., Nicu, R., Ciolacu, D.E. and Ciolacu, F., Insight into the Latest Medical Applications of Nanocellulose. *Materials*, 16, 12, 4447, 2023.
93. Paajanen, A. and Vaari, J., High-temperature decomposition of the cellulose molecule: A stochastic molecular dynamics study. *Cellulose*, 24, 2713–2725, 2017.
94. Abitbol, T., Rivkin, A., Cao, Y., Nevo, Y., Abraham, E., Ben-Shalom, T., Lapidot, S., Shoseyov, O., Nanocellulose, a tiny fiber with huge applications. *Curr. Opin. Biotechnol.*, 39, 76–88, 2016.
95. Seabra, A.B., Bernardes, J.S., Fávaro, W.J., Paula, A.J., Durán, N., Cellulose nanocrystals as carriers in medicine and their toxicities: A review. *Carbohydr. Polym.*, 181, 514–527, 2018.
96. Alexandrescu, L., Syverud, K., Gatti, A., Chinga-Carrasco, G., Cytotoxicity tests of cellulose nanofibril-based structures. *Cellulose*, 20, 1765–1775, 2013.
97. Harper, B.J., Clendaniel, A., Sinche, F., Way, D., Hughes, M., Schardt, J., Simonsen, J., Stefaniak, A.B., Harper, S.L., Impacts of chemical modification on the toxicity of diverse nanocellulose materials to developing zebrafish. *Cellulose*, 23, 1763–1775, 2016.
98. Roman, M., Toxicity of cellulose nanocrystals: A review. *Ind. Biotechnol.*, 11, 25–33, 2015.
99. Hosseinidoust, Z., Alam, M.N., Sim, G., Tufenkji, N., van de Ven, T.G.M., Cellulose nanocrystals with tunable surface charge for nanomedicine. *Nanoscale*, 7, 16647–16657, 2015.
100. DeLoid, G.M., Cao, X., Molina, R.M., Silva, D.I., Bhattacharya, K., Ng, K.W., Loo, S.C.J., Brain, J.D., Demokritou, P., Toxicological effects of ingested nanocellulose in *in vitro* intestinal epithelium and *in vivo* rat models. *Environ. Sci.: Nano*, 6, 2105–2115, 2019.

101. Ventura, C., Pinto, F., Lourenço, A.F., Ferreira, P.J.T., Louro, H., Silva, M.J., On the toxicity of cellulose nanocrystals and nanofibrils in animal and cellular models. *Cellulose*, 27, 5509–5544, 2020.
102. Zheng, L., Zhang, S., Ying, Z., Liu, J., Zhou, Y., Chen, F., Engineering of aerogel-based biomaterials for biomedical applications. *Int. J. Nanomed.*, 15, 2363–2378, 2020.
103. Husain, F.M., Khan, A., Khan, R.A., Siddique, J.A., Oves, M., Khan, A.A.P., Ansari, M.O., Cancar, H.D., Bio-based aerogels and their environment applications: An overview, in: *Advances in Aerogel Composites for Environmental Remediation*, pp. 347–356, 2021.
104. De France, K.J., Hoare, T., Cranston, E.D., Review of hydrogels and aerogels containing nanocellulose. *Chem. Mater.*, 29, 4609–4631, 2017.
105. Liu, Z., Zhang, S., He, B., Wang, S., Kong, F., Synthesis of cellulose aerogels as promising carriers for drug delivery: A review. *Cellulose*, 28, 2697–2714, 2021.
106. García-González, C.A., Alnaief, M., Smirnova, I., Polysaccharide-based aerogels—Promising biodegradable carriers for drug delivery systems. *Carbohydr. Polym.*, 86, 1425–1438, 2011.
107. Gupta, P., Vermani, K., Garg, S., Hydrogels: From controlled release to pH-responsive drug delivery. *Drug Discovery Today*, 7, 569–579, 2002.
108. Ahmed, E.M., Hydrogel: Preparation, characterization, and applications: A review. *J. Adv. Res.*, 6, 105–121, 2015.
109. Caló, E. and Khutoryanskiy, V.V., Biomedical applications of hydrogels: A review of patents and commercial products. *Eur. Polym. J.*, 65, 252–267, 2015.
110. Akhtar, M.F., Hanif, M., Ranjha, N.M., Methods of synthesis of hydrogels… A review. *Saudi Pharm. J.*, 24, 554–559, 2016.
111. Curvello, R., Raghuwanshi, V.S., Garnier, G., Engineering nanocellulose hydrogels for biomedical applications. *Adv. Colloid Interface Sci.*, 267, 47–61, 2019.
112. Yu, S., Zhang, X., Tan, G., Tian, L., Liu, D., Liu, Y., Yang, X., Pan, W., A novel pH-induced thermosensitive hydrogel composed of carboxymethyl chitosan and poloxamer cross-linked by glutaraldehyde for ophthalmic drug delivery. *Carbohydr. Polym.*, 155, 208–217, 2017.
113. Xu, L., Qiu, L., Sheng, Y., Sun, Y., Deng, L., Li, X., Bradley, M., Zhang, R., Biodegradable pH-responsive hydrogels for controlled dual-drug release. *J. Mater. Chem. B*, 6, 510–517, 2018.
114. Jin, C., Song, W., Liu, T., Xin, J., Hiscox, W.C., Zhang, J., Liu, G., Kong, Z., Temperature and pH responsive hydrogels using methacrylated lignosulfonate cross-linker: Synthesis, characterization, and properties. *ACS Sustain. Chem. Eng.*, 6, 1763–1771, 2018.

115. Guvendiren, M., Lu, H.D., Burdick, J.A., Shear-thinning hydrogels for biomedical applications. *Soft Matter*, 8, 260–272, 2012.
116. Sun, Y., Nan, D., Jin, H., Qu, X., Recent advances of injectable hydrogels for drug delivery and tissue engineering applications. *Polym. Test.*, 81, 106283, 2020.
117. Overstreet, D.J., Dutta, D., Stabenfeldt, S.E., Vernon, B.L., Injectable hydrogels. *J. Polym. Sci. Part B: Polym. Phys.*, 50, 881–903, 2012.
118. Dimatteo, R., Darling, N.J., Segura, T., *In situ* forming injectable hydrogels for drug delivery and wound repair. *Adv. Drug Delivery Rev.*, 127, 167–184, 2018.
119. Kumar, S. and Bajaj, A., Advances in self-assembled injectable hydrogels for cancer therapy. *Biomater. Sci.*, 8, 2055–2073, 2020.
120. Supramaniam, J., Adnan, R., Kaus, N.H.M., Bushra, R., Magnetic nanocellulose alginate hydrogel beads as potential drug delivery system. *Int. J. Biol. Macromol.*, 118, 640–648, 2018.
121. Kumar, A., Sood, A., Han, S.S., Potential of magnetic nano cellulose in biomedical applications: Recent advances. *Biomater. Polym. Horiz.*, 1, 32–47, 2022.
122. Lin, F., Zheng, J., Guo, W., Zhu, Z., Wang, Z., Dong, B., Lin, C., Huang, B., Lu, B., Smart cellulose-derived magnetic hydrogel with rapid swelling and deswelling properties for remotely controlled drug release. *Cellulose*, 26, 6861–6877, 2019.
123. Rahimi, M., Shojaei, S., Safa, K.D., Ghasemi, Z., Salehi, R., Yousefi, B., Shafiei-Irannejad, V., Biocompatible magnetic tris (2-aminoethyl) amine functionalized nanocrystalline cellulose as a novel nanocarrier for anticancer drug delivery of methotrexate. *New J. Chem.*, 41, 2160–2168, 2017.
124. Hasan, N., Rahman, L., Kim, S.-H., Cao, J., Arjuna, A., Lallo, S., Jhun, B.H., Yoo, J.-W., Recent advances of nanocellulose in drug delivery systems. *J. Pharm. Invest.*, 50, 553–572, 2020.
125. Emaraa, L.H., El-Ashmawya, A.A., Tahaa, N.F., El-Shaffeib, K.A., Mahdeyb, E.-S.M., Elkhollyc, H.K., Nano-crystalline cellulose as a novel tablet excipient for improving solubility and dissolution of meloxicam. *J. Appl. Pharm. Sci.*, 6, 032–043, 2016.
126. Pötzinger, Y., Kralisch, D., Fischer, D., Bacterial nanocellulose: The future of controlled drug delivery? *Ther. Deliv.*, 8, 753–761, 2017.
127. Sarkar, G., Orasugh, J.T., Saha, N.R., Roy, I., Bhattacharyya, A., Chattopadhyay, A.K., Rana, D., Chattopadhyay, D., Cellulose nanofibrils/chitosan based transdermal drug delivery vehicle for controlled release of ketorolac tromethamine. *New J. Chem.*, 41, 15312–15319, 2017.
128. Silva, N.H.C.S., Rodrigues, A.F., Almeida, I.F., Costa, P.C., Rosado, C., Neto, C.P., Silvestre, A.J.D., Freire, C.S.R., Bacterial cellulose membranes as transdermal delivery systems for diclofenac: *In vitro* dissolution and permeation studies. *Carbohydr. Polym.*, 106, 264–269, 2014.

129. Guo, T., Pei, Y., Tang, K., He, X., Huang, J., Wang, F., Mechanical and drug release properties of alginate beads reinforced with cellulose. *J. Appl. Polym. Sci.*, 134, 44495, 2017.
130. Jackson, J.K., Letchford, K., Wasserman, B.Z., Ye, L., Hamad, W.Y., Burt, H.M., The use of nanocrystalline cellulose for the binding and controlled release of drugs. *Int. J. Nanomed.*, 6, 321–330, 2011.
131. Ntoutoume, G.M.A.N., Granet, R., Mbakidi, J.P., Brégier, F., Léger, D.Y., Fidanzi-Dugas, C., Lequart, V., Joly, N., Liagre, B., Chaleix, V., Development of curcumin–cyclodextrin/cellulose nanocrystals complexes: New anticancer drug delivery systems. *Bioorganic Med. Chem. Lett.*, 26, 941–945, 2016.
132. Weng, L., Rostamzadeh, P., Nooryshokry, N., Le, H.C., Golzarian, J., *In vitro* and *in vivo* evaluation of biodegradable embolic microspheres with tunable anticancer drug release. *Acta Biomater.*, 9, 6823–6833, 2013.
133. You, J., Cao, J., Zhao, Y., Zhang, L., Zhou, J., Chen, Y., Improved mechanical properties and sustained release behavior of cationic cellulose nanocrystals reinforeced cationic cellulose injectable hydrogels. *Biomacromolecules*, 17, 2839–2848, 2016.
134. Fakhri, A., Tahami, S., Nejad, P.A., Preparation and characterization of Fe3O4-Ag2O quantum dots decorated cellulose nanofibers as a carrier of anticancer drugs for skin cancer. *J. Photochem. Photobiol. B: Biol.*, 175, 83–88, 2017.
135. Dong, S., Cho, H.J., Lee, Y.W., Roman, M., Synthesis and cellular uptake of folic acid-conjugated cellulose nanocrystals for cancer targeting. *Biomacromolecules*, 15, 1560–1567, 2014.
136. Ernsting, M.J., Tang, W.-L., MacCallum, N., Li, S.-D., Synthetic modification of carboxymethylcellulose and use thereof to prepare a nanoparticle forming conjugate of docetaxel for enhanced cytotoxicity against cancer cells. *Bioconjugate Chem.*, 22, 2474–2486, 2011.
137. Roy, A., Murakami, M., Ernsting, M.J., Hoang, B., Undzys, E., Li, S.-D., Carboxymethylcellulose-based and docetaxel-loaded nanoparticles circumvent P-glycoprotein-mediated multidrug resistance. *Mol. Pharmaceutics*, 11, 2592–2599, 2014.
138. Dong, S., Hirani, A.A., Colacino, K.R., Lee, Y.W., Roman, M., Cytotoxicity and cellular uptake of cellulose nanocrystals. *Nano Life*, 2, 1241006, 2012.
139. Huang, L., Chen, X., Nguyen, T.X., Tang, H., Zhang, L., Yang, G., Nanocellulose 3D-networks as controlled-release drug carriers. *J. Mater. Chem. B*, 1, 2976–2984, 2013.
140. Tan, T.H., Lee, H.V., Dabdawb, W.A.Y., Abd Hamid, S.B.B.O.A., A review of nanocellulose in the drug-delivery system, in: *Materials for Biomedical Engineering*, pp. 131–164, 2019.
141. Khan, S., Siddique, R., Huanfei, D., Shereen, M.A., Nabi, G., Bai, Q., Manan, S., Xue, M., Ullah, M.W., Bowen, H., Perspective applications and associated challenges of using nanocellulose in treating bone-related diseases. *Front. Bioeng. Biotechnol.*, 9, 616555, 2021.

142. Gorgieva, S., Vivod, V., Maver, U., Gradišnik, L., Dolenšek, J., Kokol, V., Internalization of (bis) phosphonate-modified cellulose nanocrystals by human osteoblast cells. *Cellulose*, 24, 4235–4252, 2017.
143. Tazi, N., Zhang, Z., Messaddeq, Y., Almeida-Lopes, L., Zanardi, L.M., Levinson, D., Rouabhia, M., Hydroxyapatite bioactivated bacterial cellulose promotes osteoblast growth and the formation of bone nodules. *AMB Express*, 2, 1–10, 2012.
144. Arianto, R.J.D., Properties of cellulose and modified cellulose-alginate for rifampicin drug delivery, *IOP Conf. Ser.: Mater. Sci. Eng.* IOP Publishing, 980, 012016.
145. Liu, Y., Sui, Y., Liu, C., Liu, C., Wu, M., Li, B., Li, Y., A physically crosslinked polydopamine/nanocellulose hydrogel as potential versatile vehicles for drug delivery and wound healing. *Carbohydr. Polym.*, 188, 27–36, 2018.
146. Alkhatib, Y., Dewaldt, M., Moritz, S., Nitzsche, R., Kralisch, D., Fischer, D., Controlled extended octenidine release from a bacterial nanocellulose/poloxamer hybrid system. *Eur. J. Pharm. Biopharm.*, 112, 164–176, 2017.
147. Tong, W.Y., bin Abdullah, A.Y.K., binti Rozman, N.A.S., bin Wahid, M.I.A., Hossain, M.S., Ring, L.C., Lazim, Y., Tan, W.-N., Antimicrobial wound dressing film utilizing cellulose nanocrystal as drug delivery system for curcumin. *Cellulose*, 25, 631–638, 2018.
148. Orasugh, J.T., Dutta, S., Das, D., Pal, C., Zaman, A., Das, S., Dutta, K., Banerjee, R., Ghosh, S.K., Chattopadhyay, D., Sustained release of ketorolac tromethamine from poloxamer 407/cellulose nanofibrils graft nanocollagen based ophthalmic formulations. *Int. J. Biol. Macromol.*, 140, 441–453, 2019.
149. Tummala, G.K., Hydrogels of poly (vinyl alcohol) and nanocellulose for ophthalmic applications: Synthesis, characterization, biocompatibility and drug delivery studies, 2018.
150. Ahn, J.H., Lim, H.W., Hong, H.R., The clinical application and efficacy of sodium hyaluronate–carboxymethylcellulose during tympanomastoid surgery. *Laryngoscope*, 122, 912–915, 2012.
151. Mencucci, R., Boccalini, C., Caputo, R., Favuzza, E., Effect of a hyaluronic acid and carboxymethylcellulose ophthalmic solution on ocular comfort and tear-film instability after cataract surgery. *J. Cataract Refract. Surg.*, 41, 1699–1704, 2015.
152. Valerieva, A., Popov, T.A., Staevska, M., Kralimarkova, T., Petkova, E., Valerieva, E., Mustakov, T., Lazarova, T., Dimitrov, V., Church, M.K., Effect of micronized cellulose powder on the efficacy of topical oxymetazoline in allergic rhinitis. *Allergy Asthma Proc.*, 36, e134–139.
153. Koivuniemi, R., Hakkarainen, T., Kiiskinen, J., Kosonen, M., Vuola, J., Valtonen, J., Luukko, K., Kavola, H., Yliperttula, M., Clinical study of nanofibrillar cellulose hydrogel dressing for skin graft donor site treatment. *Adv. Wound Care*, 9, 199–210, 2020.

154. Ning, L., You, C., Zhang, Y., Li, X., Wang, F., Polydopamine loaded fluorescent nanocellulose–agarose hydrogel: A pH-responsive drug delivery carrier for cancer therapy. *Compos. Commun.*, 26, 100739, 2021.
155. Saigal, N., Baboota, S., Ahuja, A., Ali, J., Microcrystalline cellulose as a versatile excipient in drug research. *J. Young Pharm.*, 1, 6, 2009.

10

Preparation of Antibacterial Nanocomposite Materials Using Nanocellulose

Moumita Bishai

Department of Botany, Gurudas College, Narkeldanga, Kolkata, West Bengal, India

Abstract

On account of the use of petroleum-based goods during the past few decades, environmental and ecological concerns have come under more and more scrutiny. As a result, there is a lot of interest in the synthesis of ultrafine compounds and useful materials using natural resources. Natural cellulose is the source of nanocellulose, a novel and promising natural substance that has emerged as the most environmentally acceptable and commercially viable substance available today for reducing environmental pollution. Nanocellulose has a wide range of uses thanks to its exceptional physicochemical and biological characteristics. They lack natural antibacterial qualities, though. Therefore, the inclusion of functional properties to prepare nanoscale composites is a current need to effectively formulate materials for therapeutic purposes. This chapter mainly introduces the preparation methods of various nanocellulose-based antibacterial nanocomposites and important findings related to the production process. First, briefly introduce the source, extraction process, and properties of nanocellulose. We continue to better understand the complexities associated with the production process. This knowledge on the design of nanocomposites based on nanocellulose will allow the scientific community to learn more about its future applications in the biomedical field.

Keywords: Nanocellulose, cellulose, biomaterial, biomass, nanocomposite, antibacterial, reinforcement, biomedical applications

Email: onlymoumita06@gmail.com

Mainak Mukhopadhyay and Debalina Bhattacharya (eds.) Nanocellulose: A Biopolymer for Biomedical Applications, (267–296) © 2024 Scrivener Publishing LLC

10.1 Introduction

Infectious diseases caused by antibiotic-resistant bacteria have become a global public health challenge, leading to unnecessary illness and death in society. Therefore, a new process has been devised to kill harmful microorganisms in an environmentally friendly way. The investigation of naturally fibrous nanomaterials has grown significantly in scope in order to achieve this goal. New antibacterial agents may be created using this nanoscale material's small weight, huge surface area, and great mechanical strength. The rapid advancement of nanotechnology through the use of renewable energy has engrossed a lot of interest and focus in the creation of numerous materials, goods, and applications.

One of the most sought-after nanomaterials for its many uses is nanocellulose, which is made from naturally renewable cellulose sources such agricultural waste, lignocellulosic biomass, bacteria, and animals [1]. Because they are one of the most prevalent, ecologically friendly, and biocompatible materials in the world, as well as being naturally occurring and renewable macromolecules on Earth, cellulose and its derivatives have garnered a lot of interest. According to reports, the worldwide market for cellulose and its derivatives will reach over USD 1.08 billion by 2020 as a result of rising demand from the textile, food, and pharmaceutical industries [1].

However, the widespread application of nanocellulose in biomedical research is often limited by its lack of bactericidal properties. The development of suitable decomposable and biocompatible surfaces with long-lasting antimicrobial activity at relatively low cost has not yet been achieved [2]. This shortcoming can be overcome by developing reinforced nanocomposites inside the nanocellulose, which not only improves the basic properties of the composite materials, but also maintains the original properties of the nanocellulose. Nanocellulose has achieved positive results in various sustainable industries such as packaging, pharmaceuticals, and electronic devices [2]. Recently, nanocomposites based on nanocellulose have been widely studied and welcomed by researchers.

Considering the above discussion, this article aims to present recent advancements in the extraction of nanocellulose from cellulose, including its type, source, and extraction method. Finally, the various manufacturing processes associated with nanocomposites based on nanocellulose and the errors associated with them are discussed. Finally, this chapter should pave the way for the creation of new nanocomposites from nanocellulose for various cutting-edge applications.

10.2 Nanocellulose

A high-molecular-weight, homogeneous, unbranched biopolymer with repeating units of D-glucose connected by 1–4 glycosidic linkages is cellulose [3]. The cellulose chain's one end is transformed into a hemiacetal functional group. The amount of crystallinity in cellulose, which is semi-crystalline, varies depending on the source, extraction technique, and pretreatment.

Chemical and thermal processes can transform the four primary polymorphic forms of cellulose—Cellulose I, II, III, and IV—into one another. Natural cellulose comes in two isomorphs, I and II, and is called cellulose I. Regenerated cellulose, also known as cellulose II, is the most stable crystalline form and is produced through the mercerization process [4]. The fundamental distinction between the two types of cellulose is how their atoms are ordered, with cellulose II having an antiparallel structure and cellulose I having a parallel arrangement. Typically, cellulose I and cellulose II are each treated separately with ammonia to create cellulose III, and cellulose III is then heated to 260°C to create cellulose IV [5].

As indicated in Figure 10.1, cellulose is divided into five categories based on where it comes from: wood-based cellulose (WC), plant-based cellulose (PC), bacterial-based cellulose (BC), algal cellulose, and tunicate cellulose.

The primary component of plant fibers found in plant cells is cellulose, which is surrounded by lignin and hemicellulose. They are grouped in fibrils, which combine to create microfibrils of a few nanometers in diameter. The interaction of cellulose with various macroscopic and nanoscale

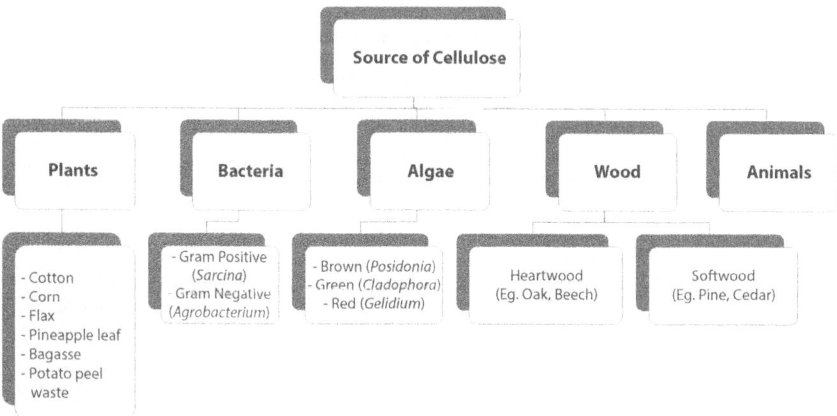

Figure 10.1 Sources of cellulose.

materials has been studied top-down [5]. Aggregates of cellulose fibrils exhibit both crystalline and amorphous structures. Ordered stacking chain molecules that support cellulose's high stiffness and strength are present in the crystalline component.

While the amorphous part contributes to the flexibility of the material. Cellulosic fibers are 3–100 mm in diameter and generally lignocellulosic biomass is 1–4 mm in length in aggregates of ordered and disordered domains [5]. Nanocellulose is a natural fiber that can be extracted from cellulose. Their characteristics are related to their type, origin, and age. It is considered to be a useful future class of materials. Of particular concern is the size of nanocellulose fibers, which are less than 100 nm in diameter and several microns in length [6]. They are biodegradable nanofibers with light weight, low density, and insoluble strength. Additionally, nanocellulose is transparent and filled with reactive side chains of hydroxyl groups that can be functionalized for various surface properties [6].

Nanocellulose obtainable from cellulose is mainly divided into three categories [3]:

(a) Cellulose nanocrystals (CNC)
(b) Cellulose nanofibrils (CNF)
(c) Bacterial cellulose (BC or microbial cellulose)

The production process is shown in Figure 10.2.

Natural cellulosic materials can be thought of as cellulose nanofibrils (CNF). Using mechanical techniques such as ultra-high friction milling, high-pressure homogenization, wet milling, cryogenic freezing, and sonication, together with enzymatic and chemical pretreatments, cellulose fibers are destroyed along the axis throughout the extraction process [7, 8]. The natural source of the cellulosic material affects the fundamental

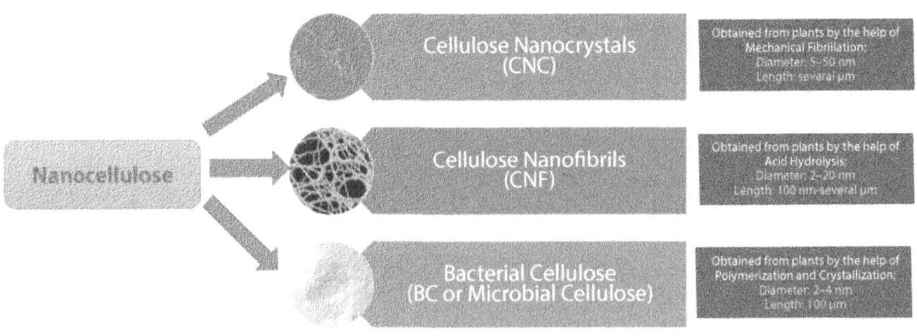

Figure 10.2 Different classes of nanocellulose.

characteristics of the CNFs that are produced. Generally speaking, CNFs made from primary cell wall fibers are longer and thinner than CNFs made from secondary cell walls.

Cellulose nanocrystals (CNCs), lovely crystal domains, are whisker-like molecules that are separated during the cellulose manufacturing process [7, 8]. Since more of the amorphous domains are destroyed during acid hydrolysis, CNCs are thought to be harder than CNFs because they have more crystallinity. Typically, pretreatment procedures (alkali and bleaching treatments) are utilized in the techniques used to remove CNC particles. The supernatant is then cleaned with deionized water (centrifugation and sonication) following the application of controlled reaction conditions such as agitation, temperature, and acid hydrolysis. To get pure CNC material, the resultant slurry can be dried using several methods.

In the presence of carbon sources and nitrogen, certain bacterial species, including Gluconacetobacter species (such as *Acetobacter xylinum*), produce bacterial nanocellulose (BNC), a different form of nanocellulose from CNF and CNC particles. after several days of aqueous medium culture, synthesized [5]. Through a manufacturing assembly process, these bacteria can transform glucose into cellulosic material. BNC is non-dispersible in water because of its natural cellulose I crystal structure, which has a greater molar mass, a higher degree of polymerization, and better crystallinity. It can also form extremely thin network structures [8, 9].

Nanoscale cellulose created by bacteria is lighter, non-toxic, and more controllable by modifying bacterial culture conditions than nanoscale cellulose derived from natural plants, which needs chemical treatment to break down lignin and hemicellulose components [8, 9]. BNC is frequently employed in the field of biomedicine because of its superior qualities, including hydrophilic action and good interpenetration with host tissues. The rising interest in tissue-engineered materials for wound care and regenerative medicine is also drawing attention to biomedical materials [8].

The extraction of cellulose microfibrils involves three main steps [7]:

1. Conventional pulping step—this process is used to remove lignin and hemicellulose;
2. Mechanical separation—separation of fibers into microfibrils by homogenization; and
3. Acid hydrolysis.

For the extraction of nanocellulose, whatever its class, several methods are available. At first, a dewaxing of the fibers is carried out to eliminate the impurities. This process is followed by a pretreatment including alkalization and bleaching. In this method, the hemicellulose part of the lignin is removed. The next step performed is disaggregation, which is performed in three ways [8]:

1. Chemical method of acid hydrolysis;
2. Biological method of enzymatic hydrolysis; and
3. Mechanical methods, including grinding, high-pressure homogenization, low-temperature pressing of liquid nitrogen, etc.

This process essentially increases the internal surface area, increases the accessibility of hydroxyl groups, modifies the crystallinity, and breaks the hydrogen bonds of cellulose to improve the reactivity of plant fibers [8]. However, depending on the desired fiber size, the cellulose extraction methods vary.

Cellulose separation has been achieved using alkaline methods, enzymatic techniques, sonication, or dilute acid pretreatment [8].

10.3 Functionalization of Nanocellulose

Nanocellulose can assemble in an array of situations owing to its high surface energy and intermolecular hydrogen bonding interactions. Therefore, nanocellulose's surface characteristics and interfacial compatibility are essential for its functionalization and useful applications. The following are the primary obstacles related to nanocelluloses [9]:

1. They have a large number of OH groups, while
2. They are highly hydrophilic.

Therefore, it is imperative to change nanocellulose in order to increase its functionality and broaden the range of applications for natural fibers. An efficient technique to increase the dispersion of nanocellulose is to enhance surface modification, based on the characteristics of certain carriers, to adjust surface hydrophilicity and improve compatibility. Recently, scientists modified the surface and internal structure of nanocellulosic materials using a variety of approaches. Chemical, physical, and enzymatic approaches are only a few of the surface functionalization possibilities available for nanocellulose [9].

Since then, new antibiotics like penicillin have been created and have been used to treat infectious infections in both people and animals.

The environment contains a number of antibacterial substances that inhibit different microbes. It has been demonstrated that a number of metal or metal oxide nanoparticles, including gold, silver, copper, copper oxide, zinc oxide, titanium dioxide, magnesium oxide, and other nanocomposites, are excellent candidates for antimicrobial composites or nanocomposites. Additionally, numerous antibacterial polymers, including the effective antibacterial agent chitosan and its derivatives, can be employed in antibacterial applications.

10.4 Nanocellulose-Based Nanocomposite

Inorganic nanoparticles and nanomaterials, as well as natural and manmade polymers, have all been used to create nanocomposites. Nanocomposites are classified as either organic or inorganic depending on the characteristics of the reinforcing elements. These two primary groups are further separated into huge solid particles, clays, metal oxides, polymers, metals, and metal nanocomposites. Numerous nanomaterials have been used to create nanocomposites thus far, including metal nanoparticles (Ag, Au, Pd, Pt, and Ni), metal oxide nanoparticles (ZnO, CuO, MgO, FeO, TiO_2, Al_2O_3, and CeO), nanoparticles from minerals (SiO_2, $CaCO_3$, and montmorillonite), and carbon nanomaterials (graphene and carbon nanotubes). Additionally, nanocomposites can be created from synthetic polymers like 2-aminoethyl acrylate and biopolymers like chitosan, alginate, hyaluronic acid, starch, gelatin, collagen, keratin, polylactic acid, polyhydroxyalkanoate, and polyvinyl alcohol.

Aldehyde groups and quaternary ammonium groups are functional groups having antibacterial characteristics. By impregnating nanocellulose with solutions of negatively charged polysaccharides such as hyaluronic acid, sodium alginate, or carrageenan, followed by positively charged chitosan, a number of novel polysaccharide-based nanocomposites have been created. In addition, biomolecule nanocomposites including cells, antibiotics, enzymes, hormones, peptides, and amino acids were produced. All of these different kinds of nanocellulose materials showed effective antibacterial properties and acceptable biocompatibility.

10.5 Properties of the Nanocomposites

Nanocomposites are currently used in many fields. With the development of nanotechnology and recent concerns about environmental issues, the use of bio-based nanomaterials has received more attention. In this regard, nanocellulose has attracted more attention due to its promising properties such as biodegradability, renewal, and lower price. Nanocellulose-based materials are receiving considerable attention as potential nanofillers for nanocomposites due to their nanoscale size (very high surface-to-volume ratio), high aspect ratio, and impressive mechanical properties (or nanometric strength), effectively responding to the desired nanocomposite. Nanoscale cellulosic materials can be isolated from various cellulosic resources using different procedures [10].

However, there are still some issues to overcome, and the main challenges in this area are related to the effective isolation of nanoscale cellulosic materials from natural resources. The incompatibility of nanocellulose with most polymers is also a key issue for its application in nanocomposites. Additionally, the drying process of nanocellulose for polymer composites is another challenge. The last issue is related to the search for a process to achieve higher yields in the separation of nanocellulose. All of these challenges and drawbacks have been a driving force to discover more efficient processes and technologies to produce nanocellulose for nanocomposites and to invent new applications.

10.6 Nanocellulose-Based Nanocomposite Processing

Nanocellulose has garnered a lot of interest as a possible reinforcing material for different polymeric materials in numerous cutting-edge applications during the past few decades. There are several challenges, most of which are connected to the method for adding nanocellulose and how to achieve an adequate dispersion of this nanocellulose in the matrix. The procedure used to create nanocomposites relies on the characteristics of the matrix and the reinforcement, but it must allow for an even distribution of the reinforcement throughout the matrix; therefore, each processing method must be specifically tailored to the situation. In a nutshell, the qualities of the nanocellulose and the nanocomposite production process are crucial for a superior finished product.

10.6.1 Solution Casting

Solution casting is one of the most popular methods for creating nanocomposites with a nanocellulose basis. The preparation is based on the dispersion of cellulose nanoparticles in a solvent, frequently using sonication, in order to obtain an efficient dispersion of the nanofillers. The polymer was already dissolved in the same solvent, and after adding the solvent-dissolved nanocellulose suspension, the combination was dried until the solvent had fully evaporated. The advantage of this approach is that it guarantees the dispersion of the nanocelluloses. Nanocellulose has been successfully added to hydrophobic and hydrophilic polymers, such as polyethylene oxide, polylactic acid, and polyhydroxyalkanoate, by solution casting [9].

10.6.2 Melt Intercalation

It is a top-down strategy that includes adding nanoparticles to molten thermoplastic polymers to improve interactions between the two types of materials. The intercalation is made possible by the polymer chains penetrating the reinforcement. A nanocomposite is created by annealing the mixture above the polymers' glass transition temperature. Due to its extensive potential for usage in the polymer sector, where it may be employed with extruders or to create composite equipment like mixers, it has since gained a reputation as the most advantageous and economical method for synthesizing nanocomposites. The matrix-filler interactions are improved by the melt intercalation approach since it is more flexible, does not require solvents or chemical processes, and lowers interfacial tension. Utilizing a straightforward and ecologically friendly fusion intercalation approach, matrix-enhanced interactions have been improved. Based on the compatibility, intercalated nanocomposites may be created. In essence, this method is more flexible and does not involve the use of solvents or chemical processes [32].

10.6.3 Impregnation

The fundamental method for creating structural composites consists of impregnating the fiber reinforcements with liquid resins, then strengthening or hardening the fiber–resin combination at a specified temperature and pressure. The pre-dried fibers are continuously infused with resin during the impregnation procedure to guarantee that they are completely saturated. Thin layers are produced by this method. To accomplish comprehensive and reliable fertilization, a variety of techniques are employed [33]. The fibers are impregnated with a resin solution during filament winding.

In the manual draping technique, the material used has already been impregnated in an environment controlled by the material supplier. During automatic fiber insertion, this step is performed on the dipping line. The fertilization line produces prepregs. The prepreg is a composite material whose reinforcing fibers are pre-impregnated with a matrix of thermoplastic or thermosetting resin in a certain proportion [33].

10.6.4 *In Situ* Polymerization

By using *in situ* polymerization, cellulose nanoparticles can be evenly distributed throughout the matrix and strongly bonded to the filler. Monomer solutions or pure nanocellulose monomers are mixed and subsequently polymerized in the presence of dispersed nanocellulose. Since monomers are frequently fluid, the impregnation can be carried out without appreciably changing the alignment of the nanocelluloses. In this method, the monomer functions as both a dispersion of nanocellulose and a matrix precursor for *in situ* polymerization. The material needs little finishing work after curing before processing.

While adding a curing agent or raising the temperature can polymerize thermoplastics, adding curing agents or peroxides can start the polymerization of thermosetting materials like epoxy resins or unsaturated polyesters [34]. The matrix and nanocellulose of the finished nanocomposite are believed to establish covalent bonds, according to numerous research on the *in-situ* polymerization technique. *In situ* polymerization, as opposed to melt processing procedures and film casting technologies, allows for the direct integration of well-dispersed nanoparticles into bulk polymer composites. *In situ* polymerization can also be utilized to create nanocomposites because melt processing and solution casting are unsuited for the polymers due to their poor temperature stability and solvent insolubility, respectively [34].

10.6.5 Coating

The coating forms a thin layer that wraps around the outside or in between two substrates to produce a composite structure. The thickness ranges from tenths of a nanometer to a micrometer. A slurry is applied during the coating process on the surface of various backing/carrier materials. Packaging made of paper and paperboard is typically coated to act as a barrier against grease, gases, and moisture. As the additional liquid medium evaporates during the coating process, the particles form a thin layer that adheres to the surface of the substance. Due to nanocellulose's high water dispersibility, thin coatings of it can be made on their own or in combination with other polymers. Different coating methods, such as aqueous, solvent-based, extrusion, and electrospinning coatings, can be used to apply nanocellulose-based substrates to cellulose-based materials or synthetic films [32].

10.6.6 Freeze Drying

It is possible to create nanocomposite aerogels from cellulose using the freeze-drying process because it eliminates the water that retains the nanocomposite in aerogel suspension or solution. This method uses sublimation to freeze the sample first while removing small molecule solvents. The air gel particle distribution and pore size may be modified during the freeze-drying process. The direction and form of the aerogel can also be improved by changing the size of the utilized container. The ice modeling approach is mostly used to produce aerogels by watching how crystal development results in the construction of nanostructures [9].

10.6.7 3D Printing

There are two ways to incorporate nanomaterials into polymeric host matrices using 3D printing. This can be done to modify the properties of conventional materials, create custom geometries, shorten lead times between designs, and improve part integration. Nanomaterials may be added automatically or manually, and matrix materials can be 3D printed with interruptions for batch printing. An alternative method is to first include nanoparticles into the host matrix, and then 3D prints the resulting combination to produce nanocomposites [35]. Newly created 3D printing technologies can be utilized to produce biomaterials for biomedical applications including tissue engineering and regenerative medicine. In order to create hierarchical 3D nanostructures in nanocomposites, the approach uses melt extrusion to create diverse biopolymers, such as polysaccharides and proteins, as bio-inks and layer-by-layer printing of these biopolymers. For 3D printing, cellulose nanoparticles can be employed as substrates or as elements of ink [36].

10.6.8 Electrospinning

Another method using the bacterial breakdown of cellulose is electrospinning. The researchers used ionic liquid 1-allyl-3-methylimidazolium chloride to dissolve pellets of freeze-dried bacterial cellulose under agitation at 70°C. DMSO was added to the solution to adjust the viscosity at room temperature. After that, the MWCNTs were added to a concentration of 0.02% and stirred before electrospinning.

These electrospun fibers have been reported to produce a composite with a smooth and continuous surface with well-dispersed multi-walled carbon nanotubes in bacterial cellulose [37]. In addition to dissolving bacterial cellulose, other electrospun composites have been developed using sulfuric acid to hydrolyze bacterial cellulose to obtain nanowhiskers. Before electrospinning occurs, the bacterial cellulose nanowhiskers are dispersed in polyethylene oxide (PEO) dissolved in water. Electrospinning has also been used to develop composites using differently treated bacterial cellulose nanowhiskers. The electrospinning technique was particularly successful, leading to good dispersion of nanowhiskers with simultaneous increases in modulus of elasticity and tensile strength as well as biomedical properties [38, 39].

10.6.9 *In Situ* Composites

Adding additives to the growth medium that are not specifically required for cell growth or cellulose production may affect the cellulose produced. Due to the biological properties of bacterial cellulose, the *in situ* method involves the addition of polymers to the cellulosic medium to bind these materials, thus providing an alternative method for making composites. Composite materials can be made simply by incorporating a host polymer into a medium, but further processing developments can improve the contact and interaction between the two materials. Identifying materials for composites with bacterial cellulose in this way could improve the development of such composites [32].

10.6.10 Other Methods

Various nanocomposites have been prepared by completely different methods. Quiro *et al.* (2010) focused on compression of bacterial cellulose bands between two PLA films. This work identified differences between bacterial cellulose composites cultured over different time periods [1].

Each of the treatment techniques explained above has its own advantages and disadvantages.

It is clear from the research that nanocellulose-based materials need to be thoroughly processed, as this affects their applicability. However, solution casting, for example, offers better thickness uniformity, as well as better physical properties and flexibility. However, to be accepted, the polymer must be dissolved in water or a volatile solvent and produce a stable solution with the appropriate minimum solids content and viscosity. Also in melt intercalation, the use of high temperatures destroys the properties of nanocellulose, which is not ideal for temperature-sensitive polymers.

This part requires further research to get a solution. Agglomeration *in situ* polymerization requires additional curing chemicals. In coating technology, there is a possibility of high consumption of coating material, which makes it difficult to control coating thickness. Therefore, further research should be conducted to address these challenges in the development of nanocellulose-based nanocomposites. As with all nanoparticles, one of the fundamental concerns is to ensure that they are evenly distributed in the polymer matrix. However, the inherent incompatibility of cellulose with most polymer matrices as well as thermal stability and hydrophobicity issues make it difficult to determine optimal conditions.

10.7 Antibacterial Properties of the Nanocomposite Material

Nanocellulose-based nanocomposites have recently made significant strides in biological applications, including tissue engineering, drug delivery, cancer therapy, and bandages for wounds [3]. However, the majority of these materials have drawbacks since they exhibit poor mechanical and antibacterial capabilities. To increase the bactericidal effectiveness of these nanocomposites in biomedical applications, various investigations have recently been carried out. Table 10.1 lists a summary of development of nanocellulose-based nanocomposites following different fabrication techniques for diverse biomedical applications.

10.8 Conclusions and Future Trends

Due to its distinct ultrafine network nanostructure, high mechanical strength, high purity, and high adsorption capacity, nanocellulose is a significant natural biopolymer with many uses. These nanocellulose-based biocomposites are now being produced in greater quantities, and they are being used more often in biomedicine, business, and technology. Nanocomposites for vascular tissue engineering and other therapies, as well as new nanocomposites for tissue engineering and drug delivery systems, have been obtained.

However, because of high manufacturing costs and low yields, commercial production of these materials has been constrained. This is demonstrated by the high cost of making nanocellulose-based nanocomposites for commercial purposes, whose large-scale manufacture is not at all feasible. Therefore, more study is required to generate a novel, cost-effective medium employing streams of industrial waste and by-products, obtain new, high-yielding strains through genetic and metabolic engineering, and improve cellular processes.

Table 10.1 A summary of development of nanocellulose-based nanocomposites following different fabrication techniques for diverse biomedical applications.

Fabrication process	Reinforced material	Remark	Reference
In situ impregnation	Titanium dioxide nanoparticles	TiO_2 nanocomposite showed considerable promise for use in medicine. They gave the nanocomposites antibacterial characteristics, which lowers the risk of microbial infections throughout the healing process. The remarkable cell multiplication, biocompatibility, and non-toxic behavior demonstrate the viability of tissue regeneration during wound healing. TiO_2-NPs were made using sodium dodecyl sulfate solution at 1%. TiO_2 nanocomposite was created during a 5-day period at 30°C and pH 5.0. The results of FE-SEM, FT-IR spectroscopy, and XRD investigation are astounding. The antibacterial test generated outstanding findings, showing that *E. coli* inhibition zones were formed using several bactericidal components.	[11–14]
	CuO particle	Within the cellulose matrix, *in situ* generation of copper oxide nanoparticles took place. With regard to bacteria, the cellulose/CuO nanocomposite showed good antibacterial action.	[15]

(*Continued*)

Table 10.1 A summary of development of nanocellulose-based nanocomposites following different fabrication techniques for diverse biomedical applications. (*Continued*)

Fabrication process	Reinforced material	Remark	Reference
	Ag nanoparticles	For the treatment of third-degree burn wounds, bacterial cellulose (BC)-based nanocomposite dressing materials (BC-PDAg) with polydopamine (PD) coating were created. Significant antimicrobial action was shown against bacteria specific to burn wounds. The BC-PDAg nanocomposite is biocompatible and encourages cell growth, according to research on *in vitro* cytotoxicity and proliferation. Further tests on female albinos revealed that BC-PDAg promoted re-epithelization and collagen deposition during wound healing, as shown by histological examination. BC-PDAg enhances healing by controlling the expression of inflammatory, angiogenesis, and growth factor genes, according to a study on molecular gene expression. The overall effectiveness of BC-PDAg nanocomposite points to its usage as a cutting-edge skin regenerating technique in contemporary medicine.	[16]

(*Continued*)

Table 10.1 A summary of development of nanocellulose-based nanocomposites following different fabrication techniques for diverse biomedical applications. (*Continued*)

Fabrication process	Reinforced material	Remark	Reference
	Gelatin and selenium nanoparticles	A unique multifunctional wound dressing without antibiotics that not only efficiently prevents wound infection but also speeds up wound healing is desperately needed. For use in the treatment of wounds, a variety of multifunctional nanocomposite hydrogels based on bacterial cellulose (BC), gelatin (Gel), and selenium nanoparticles (SeNPs) are created. These hydrogels exhibit outstanding antibacterial, antioxidant, and anti-inflammatory properties. The BC/Gel/SeNPs nanocomposite hydrogels have great antibacterial efficacy against both common bacteria (*E. coli* and *S. aureus*) and their drug-resistant cousins. They also display good biocompatibility and sustainability. The BC/Gel/SeNPs hydrogels also demonstrate excellent skin wound healing performance in a rat full-thickness defect model, as demonstrated by noticeably improved wound closure, granulation tissue formation, collagen deposition, angiogenesis, and fibroblast activation and differentiation. In addition, inflammation was markedly reduced. When used as a wound dressing in a medical setting, the multifunctional BC/Gel/SeNPs nanocomposite hydrogel shows promise for reducing wound infection risk and promoting skin regeneration.	[17]

(*Continued*)

Table 10.1 A summary of development of nanocellulose-based nanocomposites following different fabrication techniques for diverse biomedical applications. (*Continued*)

Fabrication process	Reinforced material	Remark	Reference
Ex situ addition	Poly(vinyl alcohol), silk sericin, azithromycin	The difficulty of choosing a suitable antimicrobial dressing with antibiofilm activity owing to antimicrobial resistance in biofilms limits the therapy of chronic wound infection. In this instance, genipin-crosslinked silk sericin (SS), a bacterial cellulose/poly(vinyl alcohol) nanocomposite, and azithromycin (AZM) (SSga) are used to *ex situ* modify the nanocomposite. The composite performed well as a material for wound dressing in limiting the growth, spread, and elimination of microbial biofilm. Studies involving bacteria and fungus show that the presence of AZM significantly inhibits and disrupts biofilms. The composite is confirmed to be harmless to HaCaT and NIH_3T_3 cells by *in vitro* biocompatibility testing, whereas the addition of SS promotes cell growth. The created nanocomposite quickens the healing of wounds in the diseased mouse model, making it a potentially useful biomaterial for treating wounds.	[18]

(*Continued*)

Table 10.1 A summary of development of nanocellulose-based nanocomposites following different fabrication techniques for diverse biomedical applications. (*Continued*)

Fabrication process	Reinforced material	Remark	Reference
	Chitosan and diamond nanoparticles	A mitochondrial activity experiment employing L929 fibroblast cells revealed that the nanocomposite films are biocompatible after 24 h of incubation. Flexible and translucent chitosan-based bacterial cellulose nanocomposite films containing diamond nanoparticles (NDs) were employed to evaluate this. Numerous lamellipodia and evidence of cell–cell contact point to the possibility of using the produced films as a flexible platform for wound dressing. Nanocomposite films have the ability to promote cellular behavior *in vitro*, according to studies on cell adhesion and interactions.	[19]

(*Continued*)

Table 10.1 A summary of development of nanocellulose-based nanocomposites following different fabrication techniques for diverse biomedical applications. (*Continued*)

Fabrication process	**Reinforced material**	**Remark**	**Reference**
	Chitosan and collagen	By combining collagen (COL) and chitosan (CS) with oxidized bacterial cellulose (OBC), a new hemostatic nanocomposite (OBC/COL/CS) was created. The OBC/COL/CS nanocomposite demonstrated good *in vivo* biodegradation and broad range antibacterial activities. Greater procoagulant and blood-clotting abilities, increased erythrocyte and platelet adhesion, and less blood loss were all displayed by OBC/COL/CS. The findings imply that the OBC/COL/CS has a high potential for application as an absorbable hemostat for the control of internal bleeding. It is a quick and effective procoagulant agent with good antibacterial characteristics.	[20]
	Poly(2-hydroxyethyl methacrylate) and silver nanoparticle	A transparent wound dressing consisting of bacterial cellulose (BC) and poly(2-hydroxyethyl methacrylate) (PHEMA) hydrogel coated with silver (Ag) nanoparticles with antibacterial properties produced inhibition zones against *E. coli* and *S. aureus*, respectively. The PHEMA/BCWs/Ag hydrogel facilitated the growth of NIH3T3 fibroblast, showing their low toxicity. PHEMA/BCWs/Ag hydrogel is a potential transparent wound dressing material for skin repair.	[21]

(*Continued*)

Table 10.1 A summary of development of nanocellulose-based nanocomposites following different fabrication techniques for diverse biomedical applications. (*Continued*)

Fabrication process	Reinforced material	Remark	Reference
	Montmorillonite (MMT)	Various composites of MMT were created by impregnating cellulose fiber with suspensions of medicinal clay. Examined were the composites' antibacterial effects on *Staphylococcus aureus* and *E. coli.* They clearly showed zones of inhibition against *S. aureus* and *E. coli*. The research aided in the creation of nanocomposite sterile wound dressings and regeneration materials with antibacterial characteristics for medicinal uses.	[22]
	Curcumin	By combining the therapeutic properties of curcumin with the distinctive structural, physicochemical, and biological characteristics of bacterial cellulose (BC), curcumin-loaded bacterial cellulose (BC-Cur) nanocomposite was created as a possible wound dressing for partial thickness burns. The BC-Cur nanocomposite demonstrated improved fibroblast cell adhesion and proliferation following the successful impregnation of curcumin into the BC matrix. Curcumin was examined for its antibacterial properties against *E. coli*, *P. aeruginosa*, *S. typhimurium*, and *S. aureus*. Partial-thickness burns in Balbc mice underwent wound healing study, which revealed an increased rate of wound closure.	[23]

(*Continued*)

Table 10.1 A summary of development of nanocellulose-based nanocomposites following different fabrication techniques for diverse biomedical applications. (*Continued*)

Fabrication process	Reinforced material	Remark	Reference
		In the BC-Cur nanocomposite group, histological investigations revealed healthy granulation tissues, fine re-epithelialization, vascularization, and resurfacing of the wound bed. Combining BC with biomaterial made from curcumin might serve as a foundation for the creation of viable substitutes for the traditional dressing method used to treat burns.	
	Tea extracts Acids (polycaffeic, tannic, ferulic)	The cellulose fiber can be impregnated with tea extracts and other acids, increasing the membrane's permeability and enabling silver ions and ROS assault. Additionally, it has antibacterial and antifungal action and offers decreased or no toxicity to human cells.	[24]

(*Continued*)

Table 10.1 A summary of development of nanocellulose-based nanocomposites following different fabrication techniques for diverse biomedical applications. (*Continued*)

Fabrication process	Reinforced material	Remark	Reference
	Carbon nanotubes and poly(ethylene imine)	High-density phage immobilization is made possible by carboxylated multiwalled carbon nanotubes (c-MWCNTs) coated with BC. A positive charge that results from surface modification of BC/c-MWCNTs with polyethyleneimine (PEI) makes orientated phage immobilization easier. By generating a distinct lytic zone and inhibiting bacterial growth, the produced nanocomposite demonstrated anti-staphylococcal efficacy. The biosensor demonstrated strong *S. aureus* selectivity in both pure and mixed cultures of non-host bacteria.	[25]
	Quaternized chitosan	Quaternized chitosan (also known as hydroxypropyltrimethylammonium chloride chitosan, or HACC), an antimicrobial substance that dissolves in water, was developed. With the addition of HACC, the antibacterial activities of *S. aureus* and methicillin-resistant *S. aureus* may be improved. HACC also demonstrates biocompatibility, making it a suitable choice for wound dressings in clinical settings.	[26]

(*Continued*)

Table 10.1 A summary of development of nanocellulose-based nanocomposites following different fabrication techniques for diverse biomedical applications. (*Continued*)

Fabrication process	Reinforced material	Remark	Reference
	ZnO nanoparticle	Zinc oxide (ZnO) nanoparticles are added to a composite of poly(3-hydroxybutyrate) and nanocellulose for improved characteristics and antibacterial activity against *E. coli* and *S. aureus*. Nanocellulose treated with ZnO plasma completely stops *S. aureus* from growing.	[27]
3D printing method	Alginate/hyaluronic acid with CNF	To record the temperature of the wound, an electrode printed in silver ink has been inserted onto the bandage. Additionally, using the CNF for stem cell culture and alginate/hyaluronic acid for 3D printing cartilage, it was possible to integrate *in situ* physiological conditions. The 3D printing of CNC-based hydrogel ink with sodium alginate and gelatin aims to create a porous cubic scaffold with a homogeneous pore structure. The created 3D scaffolds demonstrated porosity and compression modulus values that were comparable to those required for applications involving cartilage regeneration. Thus, utilizing this approach, scaffolds may be created that are tailored to the individual needs of patients.	[28]

(*Continued*)

Table 10.1 A summary of development of nanocellulose-based nanocomposites following different fabrication techniques for diverse biomedical applications. (*Continued*)

Fabrication process	Reinforced material	Remark	Reference
Electron beam irradiation	Poly(acrylic acid)-graphene oxide	Using electron beam irradiation, a composite hydrogel made of bacterial nanocellulose (BNC)-grafted poly(acrylic acid) (AA)-graphene oxide (BNC/P (AA)/GO) has been created. The swelling capacity, water retention ratio, bioadhesion, and water absorption properties of composite hydrogel were found to be improved while maintaining their mechanical properties after assessment of their bioadhesiveness, mechanical properties, water vapor transmission rates (WVTRs), biocompatibility, and water retention abilities. Additionally, the BNC/P(AA)/GO composite hydrogels increased cell adhesion and proliferation and were biocompatible. According to the study, BNC/P(AA)/GO composite hydrogel may have uses as a material for treating wounds.	[29]

(*Continued*)

Table 10.1 A summary of development of nanocellulose-based nanocomposites following different fabrication techniques for diverse biomedical applications. (*Continued*)

Fabrication process	Reinforced material	Remark	Reference
Ultrasonic irradiation	Magnetite and hydroxyapatite	In a prior study, BC with HAp and Fe_3O_4 (Ferrate) nanoparticles were used to create nanocomposite scaffolds. It was found that the nanoparticles were evenly dispersed throughout the BC matrix. The swelling capacity of the BC was reduced and the matrix's 80% porosity level was maintained by adding nanoparticles to the BC matrix. The outcomes demonstrated that the created nanocomposite scaffold was non-toxic to mouse fibroblast L929 cells and biocompatible with the osteoblast MC3T3-E1 cell line. When utilized to restore bone tissue, nanoparticle integration also led to a decrease in the scaffold's crystallinity, which aided in the material's prompt biodegradation. The study's findings support the use of the composite scaffold for bone regeneration.	[30]
Freeze drying	Sodium compound	Sodium periodate produced 2,3-dialdehyde nanofibrillated cellulose (DANFC) using the freeze-drying technique. Positive findings were seen in the antibacterial activity of DANFC against both *S. aureus* and methicillin-resistant *S. aureus* (MRSA). Additionally, DANFC has high biocompatibility with mammalian cells and has the ability to accelerate wound healing.	[31]

(*Continued*)

Table 10.1 A summary of development of nanocellulose-based nanocomposites following different fabrication techniques for diverse biomedical applications. (*Continued*)

Fabrication process	Reinforced material	Remark	Reference
	2,2,6,6-Tetramethylpiperidinyloxy (TEMPO)	Under aqueous conditions, TEMPO may bond on the surface of nanocellulose, which causes the hydroxyl group at the C6 position of the nanocellulose to change into aldehyde and carboxyl functional groups. A potential option for treating shrimp bacteriosis was created by depositing silver nanoparticles over TEMPO-oxidized BC. This composite material is known as TEMPO-BC/silver.	[32]
Solution casting	Graphene oxide	Solution casting was used to make cellulose nanocrystals (CNCs) and reduced graphene oxide (rGO) with success. They demonstrated the PLA/CNC/rGO nanocomposite film's unique antibacterial activity against Gram-positive *S. aureus* and Gram-negative *E. coli* bacterial strains, respectively. Additionally, the fibroblast cell line (NIH-3T3) demonstrated low cytotoxicity after being exposed to nanocomposite film in an *in vitro* cell-based cytotoxicity experiment. Consequently, the nanocomposite film that has been created has enormous potential for use in both biomedical and food packaging applications.	[33]
Other method	Antibiotic (Allicin)	Fabricated allicin-conjugated nanocellulose was used to test the materials' antibacterial efficacy. *Staphylococcus aureus* was very resistant to the results' antibacterial action.	[34]

Three other significant hurdles for researchers include enhancing the performance of nanocomposites based on nanocellulose, lowering manufacturing costs, and developing appropriate industrial production lines. Fortunately, research is underway in each of these fields, and several new applications are anticipated in the coming years. The creation of novel techniques for the modification and functionalization of materials based on nanocellulose is also given priority. In the future, it may be possible to combine these methods in new and interesting ways, opening up possibilities for the production of materials that will improve living quality and the environment for people.

References

1. Quero, F., Nogi, M., Yano, H., Abdulsalami, K., Holmes, S.M., Sakakini, B.H., Eichhorn, S.J., Optimization of the mechanical performance of bacterial cellulose/poly(l-lactic) acid composites. *ACS Appl. Mater. Interfaces*, 2, 1, 321–330, 2010.
2. Yang, X., Wang, X., Liu, H., Zhao, Y., Jiang, S., Liu, L., Impact of dimethyl sulfoxide treatment on morphology and characteristics of nanofibrillated cellulose isolated from corn husks. *Bioresources*, 12, 95–106, 2016.
3. Tavakolian, M., Jafari, S.M., van de Ven, T.G.M., A Review on surface functionalized cellulosic nanostructures as biocompatible antibacterial materials. *Nano-Micro Lett.*, 12, 73, 2020, https://doi.org/10.1007/s40820-020-0408-4.
4. Joseph, B., Sagarika, V.K., Sabu, C., Kalarikkal, N., Thomas, S. Cellulose nanocomposites: Fabrication and biomedical applications. *J. Bioresour. Bioprod.*, 5, 223–237, 2020.
5. Norizan, M.N., Shazleen, S.S., Alias, A.H., Sabaruddin, F.A., Asyraf, M.R.M., Zainudin, E.S., Abdullah, M., Samsudin, M.S., Kamarudin, S.H., Norrrahim, M.N.F., Nanocellulose-based nanocomposites for sustainable applications: A review. *Nanomater.*, 12, 3483, 2022, https://doi.org/10.3390/nano12193483.
6. Farooq, A., Patoary, M.K., Zhang, M., Mussana, H., Li, M., Naeem, M., Mushtaq, M., Farooq, M., Liu, L., Cellulose from sources to nanocellulose and an overview of synthesis and properties of nanocellulose/zinc oxide nanocomposite materials-review. *Int. J. Biol. Macromol.*, 154, 1050–1073, 2020.
7. Dhali, K., Ghasemlou, M., Daver, F., Cass, P., Adhikari, B., A review of nanocellulose as a new material towards environmental sustainability. *Sci. Total Environ.*, 775, 145871, 2021.
8. Pinto, E., Aggrey, W.N., Boakye, P., Amenuvor, G., Sokama-Neuyam, Y.A., Fokuo, M.K., Karimaie, H., Sarkodie, K., Adenutsi, C.D., Erzuah, S., Rockson, M.A.D., Cellulose processing from biomass and its derivatization into carboxymethylcellulose: A review. *Sci. Afr.*, 15, e01078, 2022, https://doi.org/10.1016/j.sciaf.2021.e01078.

9. Rojas, J., Bedoya, M., Ciro, Y., Current trends in the production of cellulose nanoparticles and nanocomposites for biomedical applications, in: *Cellulose - Fundamental Aspects and Current Trends,* M. Poletto, H. Luiz, O. Junior (Eds.), InTech, London, UK, 2015.
10. Patel, D.K., Dutta, S.D., Lim, K.T., Nanocellulose-based polymer hybrids and their emerging applications in biomedical engineering and water purification. *RSC Adv.*, 9, 19143–19162, 2019.
11. Revin, V.V., Liyaskina, E.V., Parchaykina, M.V., Kuzmenko, T.P., Kurgaeva, I.V., Revin, V.D., Ullah, M.W., Bacterial cellulose-based polymer nanocomposites: A review. *Polymers,* 14, 4670, 2022, https://doi.org/10.3390/polym14214670.
12. Toro, R., Adel, A.M., Caro, T., Brunetti, B., Al-Shemy, M.T., Caschera, D., A facile one-pot approach to the fabrication of nanocellulose–titanium dioxide nanocomposites with promising photocatalytic and antimicrobial activity. *Mater.*, 15, 5789, 2022, https://doi.org/10.3390/ma15165789.
13. Khan, S., Ul-Islam, M., Khattak, W.A., Ullah, M.W., Park, J.K., Bacterial cellulose-titanium dioxide nanocomposites: nanostructural characteristics, antibacterial mechanism, and biocompatibility. *Cellul.*, 22, 565–579, 2015.
14. Ullah, M.W., Ul-Islam, M., Khan, S., Kim, Y., Jang, J.H., Park, J.K., *In situ* synthesis of a bio-cellulose/titanium dioxide nanocomposite by using a cell-free system. *RSC Adv.*, 6, 22424–22435, 2016.
15. Muthulakshmi, L., Rajini, N., Nellaiah, H., Kathiresan, T., Jawaid, M., Varada, A., Preparation and properties of cellulose nanocomposite films with *in situ* generated copper nanoparticles using *Terminalia catappa* leaf extract. *Int. J. Biol. Macromol.*, 95, 1064–1071, 2017, http://dx.doi.org/10.1016/j.ijbiomac.2016.09.114.
16. Jiji, S., Udhayakumar, S., Maharajan, K., Rose, C., Muralidharan, C., Kadirvelu, K., Bacterial cellulose matrix with *in situ* impregnation of silver nanoparticles via catecholic redox chemistry for third degree burn wound healing. *Carbohydr. Polym.*, 245, 116573, 2020, https://doi.org/10.1016/j.carbpol.2020.116573.
17. Wang, Y., Yu, Y., Duan, Y., Wang, Q., Cong, X., He, Y., Gao, C., Hafeez, M., Jan, S., Rasheed, S.M., Cheng, S., Wang, Z., Enhancing the activity of carboxymethyl cellulase enzyme using highly stable selenium nanoparticles biosynthesized by *Bacillus paralicheniformis* Y4. *Molecules*, 27, 4585, 2022, https://doi.org/10.3390/molecules27144585.
18. Bakadia, B.M., Boni, B.O.O., Ahmed, A.A.Q., Zheng, R., Shi, Z., Ullah, M.W., Lamboni, L., Yang, G., *In situ* synthesized porous bacterial cellulose/poly(vinyl alcohol)-based silk sericin and azithromycin release system for treating chronic wound biofilm. *Macromol. Biosci.*, 10, 2200201, 2022, https://doi.org/10.1002/mabi.202200201.
19. Ostadhossein, F., Mahmoudi, N., Morales-Cid, M., Tamjid, E., Navas-Martos, F.J., Soriano-Cuadrado, B., Paniza, J.M.L., Simchi, A., Development of chitosan/bacterial cellulose composite films containing nanodiamonds as

a potential flexible platform for wound dressing. *Materials*, 8, 6401–6418, 2015.

20. Yuan, H., Chen, L., Hong, F.F., A biodegradable antibacterial nanocomposite based on oxidized bacterial nanocellulose for rapid hemostasis and wound healing. *ACS Appl. Mater. Interfaces*, 12, 3, 3382–3392, 2020.
21. Di, Z., Shi, Z., Ullah, M.W., Li, S., Yang, G., A transparent wound dressing based on bacterial cellulose whisker and poly (2-hydroxyethyl methacrylate). *Int. J. Biol. Macromol.*, 105, 1, 638–644, 2017, https://doi.org/10.1016/j.ijbiomac.2017.07.075.
22. Ul-Islam, M., Khattak, W.A., Ullah, M.W., Khan, S., Park, J.K., Synthesis of regenerated bacterial cellulose-zinc oxide nanocomposite films for biomedical applications. *Cellulose*, 21, 433–447, 2014.
23. Sajjad, W., He, F., Ullah, M.W., Ikram, M., Shah, S.M., Khan, R., Khan, T., Khalid, A., Yang, G., Wahid, F., Fabrication of bacterial cellulose-curcumin nanocomposite as a novel dressing for partial thickness skin burn. *Front. Bioeng. Biotechnol.*, 8, 553037, 2020, https://doi.org/10.3389/fbioe.2020.553037.
24. Kukushkina, E.A., Hossain, S.I., Sportelli, M.C., Ditaranto, N., Picca, R.A., Cioffi, N., Ag-based synergistic antimicrobial composites- A critical review. *Nanomater.*, 11, 1687, 2021, https://doi.org/10.3390/nano11071687.
25. Farooq, M., Ullah, M.W., Yang, Q., Aziz, A., Xu, J., Zhou, L., Wang, S., High-density phage particles immobilization in surface-modified bacterial cellulose for ultra-sensitive and selective electrochemical detection of Staphylococcus aureus. *Biosens. Bioelectron.*, 157, 112163, 2020.
26. Ao, H., Jiang, W., Nie, Y., Zhou, C., Zong, J., Liu, M., Liu, X., Wan, Y., Engineering quaternized chitosan in the 3D bacterial cellulose structure for antibacterial wound dressings. *Polym. Test.*, 86, 106490, 2020.
27. Panaitescu, D.M., Ionita, E.R., Nicolae, C.A., Gabor, A.R., Ionita, M.D., Trusca, R., Lixandru, B.E., Codita, I., Dinescu, G., Poly(3-hydroxybutyrate) modified by nanocellulose and plasma treatment for packaging applications. *Polym.*, *10*, 1249, 2018, https://doi.org/10.3390/polym10111249.
28. Sultan, S., Abdelhamid, H.N., Zou, X., Mathew, A.P., Cello MOF: Nanocellulose enabled 3D printing of metal-organic frameworks. *Adv. Funct. Mater.*, 29, 2, 1805372, 2019.
29. Chen, X.Y., Low, H.R., Loi, X.Y., Merel, L., Mohd Cairul Iqbal, M.A., Fabrication and evaluation of bacterial nanocellulose/poly(acrylic acid)/graphene oxide composite hydrogel: Characterizations and biocompatibility studies for wound dressing. *J. Biomed. Mater. Res.*, 107, 6, 2140–2151, 2019, https://doi.org/10.1002/jbm.b.34309.
30. Torgbo, S. and Sukyai, P., Fabrication of microporous bacterial cellulose embedded with magnetite and hydroxyapatite nanocomposite scaffold for bone tissue engineering. *Mater. Chem. Phys.*, 237, 121868, 2019, https://doi.org/10.1016/j.matchemphys.2019.121868.

31. Mao, L., Wang, L., Zhang, M., Ullah, M.W., Liu, L., Zhao, W., Li, Y., Ahmed, A.A.Q., Cheng, H., Shi, Z., Yang, G., *In situ* synthesized selenium nanoparticles-decorated bacterial cellulose/gelatin hydrogel with enhanced antibacterial, antioxidant, and anti-inflammatory capabilities for facilitating skin wound healing. *Adv. Healthcare Mater.*, 2100402, 16, 2017.
32. Elayaraja, S., Zagorsek, F., Li, K., Xiang, J., *In situ* synthesis of silver nanoparticles into TEMPO-mediated oxidized bacterial cellulose and their antivibriocidal activity against shrimp pathogens. *Carbohydr. Polym.*, 166, 329–337, 2017.
33. Pal, N., Dubey, P., Gopinath, P., Pal, K., Combined effect of cellulose nanocrystal and reduced graphene oxide into poly-lactic acid matrix nanocomposite as a scaffold and its anti-bacterial activity. *Int. J. Biol. Macromol.*, 95, 94–105, 2017, https://doi.org/10.1016/j.ijbiomac.2016.11.041.
34. Jafary, R., Mohammad, M.K., Seyedhossein, H., Ali, J., Antibacterial property of cellulose fabric finished by allicin-conjugated nanocellulose. *J. Text. Inst.*, 106, 1–7, 2014.
35. Qiu, K. and Netravali, A.N., A review of fabrication and applications of bacterial cellulosebased nanocomposites. *Polym. Rev.*, 54, 4, 598–626, 2014.
36. Norrrahim, M.N.F., Nurazzi, N.M., Jenol, M.A., Farid, M.A.A., Janudin, N., Ujang, F.A., Yasim-Anuar, T.A.T., Najmuddine, S.U.F.S., Ilyasf, R.F., Emerging development of nanocellulose as an antimicrobial material: An overview. *Mater. Adv.*, 2, 3538–3551, 2021.
37. Li, J., Cha, R., Mou, K., Zhao, X., Long, X., Luo, H., Zhou, F., Jiang, X., Nanocellulose-based antibacterial materials. *Adv. Healthcare Mater.*, 7, 20, 1800334, 2018.
38. Finny, A.S., Popoola, O., Andreescu, S., 3D-printable nanocellulose-based functional materials: Fundamentals and applications. *Nanomaterials (Basel)*, 11, 2358, 2021.
39. Jiang, Z., Chen, D., Yu, Y., Miao, J., Liu, Y., Zhang, L., Composite fibers prepared from multi-walled carbon nanotubes/cellulose dispersed/dissolved in ammonium/dimethyl sulfoxide mixed solvent. *RSC Adv.*, 7, 2186–2192, 2017.

11

Application of Nanocellulose for Treatment of Renal Failure

Prapti Chakraborty* **and Priyanka Ray**

Guru Nanak Institute of Pharmaceutical Science & Technology, Kolkata, India

Abstract

Nanocellulose compositions are becoming more popular because of their adaptable qualities. High strength, a large surface area, variable surface chemistry, and the provision of novel theranostic techniques in clinical settings are only a few of its special qualities. Due to the widespread inadequacy of current treatments, renal illness places a significant cost on worldwide public health. Future prevention and progression of kidney disease depend on early detection and appropriate treatment. The pharmacokinetic characteristics, targetability, and bioavailability of pharmaceuticals are improved by the nanocelluloses, which differ in particle size, charge, shape, and density of targeting ligands. As a result, the development of nanocelluloses in medicine has brought about fresh approaches to the potential diagnosis and care of renal disorders. There is a promise for a more effective method of treating renal disorders thanks to the growing body of research on the use of nanomaterials in kidney therapy. In the field of therapeutic medicine, the convergence of biotechnology, materials science, and nanotechnology has quickly gained speed. Exponential growth is being seen as the foundation of this new field is being established, and this could greatly enhance human life quality. A number of nanomaterial-based methods for treating and regrowing kidney tissue have been looked into. This chapter provides in-depth descriptions of innovative approaches for treating renal failure employing nanomaterials, mesenchymal stem cell-derived nanovesicles, carbon nanotubes, nanofibrous membranes, and nanomaterial-based adsorbents and membranes. A growing number of nanomaterials have the potential to affect renal disease. It covers the most recent experimental studies, clinical uses of nanocelluloses, present difficulties, and future prospects in the detection and treatment of renal illness.

Keywords: Nanocellulose, nanomaterials, kidney, renal disease

**Corresponding author*: prapti.chakraborty@gnipst.ac.n

Mainak Mukhopadhyay and Debalina Bhattacharya (eds.) Nanocellulose: A Biopolymer for Biomedical Applications, (297–318) © 2024 Scrivener Publishing LLC

11.1 Introduction

Nanocellulose is a flexible and biocompatible material that has been studied for a wide range of medical uses [1]. Because of their biocompatibility, biodegradability, and low cytotoxicity, nanocelluloses in particular hold a lot of promise as an advanced material for biomedical applications that is both affordable and practical [2, 3]. Also, because of their chemical functionality, they are simple to alter and produce useful compounds. Even though the hierarchical nanostructure of cellulose is what helps plants support weight, a lot of research has been done to make better cellulosic materials with a variety of nanostructures and functional properties [4]. Nanocelluloses are widely used in a wide range of medical procedures, such as tissue engineering, drug delivery, wound healing, diagnostics, and more [5, 6]. However, its use in renal failure is not well established, and more research is needed to determine its potential applications. A condition known as renal failure, commonly referred to as kidney failure, occurs when the kidneys are unable to filter waste and extra fluid from the circulation [7]. This can lead to a buildup of toxins in the body and can be life-threatening if left untreated. One potential application of nanocellulose in renal failure could be in the development of dialysis membranes. Dialysis is a common treatment for renal failure, in which a machine filters waste and excess fluids from the blood. However, current dialysis membranes have limitations in terms of efficiency and biocompatibility. Nanocellulose has been shown to have excellent filtration properties, and its biocompatibility makes it an attractive material for use in dialysis membranes [8]. Additionally, the nanoscale size of the cellulose fibers could potentially allow for more efficient filtration of toxins from the blood. The creation of scaffolds for kidney tissue engineering may be another possible use for nanocellulose in renal failure [9]. A new area called tissue engineering uses a combination of cells, biomaterials, and biochemical signals to create functional tissue. Since it has been shown to stimulate cell growth and proliferation, nanocellulose is a promising material for use in tissue engineering applications [10]. By combining nanocellulose scaffolds with kidney cells, it may be possible to create functional kidney tissue for transplantation or other therapeutic purposes. However, it is important to note that these potential applications of nanocellulose in renal failure are still in the early stages of research, and more studies are needed to fully understand their feasibility and effectiveness. Renal disease, also known as kidney disease, refers to any

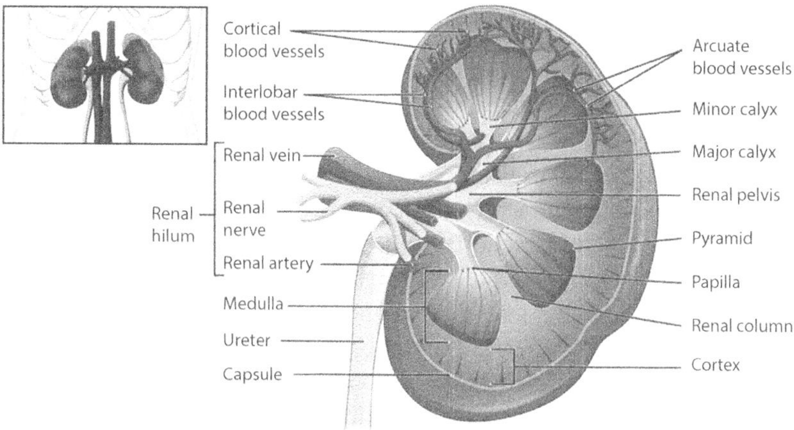

Figure 11.1 Anatomy of the human kidney [21].

condition that affects the function of the kidneys [11]. The kidneys are critical organs that regulate several body processes, including blood pressure and the formation of red blood cells, as well as filter waste and extra fluid from the blood [12].

Treatment for renal disease varies depending on the type and severity of the condition [13]. In some instances, medications and lifestyle modifications like eating well and exercising can help control symptoms and limit the development of the condition [14]. In more severe cases, dialysis or kidney transplantation may be necessary. It is important to consult with a healthcare professional for proper diagnosis and treatment of renal disease. The anatomy of human kidney is shown in Figure 11.1.

11.2 Renal Disease and Its Implications

A condition known as renal disease, sometimes known as kidney disease, occurs when both kidneys are unable to filter the blood of waste items produced by the body [15]. Thus, it can lead to a buildup of various toxins in the body, causing a range of health problems. Renal disease affects millions of people throughout the world and it is a prime cause of morbidity and death. In this article, we will discuss various renal disease and its implications [16, 17].

11.2.1 The Structure and Function of the Kidneys

Essentially two bean-shaped organs, the kidneys are found on either side of the spine, just below the ribs. Filtering waste products from the blood and excreting them through the urine is the kidney's main job. Additionally, the kidneys control the body's blood pressure, electrolyte levels, and fluid balance [18].

The kidney's nephron, which is made up of a tubule and a glomerulus, is both its structural and functional unit. A system of tiny blood tubes called the glomerulus removes waste from the blood. Following this, the tubule reabsorbs vital elements like glucose and amino acids and excretes waste materials like urea and creatinine in the form of urine [19, 20].

Treatment for renal disease depends on what kind of disease it is and how bad it is [13]. In some instances, medications and lifestyle modifications like eating well and exercising can help control symptoms and limit the development of the condition [14]. In more severe cases, dialysis or kidney transplantation may be necessary. It is important to consult with a healthcare professional for the proper diagnosis and treatment of renal disease.

11.2.2 Types of Renal Disease

There are various types of renal disease, each with its own causes and implications. Some of the most common types of renal disease are as follows:

11.2.2.1 Acute Kidney Injury (AKI)

Acute kidney injury (AKI) is a sudden loss of kidney function that frequently results from renal disease, a blockage in the urinary tract, or a decrease in blood supply to the kidneys. AKI is a dangerous illness that can result in electrolyte imbalances and the buildup of waste products in the body [21, 22].

11.2.2.2 Chronic Kidney Disease (CKD)

Chronic kidney disease (CKD) is characterized by a steady decline in kidney function over time. High blood pressure and diabetes are the two primary causes of CKD. Therefore, CKD has the potential to cause a number of problems, such as anaemia, bone disease, and cardiovascular disease [23].

11.2.2.3 Glomerulonephritis

When the glomeruli enlarge, it can lead to glomerulonephritis, a type of kidney disease. It can hurt the kidneys and make it harder for them to filter waste out of the blood. Glomerulonephritis is mainly caused by a range of factors, including infections, autoimmune disorders, and certain medications [24, 25].

11.2.2.4 Polycystic Kidney Disease (PKD)

PKD is primarily a hereditary condition that results in kidney cyst formation. The kidney's capacity to filter waste from the blood is hampered by these cysts, which can result in a number of issues. PKD is a long-term disease that results in kidney failure over time [26].

11.2.3 Implications of Renal Disease

Renal disease can have a range of implications for the body, including the following:

11.2.3.1 Electrolyte Imbalances

The kidneys are in charge of controlling the levels of sodium, potassium, and calcium in the body. Electrolyte imbalances can happen when the kidneys are hurt, which can lead to several health problems.

11.2.3.2 Accumulation of Waste Products

The primary function of the kidneys is to filter waste materials from the blood for excretion through urine. Waste products can build up in the body when the kidneys are damaged, which can cause a number of issues [27].

11.2.3.3 Hypertension

In controlling blood pressure, both kidneys are important. Hypertension may develop as a result of renal impairment, raising the risk of cardiovascular disease.

11.2.3.4 Anemia

Erythropoietin is a hormone that is made by the kidneys and helps the body make more red blood cells. Reduced erythropoietin production as a result of renal injury can cause anemia [28, 29].

11.2.3.5 Bone Disease

Renal (kidney) disease can have implications on bone health, leading to a condition called renal bone disease or chronic kidney disease–mineral and bone disorder (CKD-MBD) [30]. The implications of renal disease on bone health include the following:

- Weakened bones: Both kidneys play a vital role in maintaining the balance of minerals, such as calcium and phosphorus, in the body. In people with renal disease, the kidneys may not be able to regulate these minerals properly, leading to weakened bones.
- Osteoporosis: People with renal disease are at an increased risk of developing osteoporosis, a condition characterized by weakened bones and a high risk of fractures.
- Bone pain: People with renal bone disease may experience bone pain or muscle weakness, which can affect their mobility and quality of life.
- Vascular calcification: In people with renal disease, calcium can build up in the walls of blood vessels, leading to vascular calcification. This can increase the risk of heart disease and stroke.
- Fractures are more likely to happen to people with renal bone disease because their bones are weaker and they are more likely to fall. The kidneys are particularly important bodily organs because they filter wastes and extra fluids from the blood, keeping the electrolyte balance of the body and generating hormones that keep blood pressure and red blood cell formation in check. However, a condition known as renal disease can develop when various diseases affect the kidneys. Renal disease has various implications on the health of an individual, including increased morbidity and mortality, reduced quality of life, and an economic burden due to the high cost of treatment [31].

11.2.4 Causes of Renal Disease

Renal disease can be caused by various factors, including the following:

- Diabetes: Renal disease is primarily brought on by diabetes. Blood arteries inside the kidneys are harmed by high blood sugar levels, which causes kidney damage and finally renal failure.

- High blood pressure: Renal disease can result from high blood pressure, often known as hypertension, which can harm kidney blood vessels and impair kidney function.
- Glomerulonephritis: Glomerulonephritis is an inflammation of the glomeruli, which are the microscopic blood arteries in the kidneys that filter blood. Different infections, autoimmune conditions, and other illnesses can all lead to glomerulonephritis.
- Polycystic kidney disease: A genetic condition known as polycystic kidney disease results in the development of kidney cysts, which eventually cause kidney failure.
- Kidney stones: If left untreated, kidney stones can cause severe blockages in the urinary tract, which can lead to renal disease [32].

11.2.5 Symptoms of Renal Disease

The various symptoms of renal disease can vary depending on the type of renal disease and the severity of the condition. However, some common symptoms include the following:

- Tiredness
- Inflammation of the legs, ankles, feet, or face
- Decreased urine output
- Foamy urine
- Blood in the urine
- High blood pressure
- Shortness of breath
- Loss of appetite
- Nausea and vomiting
- Itching
- Muscle cramps

11.2.6 Diagnosis of Renal Disease

The diagnosis of renal disease usually involves a combination of medical history, physical examination, and laboratory tests. Some common tests include the following:

- Urine tests: Detecting anomalies in the urine, such as the presence of blood, protein, or sugar, can be done with the aid of urine tests.

- Blood tests: Creatinine is a waste product that the kidneys eliminate from the blood and can be measured in blood tests. Blood creatinine levels that are higher than normal may be a sign of impaired renal function.
- Imaging tests: Imaging examinations like an ultrasound, CT scan, or MRI can aid in the detection of any abnormalities in the kidneys, such as cysts or blockages [33].

11.3 Nanocellulose—Concept and Formulation

Due to its distinctive qualities and prospective applications in a variety of industries, including the medical, environmental, and industrial sectors, nanocellulose is a promising biomaterial that has drawn a lot of interest. In addition to other things, including bacteria, algae, and plants, nanocellulose is a kind of cellulose. It is created by shrinking cellulose fibers to nanoscale dimensions utilizing a number of processes, such as acid hydrolysis, mechanical homogenization, and enzymatic hydrolysis [34].

The components of nanocellulose include cellulose nanofibrils (CNFs) or cellulose nanocrystals (CNCs), which have a large surface area and aspect ratio. Contrary to CNCs, CNFs are created by acid-hydrolyzing cellulose fibers, not by mechanically processing it. As a result, the nanoscale particles have extraordinary qualities like high tensile strength, high aspect ratio, increased surface area, low density, and biodegradability.

The outstanding mechanical qualities of nanocellulose are one of its most important benefits. In comparison to Kevlar and steel, nanocellulose is more tensilely strong. Moreover, it is low in density and lightweight, making it a superb choice for lightweight structural applications. To enhance the mechanical properties of composites, nanocellulose can also be utilized as a reinforcement material [35].

Nanocellulose could be used in optical films, coatings, and sensors because it has better optical properties than other materials. High-transparency nanocellulose films can replace traditional plastic films in a variety of applications. Moreover, they have a low coefficient of thermal expansion, which makes them perfect for applications involving high temperatures [36].

Nanocellulose is also good for the environment because it is made from renewable resources and breaks down over time. It makes a great option for sustainable applications since soil microorganisms can quickly break it down.

Nanocellulose could be used in the biomedical industry for tissue engineering, drug delivery, wound healing, and coatings for implants. Nanocellulose can be used to make scaffolds for tissue engineering that can help cells grow and repair themselves. The drug can also be put on nanocellulose particles and used as a drug carrier to get the medicine to the right place. Because it can kill bacteria, nanocellulose is a great choice for coating implants and helping wounds heal [37].

Nanocellulose can be used in the food industry to thicken, stabilize, and mix things. It can also be used to replace fat, which helps lower the number of calories in food.

In conclusion, nanocellulose is an adaptable biomaterial with special qualities and prospective uses in a range of industries. It is a prime contender for environmentally friendly and sustainable applications due to its superior mechanical and optical qualities, biodegradability, and biocompatibility. Products made from nanocellulose have the potential to revolutionize many industries and offer solutions to several societal problems [38].

The unique characteristics of nanocellulose, such as its high strength, stiffness, and surface area as well as its biocompatibility and low toxicity, have led to a substantial increase in attention in recent years [39]. Wood, cotton, and bacteria are just a few of the substances from which nanocellulose can be made.

Some of the industries that could change because of nanocellulose-based products are packaging, textiles, construction, and electronics. However, making nanocellulose requires careful thought about many things, such as where the cellulose comes from, how it will be extracted, and what qualities the final product should have [40].

11.3.1 Methods of Extraction

Wood, cotton, and bacteria are just a few of the substances from which nanocellulose can be made. The factors influencing the extraction process are the cellulose source and the desired characteristics of the nanocellulose. Mechanical, chemical, and enzymatic extraction techniques are the three basic types.

11.3.1.1 Mechanical Extraction

The physical disintegration of cellulose fibers into smaller particles is a component of mechanical extraction. The most common source of cellulose is wood pulp, which is commonly used in this process to create nanocellulose. During the procedure, the pulp is ground in a high-pressure homogenizer or a microfluidizer, which reduces the fibers to nanoparticle size. Nanocellulose usually takes the shape of long, flexible strands called cellulose nanofibrils (CNFs) or stiffer, needle-like particles called cellulose nanocrystals (CNCs).

The ease of use and lack of dangerous chemicals are two benefits of mechanical extraction. In contrast to other procedures, it can be energy-intensive and lead to a reduced yield of nanocellulose [41].

11.3.1.2 Chemical Extraction

Chemical extraction is the process of dissolving or disintegrating cellulose fibers into smaller pieces using chemicals. With the use of this technique, nanocellulose can be produced from materials other than wood pulp, such as cotton or bacterial cellulose. Using strong acids, such as sulfuric acid, to treat the cellulose causes the fibers to disintegrate into nanoscale particles. Usually, the resulting nanocellulose takes the shape of CNCs, which are crystalline structures with a high aspect ratio [42].

Chemical extraction has several benefits, including a high yield and the capacity to create CNCs with a high degree of crystallinity. However, compared to other procedures, it necessitates the employment of dangerous chemicals and could lead to a poorer purity of nanocellulose [43].

A unique class of biomaterials called nanocellulose is made from renewable resources like the cellulose found in plants. Nanocellulose materials are intensively explored for a wide range of applications in many fields such as packaging, textiles, cosmetics, biomedical engineering, and more due to their special physical, chemical, and mechanical properties [44]. We will look at several nanocellulose compositions and possible uses in this post.

11.3.2 Cellulose Nanofibers (CNFs)

By mechanically shrinking cellulose fibers to a nanoscale dimension, cellulose nanofibers (CNF) are produced. CNF is a good contender for a variety of applications due to its high mechanical strength, high aspect ratio, and enhanced surface area. CNF's characteristics can be further improved through chemical or physical modification for certain purposes. For instance, to increase the solubility of CNF in water and make it acceptable for use as a thickening agent or stabilizer in the food sector, carboxymethyl cellulose (CMC) can be grafted onto the surface of CNF [45].

11.3.3 Cellulose Nanocrystals (CNCs)

Acid hydrolysis of cellulose fibers results in the creation of rod-like particles with high aspect ratios, which are known as cellulose nanocrystals (CNC). The huge surface area, high mechanical strength, and distinctive optical features of CNC make it a potential material for a variety of uses, including in the biomedical industry as scaffolds for tissue engineering and drug delivery carriers. CNC can also be chemically altered to improve its characteristics for particular purposes. A potential substance for water filtration is CNC, which can have its surface altered using cationic polymers to enhance its capacity to adsorb heavy metal ions [46].

11.3.4 Bacterial Cellulose (BC)

A biopolymer called bacterial cellulose (BC) is created when bacteria digest glucose or other carbohydrates. The various sources of cellulose has been illustrated in Figure 11.2. BC is a potential material for a variety of

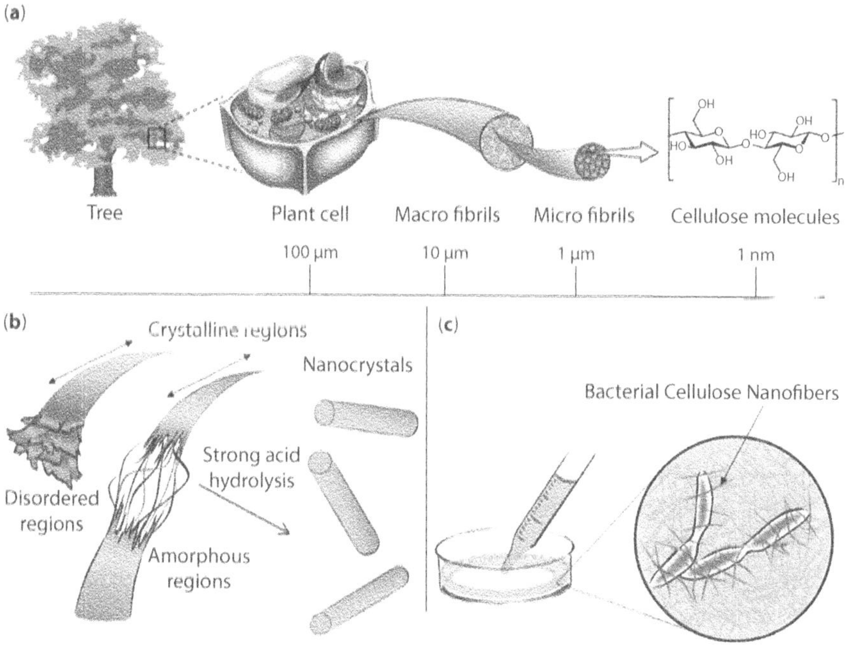

Figure 11.2 (a) Cellulose in plant and trees has a schematic structure from meter to nano scale. (b) Diagram showing the reaction between cellulose and strong acids to form nanocellulose. (c) Bionanocellulose cultured from cellulose-synthesizing bacteria [49].

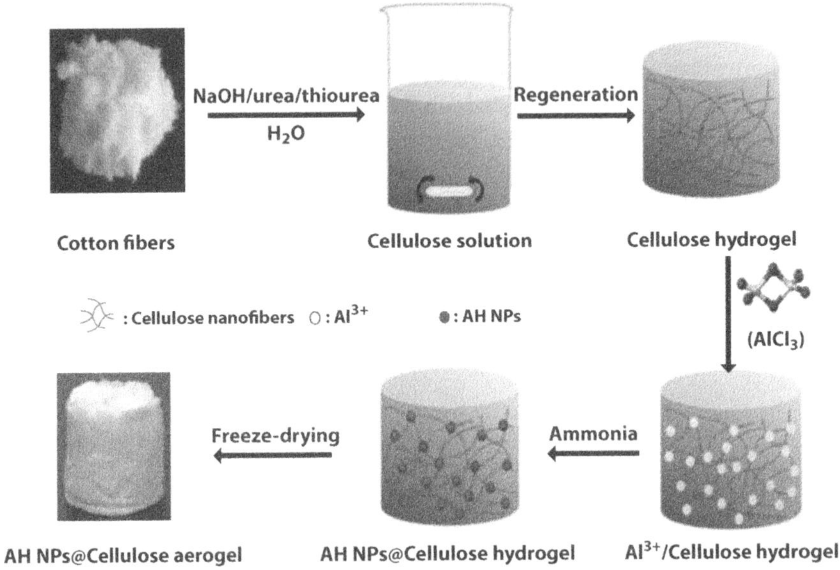

Figure 11.3 Schematic representation of nanocellulose aerogel formulation [51].

applications, including wound dressing, tissue engineering, and food packaging, thanks to its distinctive 3D network structure, improved mechanical strength, high water-holding capacity, and biocompatibility. BC can also be chemically or physically altered to improve its characteristics for particular purposes. A potential substance for wound dressings is BC, which can be altered with chitosan to enhance its antibacterial properties [47]. A schematic represenation of the nanocellulose aerogel formulation has been shown in Figure 11.3.

11.3.5 Cellulose Nanofibrils (CNFib)

By mechanically reducing cellulose fibers to a nanoscale dimension using high-pressure homogenization or micro fluidization, cellulose nanofibrils (CNFib) are produced. CNFib is a potential material for a variety of applications, such as in the food sector as a fat replacement or as a reinforcing agent in paper or composites, thanks to its high surface area, high mechanical strength, and high water holding capacity. Moreover, CNFib can be chemically or physically altered to improve its characteristics for particular applications. One potential material for food packaging is CNFib, which may be altered to enhance its antibacterial characteristics using silver nanoparticles [48].

11.3.6 Nanocellulose Aerogels

Nanocellulose aerogels are a family of materials with high porosity, low density, and increased surface area, which makes them suitable for a variety of uses, including as drug delivery vehicles in the biomedical field, insulation in the building industry, and absorbents in the oil and gas business. Many techniques, including solvent exchange, freeze-drying, and supercritical drying, can be used to create nanocellulose aerogels. By adjusting the preparation parameters, such as the nanocellulose concentration, the characteristics of nanocellulose aerogels can be tailored [50].

11.4 Kidney-Targeted Drug Delivery by Nanocellulose

Nanocellulose could be used to deliver drugs directly to the kidneys, reducing the amount of drug required and minimizing side effects [52]. One approach to achieve this is to functionalize the surface of the nanocellulose particles with specific targeting ligands that can recognize and bind to receptors on the surface of kidney cells [53].

A study showed that a model drug (doxorubicin) could be delivered to kidney cells *in vitro* by using carboxymethyl cellulose nanocrystals (CMCNCs) that were made to target folic acid. The findings demonstrated that the kidney cells efficiently absorbed the targeted CMCNCs compared to non-targeted particles, increasing drug efficacy and lowering toxicity [54, 55].

Using the unique properties of the nanomaterial to control drug release is another way nanocellulose could be used to deliver drugs to the kidneys. Nanocellulose has a large surface area and high surface energy, which can be used to adsorb drugs onto its surface and release them in a controlled manner [56]. By tuning the properties of the nanocellulose particles, such as their size and surface charge, it may be possible to achieve sustained release of drugs specifically in the kidney, providing a longer duration of therapeutic effect and reducing the frequency of dosing [57].

One area where nanocellulose-based materials may be useful is in the development of new therapies for kidney fibrosis. Kidney fibrosis is a common feature of many forms of renal disease and involves the excessive deposition of extracellular matrix proteins, leading to the loss of normal kidney function [58]. Recent studies have shown that nanocellulose-based materials can inhibit the development of kidney fibrosis by reducing the production of extracellular matrix proteins and promoting the regeneration of kidney tissue [59]. Table 11.1 represents the application of nanocellulose in renal failure.

Table 11.1 Nanocellulose formulation and its therapeutic applications in renal failure.

Nanocellulose type	Formulation	Renal treatment	Therapeutic applications	Reference
Bacterial nanocellulose	Bacterial nanocellulose scaffolds with vascular structure	Acute kidney injury treatment	Creating a mold for casting a full-sized kidney-like construct.	[71]
Plant-derived nanocellulose	Cellulose nanocrystals	Hemodialysis treatment	*In vivo*, water-soluble functionalized derivatives display special fibrillar nanomaterial characteristics for renal specific targeting.	[72]
Bacterial nanocellulose	Polypyrrole@ BNCC/ PVA nanocomposite	Hemodialysis	The dialysis-free preparation demonstrated enhanced conductivity.	[73]
Plant-derived nanocellulose	Nanocellulose-based membranes	Hemodialysis treatment	Bacterial nanocellulose-based membranes showed potential for use in hemodialysis treatment, with improved permeability and blood compatibility.	[74]
CNC	cCNC/SA double-membrane hydrogels	Chronic kidney disease treatment	Nanocellulose hydrogel showed potential for use as a therapeutic agent for chronic kidney disease, with improved anti-inflammatory and antioxidant properties.	[75]

Nanocellulose-based materials might also be useful in making new ways to give drugs to people with renal disease. Nanocellulose-based materials can be engineered to have specific properties, such as controlled release and targeted delivery, which could be useful in delivering drugs directly to the kidneys [60, 61]. This approach could potentially reduce the side effects associated with systemic drug delivery and improve the efficacy of existing treatments for renal diseases [62].

Chen *et al.* synthesized a novel, non-toxic PEG-PCL-PEI polymer, and PPP-RH-NPs. PPP-RH-NPs demonstrated a good renal distribution in the STZ-induced DN model, which improved a variety of pathological markers and reversed the worsening DN. As a result, these nanoparticles enhanced the pharmacological effects of RH on DN, providing new opportunities for its application [63].

In the research titled "Self-assembled hyaluronic acid-coated nanocomplexes for targeted delivery of curcumin to alleviate acute kidney injury", Huang *et al.* used polymeric nanocarriers made from methoxy polyethylene glycol-block poly(lactic-co-glycolic acid) (mPEG-PLGA) to encapsulate curcumin for renal-targeted drug delivery. The nanocarriers were functionalized with a kidney-targeting ligand (cinnamoyl phenethylamine) to enhance their uptake by kidney cells. The results showed that the targeted nanocarriers were taken up more efficiently by kidney cells and had a greater therapeutic effect than non-targeted nanocarriers [64].

Hachaichi *et al.* used carboxymethyl cellulose nanocrystals (CMCNCs) as carriers for renal drug delivery. The CMCNCs were functionalized with the targeting ligand folic acid to enhance their uptake by kidney cells. The authors optimized the formulation parameters of the CMCNCs to achieve sustained drug release *in vitro*. They also demonstrated the biocompatibility of the CMCNCs *in vivo* using a zebrafish model [65].

Eftekhari *et al.* discuss the potential use of nanocellulose-based biomaterials for the treatment of renal failure. The authors describe the different types of nanocellulose materials that have been studied for renal applications, including nanocellulose hydrogels and nanocellulose membranes. They also provide an overview of recent studies that have investigated the use of nanocellulose-based biomaterials for the treatment of acute kidney injury and chronic kidney disease [66].

Waheed *et al.* focused on the use of nanocellulose in hemodialysis membranes for the treatment of renal failure. They discuss the advantages of using nanocellulose, such as its biocompatibility, mechanical strength, and ability to control pore size and surface charge. They also describe the different methods that have been used to incorporate nanocellulose into hemodialysis membranes and the potential benefits of these membranes for improving patient outcomes [67].

Wang *et al.* discussed the potential of nanocellulose-based scaffolds for renal regeneration in cases of renal failure. They describe the properties of nanocellulose that make it suitable for tissue engineering, such as its biocompatibility, biodegradability, and ability to support cell attachment and proliferation. They also provide an overview of recent studies that have investigated the use of nanocellulose-based scaffolds for renal regeneration in animal models [68].

Khalil *et al.* provide a critical overview of cellulose-based micro- and nanoscale scaffold preparation materials in their review article. It also examines the connection between the fabrication method and the scaffold's characteristics [69, 70].

11.5 Conclusion

There is a great deal of promise for the use of nanocellulose to treat renal failure in the realm of regenerative medicine. Nanocellulose has been demonstrated to be a strong candidate for producing biomaterials for engineering kidney tissue due to its distinctive qualities, such as being biocompatible, biodegradable, and non-toxic. It has been demonstrated that nanocellulose-based scaffolds can promote cell growth and division, increase kidney cell survival and function, and hasten the healing of damaged renal tissue. Dialysis membranes may also be made from nanocellulose-based materials. These membranes can improve the efficiency of the dialysis procedure and lower the risk of infection. Moreover, nanocellulose-based materials have demonstrated promise as renal disease treatment drug delivery methods. Overall, research into the application of nanocellulose in the field of renal regenerative medicine is expanding quickly and showing great promise. The early results are very encouraging, but further research and clinical trials are needed to properly investigate the potential of nanocellulose-based materials for the treatment of renal failure.

References

1. Trache, D., Tarchoun, A.F., Derradji, M., Hamidon, T.S., Masruchin, N., Brosse, N., Hussin, M.H., Nanocellulose: From fundamentals to advanced applications. *Front. Chem.*, 8, 392, 2020 May 6.
2. Jasmania, L. and Thielemans, W., Preparation of nanocellulose and its potential application. *Int. J. Nanomater. Nanotechnol. Nanomed.*, 4, 2, 14–21, 2018 Oct.

3. Thomas, B., Raj, M.C., Joy, J., Moores, A., Drisko, G.L., Sanchez, C., Nanocellulose, a versatile green platform: From biosources to materials and their applications. *Chem. Rev.*, 118, 24, 11575–625, 2018 Nov 7.
4. Ewulonu, C.M., Liu, X., Wu, M., Yong, H., Lignin-containing cellulose nanomaterials: A promising new nanomaterial for numerous applications. *J. Bioresour. Bioprod.*, 4, 1, 3–10, 2019 Feb 1.
5. Wang, S., Lu, A., Zhang, L., Recent advances in regenerated cellulose materials. *Prog. Polym. Sci.*, 53, 169–206, 2016 Feb 1.
6. Wang, X., Yao, C., Wang, F., Li, Z., Cellulose-based nanomaterials for energy applications. *Small*, 13, 42, 1702240, 2017 Nov.
7. Wei, J., Yang, Z., Sun, Y., Wang, C., Fan, J., Kang, G., Zhang, R., Dong, X., Li, Y., Nanocellulose-based magnetic hybrid aerogel for adsorption of heavy metal ions from water. *J. Mater. Sci.*, 54, 8, 6709–18, 2019 Apr 30.
8. Eftekhari, A., Maleki Dizaj, S., Ahmadian, E., Przekora, A., Hosseiniyan Khatibi, S.M., Ardalan, M. *et al.*, Application of advanced nanomaterials for kidney failure treatment and regeneration. *Mater.* [Internet], 14, 11, 2939, 2021, Available from: http://dx.doi.org/10.3390/ma14112939.
9. Adewuyi, A., Otuechere, C.A., Adebayo, O.L., Anazodo, C., Pereira, F.V., Renal toxicological evaluations of sulphonated nanocellulose from Khaya sengalensis seed in Wistar rats. *Chem. Biol. Interact.*, 284, 56–68, 2018.
10. Andreoli, M.C., Dalboni, M.A., Watanabe, R., Manfredi, S.R., Canziani, M.E., Kallás, E.G., Sesso, R.C., Draibe, S.A., Balakrishnan, V.S., Jaber, B.L., Liangos, O., Impact of dialyzer membrane on apoptosis and function of polymorphonuclear cells and cytokine synthesis by peripheral blood mononuclear cells in hemodialysis patients. *Artif. Organs*, 31, 12, 887–92, 2007.
11. Adewuyi, A., Otuechere, C.A., Adebayo, O.L., Anazodo, C., Pereira, F.V., Renal toxicological evaluations of sulphonated nanocellulose from Khaya sengalensis seed in Wistar rats. *Chem.-Biol. Interact.*, 284, 56–68, 2018 Mar 25.
12. Saifi, M.A., Sangomla, S., Khurana, A., Godugu, C., Protective effect of nanoceria on cisplatin-induced nephrotoxicity by amelioration of oxidative stress and pro-inflammatory mechanisms. *Biol. Trace Elem. Res.*, 189, 145–56, 2019 May 15.
13. Long, D.A., Norman, J.T., Fine, L.G., Restoring the renal microvasculature to treat chronic kidney disease. *Nat. Rev. Nephrol.*, 8, 4, 244–50, 2012 Apr.
14. Murugesan, S., Mousa, S., Vijayaraghavan, A., Ajayan, P.M., Linhardt, R.J., Ionic liquid-derived blood-compatible composite membranes for kidney dialysis. *J. Biomed. Mater. Res. Part B: Appl. Biomater.: An Official Journal of The Society for Biomaterials, The Japanese Society for Biomaterials, and The Australian Society for Biomaterials and the Korean Society for Biomaterials*, 79, 2, 298–304, 2006 Nov.
15. Babicka, M., Woźniak, M., Dwiecki, K., Borysiak, S., Ratajczak, I., Preparation of nanocellulose using ionic liquids: 1-propyl-3-methylimidazolium chloride and 1-ethyl-3-methylimidazolium chloride. *Molecules*, 25, 7, 1544, 2020.

16. Jasmania, L., Preparation of nanocellulose and its potential application. *Int. J. Nanomater. Nanotechnol. Nanomedicine*, 4, 2, 014–21, 2018.
17. Ferraz, N., Carlsson, D.O., Hong, J., Larsson, R., Fellström, B., Nyholm, L. *et al.*, Haemocompatibility and ion exchange capability of nanocellulose polypyrrole membranes intended for blood purification. *J. R. Soc. Interface*, 9, 73, 1943–55, 2012.
18. Eftekhari, A., Maleki Dizaj, S., Ahmadian, E., Przekora, A., Hosseiniyan Khatibi, S.M., Ardalan, M. *et al.*, Application of advanced nanomaterials for kidney failure treatment and regeneration. *Materials*, 14, 11, 2939, 2021.
19. Moore, J.P. 2nd, Robling, K., Romero, C., Kiper, K., Dachavaram, S.S., Crooks, P.A., Hestekin, J.A., Oxone®-mediated TEMPO-oxidized cellulose nanomaterial ultrafiltration and dialysis mixed-matrix hollow fiber membranes. *Polym. (Basel)*, 12, 6, 1348, 2020.
20. Langari, M.M., Nikzad, M., Labidi, J., Nanocellulose-based sensors in medical/clinical applications: The state-of-the-art review. *Carbohydr. Polym.*, 304, 120509, 2023.
21. Grząbka-Zasadzińska, A., Skrzypczak, A., Borysiak, S., The influence of the cation type of ionic liquid on the production of nanocrystalline cellulose and mechanical properties of chitosan-based biocomposites. *Cellulose*, 26, 4827–4840, 2019.
22. Huang, Z.M., Zhang, Y.Z., Kotaki, M., Ramakrishna, S., A review on polymer nanofibers by electrospinning and their applications in nanocomposites. *Compos. Sci. Technol.*, 63, 2223–2253, 2003.
23. Lanza, R., Langer, R., Vacanti, J.P., Atala, A., *Principles of tissue engineering*, Academic Press, Cambridge, MA, USA, 2020.
24. Eftekhari, A., Maleki Dizaj, S., Sharifi, S., Salatin, S., Rahbar Saadat, Y., Zununi Vahed, S., Samiei, M., Ardalan, M., Rameshrad, M., Ahmadian, E., The use of nanomaterials in tissue engineering for cartilage regeneration; current approaches and future perspectives. *Int. J. Mol. Sci.*, 21, 536, 2020.
25. Parveen, S., Krishnakumar, K., Sahoo, S., New era in health care: Tissue engineering. *J. Stem Cells Regen. Med.*, 1, 8, 2006.
26. Sharma, C., Dinda, A.K., Mishra, N.C., Synthesis and characterization of glycine modified chitosan-gelatin-alginate composite scaffold for tissue engineering applications. *J. Biomater. Tissue Eng.*, 2, 133–142, 2012.
27. Liu, C., Xia, Z., Czernuszka, J., Design and development of three-dimensional scaffolds for tissue engineering. *Chem. Eng. Res. Des.*, 85, 1051–1064, 2007.
28. Eaglstein, W.H. and Falanga, V., Tissue engineering and the development of Apligraf®, a human skin equivalent. *Clin. Ther.*, 19, 894–905, 1997.
29. Chandler, L.A., Ma, C., Gonzalez, A.M., Doukas, J., Nguyen, T., Pierce, G.F., Phillips, M.L., Matrix-enabled gene transfer for cutaneous wound repair. *Wound Repair Regen.*, 8, 473–479, 2000.

30. Xiao, Y., Qian, H., Young, W.G., Bartold, P.M., Tissue engineering for bone regeneration using differentiated alveolar bone cells in collagen scaffolds. *Tissue Eng.*, 9, 1167–1177, 2003.
31. Lee, S.J., Wang, H.J., Kim, T.H., Choi, J.S., Kulkarni, G., Jackson, J.D., Atala, A., Yoo, J.J., *In situ* tissue regeneration of renal tissue induced by collagen hydrogel injection. *Stem Cells Transl. Med.*, 7, 241–250, 2018.
32. Vashi, A.V., Abberton, K.M., Thomas, G.P., Morrison, W.A., O'Connor, A.J., Cooper-White, J.J., Thompson, E.W., Adipose tissue engineering based on the controlled release of fibroblast growth factor-2 in a collagen matrix. *Tissue Eng.*, 12, 3035–3043, 2006.
33. Lee, J.-Y., Nam, S.-H., Im, S.-Y., Park, Y.-J., Lee, Y.-M., Seol, Y.-J., Chung, C.-P., Lee, S.-J., Enhanced bone formation by controlled growth factor delivery from chitosan-based biomaterials. *J. Control. Release*, 78, 187–197, 2002.
34. Ghosh, S. and Jassal, M., Use of polysaccharide fibres for modern wound dressings. *Indian J. Fibre Text. Res.*, 27, 434–450, 2002.
35. Cai, S., Liu, Y., Shu, X.Z., Prestwich, G.D., Injectable glycosaminoglycan hydrogels for controlled release of human basic fibroblast growth factor. *Biomaterials*, 26, 6054–6067, 2005.
36. Malafaya, P.B., Silva, G.A., Reis, R.L., Natural–origin polymers as carriers and scaffolds for biomolecules and cell delivery in tissue engineering applications. *Adv. Drug Deliv. Rev.*, 59, 207–233, 2007.
37. Long, D.A., Norman, J.T., Fine, L.G., Restoring the renal microvasculature to treat chronic kidney disease. *Nat. Rev. Nephrol.*, 8, 244, 2012.
38. Uzarski, J.S., Xia, Y., Belmonte, J.C., Wertheim, J.A., New strategies in kidney regeneration and tissue engineering. *Curr. Opin. Nephrol. Hypertens.*, 23, 399–405, 2014.
39. Jeevanandam, J., Barhoum, A., Chan, Y.S., Dufresne, A., Danquah, M.K., Review on nanoparticles and nanostructured materials: History, sources, toxicity and regulations. *Beilstein J. Nanotechnol.*, 9, 1050–1074, 2018.
40. Ahmadian, E., Dizaj, S.M., Eftekhari, A., Dalir, E., Vahedi, P., Hasanzadeh, A., Samiei, M., The potential applications of hyaluronic acid hydrogels in biomedicine. *Drug Res.*, 70, 6–11, 2020.
41. Ahmadian, E., Dizaj, S.M., Rahimpour, E., Hasanzadeh, A., Eftekhari, A., Halajzadeh, J., Ahmadian, H., Effect of silver nanoparticles in the induction of apoptosis on human hepatocellular carcinoma (HepG2) cell line. *Mater. Sci. Eng. C*, 93, 465–471, 2018.
42. Dizaj, S. and Adibkia, K., A short overview on the nanoparticle-based smart drug delivery systems. *J. Pharm. Pharm.*, 2, 1–2, 2015.
43. Dizaj, S.M., Can nanotechnology present new strategies to overcome COVID-19? *J. Adv. Chem. Pharm. Mater.*, 3, 258–259, 2020.
44. Fujisato, T., Sajiki, T., Liu, Q., Ikada, Y., Effect of basic fibroblast growth factor on cartilage regeneration in chondrocyte-seeded collagen sponge scaffold. *Biomaterials*, 17, 155–162, 1996.

45. Dhandayuthapani, B., Yoshida, Y., Maekawa, T., Kumar, D.S., Polymeric scaffolds intissue engineering application: A review. *Int. J. Polym. Sci.*, 1–19, 2011, 2011.
46. Uyama, S., Kaufmann, P.M., Kneser, U., Fiegel, H.C., Pollok, J.M., Kluth, D., Vacanti, J.P., Rogiers, X., Hepatocyte transplantation using biodegradable matrices in ascorbic acid-deficient rats: Comparison with heterotopically transplanted liver grafts1. *Transplantation*, 71, 1226–1231, 2001.
47. Gomes, S.R., Rodrigues, G., Martins, G.G., Roberto, M.A., Mafra, M., Henriques, C., Silva, J.C., *In vitro* and *in vivo* evaluation of electrospun nanofibers of PCL, chitosan and gelatin: A comparative study. *Mater. Sci. Eng. C*, 46, 348–358, 2015.
48. Miyashiro, D., Hamano, R., Umemura, K., A review of applications using mixed materials of cellulose, nanocellulose and carbon nanotubes. *Nanomaterials*, 10, 2, 186, 2020 Jan 21.
49. Hasirci, V., Berthiaume, F., Bondre, S., Gresser, J., Trantolo, D., Toner, M., Wise, D., Expression of liver-specific functions by rat hepatocytes seeded in treated poly (lactic-co-glycolic) acid biodegradable foams. *Tissue Eng.*, 7, 385–394, 2001.
50. Jiang, D., Rosenkrans, Z.T., Ni, D., Lin, J., Huang, P., Cai, W., Nanomedicines for renal management: From imaging to treatment. *Acc. Chem. Res.*, 53, 9, 1869–80, 2020 Aug 12.
51. Gopakumar, D.A., Pottathara, Y.B., Sabu, K.T., Khalil, H.A., Grohens, Y., Thomas, S., Nanocellulose-based aerogels for industrial applications, in: *Industrial Applications of Nanomaterials*, pp. 403–421, Elsevier, Amsterdam, Netherlands, 2019 Jan 1.
52. Ghasemi-Mobarakeh, L., Prabhakaran, M.P., Morshed, M., Nasr-Esfahani, M.-H., Ramakrishna, S., Electrospun poly (_-caprolactone)/gelatin nanofibrous scaffolds for nerve tissue engineering. *Biomaterials*, 29, 4532–4539, 2008.
53. Kazimierczak, P., Benko, A., Nocun, M., Przekora, A., Novel chitosan/agarose/hydroxyapatite nanocomposite scaffold for bone tissue engineering applications: Comprehensive evaluation of biocompatibility and osteo inductivity with the use of osteoblasts and mesenchymal stem cells. *Int. J. Nanomed.*, 14, 6615, 2019.
54. Gu, X., Liu, Z., Tai, Y.F., Zhou, L.Y., Liu, K., Kong, D., Midgley, A.C., Zuo, X.C., Hydrogel and nanoparticle carriers for kidney disease therapy: Trends and recent advancements. *Prog. Biomed. Eng.*, 4, 28–56, 2022 May 9.
55. Midgley, A.C., Wei, Y., Zhu, D., Gao, F., Yan, H., Khalique, A., Luo, W., Jiang, H., Liu, X., Guo, J., Zhang, C., Multifunctional natural polymer nanoparticles as antifibrotic gene carriers for CKD therapy. *J. Am. Soc. Nephrol.*, 31, 10, 2292–311, 2020 Oct 1.
56. Alallam, B., Choukaife, H., Seyam, S., Lim, V., Alfatama, M., Advanced drug delivery systems for renal disorders. *Gels*, 9, 2, 115, 2023 Feb 1.

57. Tan, W., Twomey, J., Guo, D., Madhavan, K., Li, M., Evaluation of nanostructural, mechanical, and biological properties of collagen–nanotube composites. *IEEE Trans. Nanobiosci*, 9, 2, 111–20, 2010 Mar 8.
58. Endes, C., Camarero-Espinosa, S., Mueller, S., Foster, E.J., Petri-Fink, A., Rothen-Rutishauser, B., Weder, C., Clift, M.J., A critical review of the current knowledge regarding the biological impact of nanocellulose. *J. Nanobiotechnol.*, 14, 1, 1–4, 2016 Dec.
59. Lin, N. and Dufresne, A., Nanocellulose in biomedicine: Current status and future prospect. *Eur. Polym. J.*, 59, 302–25, 2014 Oct 1.
60. Edwards, J., Prevost, N., French, A., Concha, M., DeLucca, A., Wu, Q., Nanocellulose-based biosensors: Design, preparation, and activity of peptide-linked cotton cellulose nanocrystals having Fluorimetric and Colorimetric elastase detection sensitivity. *Engineering*, 5, 9A, 20–28, 2013.
61. Piasecka-Zelga, J., Zelga, P., Gronkowska, K., Madalski, J., Szulc, J., Wietecha, J., Ciechańska, D., Dziuba, R., Toxicological and sensitization studies of novel vascular prostheses made of bacterial nanocellulose modified with chitosan (MBC) for application as the tissue-engineered blood vessels. *Regener. Eng. Transl. Med.*, 7, 2, 218–33, 2021 Jun.
62. Robling, K., Characterization of oxone mediated TEMPO-oxidized nano cellulose mixed-matr ix Membranes During Ultrafiltration and Hemodialysis. Biomedical Engineering Undergraduate Honors Theses Retrieved from https://scholarworks.uark.edu/bmeguht/77
63. Chen, D., Han, S., Zhu, Y., Hu, F., Wei, Y., Wang, G., Kidney-targeted drug delivery via rhein-loaded polyethyleneglycol-co-polycaprolactone-co-polyethylenimine nanoparticles for diabetic nephropathy therapy. *Int. J. Nanomed.*, 13, 3507–3527, 2018.
64. Huang, J., Guo, J., Dong, Y., Xiao, H., Yang, P., Liu, Y., Liu, S., Cheng, S., Song, J., Su, Y., Wang, S., Self-assembled hyaluronic acid-coated nanocomplexes for targeted delivery of curcumin alleviate acute kidney injury. *Int. J. Biol. Macromol.*, 226, 1192–1202, 2023 Jan 31.
65. Hachaichi, A., Kouini, B., Kian, L.K., Asim, M., Fouad, H., Jawaid, M., Sain, M., Nanocrystalline cellulose from microcrystalline cellulose of date palm fibers as a promising candidate for bio-nanocomposites: Isolation and characterization. *Materials*, 14, 18, 5313, 2021 Sep 15.
66. Eftekhari, A., Maleki Dizaj, S., Ahmadian, E., Przekora, A., Hosseiniyan Khatibi, S.M., Ardalan, M., Zununi Vahed, S., Valiyeva, M., Mehraliyeva, S., Khalilov, R., Hasanzadeh, M., Application of advanced nanomaterials for kidney failure treatment and regeneration. *Mater. (Basel)*, 14, 11, 2939, 2021.
67. Waheed, H., Farrukh, S., Hussain, A., Mukhtar, A., Mubashir, M., Saqib, S., Ullah, S., Peter, A.P., Khoo, K.S., Show, P.L., Green synthesized nano-cellulose polyethylene imine-based biological membrane. *Food Chem. Toxicol.*, 160, 112773, 2022 Feb 1.

68. Wang, C., Bai, J., Tian, P., Xie, R., Duan, Z., Lv, Q., Tao, Y., The application status of nanoscale cellulose-based hydrogels in tissue engineering and regenerative biomedicine. *Front. Bioeng. Biotechnol.*, 9, 732513, 2021 Nov 18.
69. Khalil, H.A., Jummaat, F., Yahya, E.B., Olaiya, N.G., Adnan, A.S., Abdat, M., Nam, N., Halim, A.S., Kumar, U.S., Bairwan, R., Suriani, A.B., A review on micro-to nanocellulose biopolymer scaffold forming for tissue engineering applications. *Polymers*, 12, 9, 2043, 2020 Sep 8.
70. Mandrycky, C., Phong, K., Zheng, Y., Tissue engineering toward organ-specific regeneration and disease modeling. *MRS Commun.*, 7, 3, 332–47, 2017 Sep.
71. Wong, S., Alidori, S., Mello, B.P., Almeida, B.A., Ulmert, D., Brendel, M.B., Scheinberg, D.A., McDevitt, M.R., Fibrillar pharmacology of functionalized nanocellulose. *Sci. Rep.*, 11, 1, 157, 2021 Jan 8.
72. Nirmal, N., Pillay, M.N., Mariola, M., Petruccione, F., van Zyl, W.E., Formation of dialysis-free Kombucha-based bacterial nanocellulose embedded in a polypyrrole/PVA composite for bulk conductivity measurements. *RSC Adv.*, 10, 46, 27585–97, 2020.
73. Ferraz, N., Leschinskaya, A., Toomadj, F., Fellström, B., Strømme, M., Mihranyan, A., Membrane characterization and solute diffusion in porous composite nanocellulose membranes for hemodialysis. *Cellulose*, 20, 2959–70, 2013 Dec.
74. Robling, K., Characterization of oxone mediated TEMPO-oxidized nano cellulose mixed-matrix membranes during ultrafiltration and hemodialysis.
75. Gunathilake, T.M., Ching, Y.C., Chuah, C.H., Enhancement of curcumin bioavailability using nanocellulose reinforced chitosan hydrogel. *Polymers*, 9, 64–75, 2017.

12

Use of Nanocellulose Hydrogels for Ophthalmic Applications

Gouranga Dutta[1], Debabrata Ghosh Dastidar[2]* and Abimanyu Sugumaran[3]

[1]Department of Pharmaceutics, SRM College of Pharmacy, SRM Institute of Science and Technology, SRM Nagar, Kattankulathur, Chennai, Tamil Nadu, India
[2]Department of Pharmaceutics, Guru Nanak Institute of Pharmaceutical Science & Technology, Panihati, Kolkata, West Bengal, India
[3]Department of Pharmaceutical Sciences, Assam University (A Central University), Silchar, Assam, India

Abstract

Nanocelluloses comprise two distinct cellulose particle types: crystalline cellulose nanocrystals (CNCs) and longer, intertwined cellulose nanofibers (CNFs). Their utilization in constructing interconnected networks stems from the hygroscopic properties of cellulose, its extensive specific surface area, and its relatively wide aspect ratio. Furthermore, nanocellulose can undergo chemical modifications to synthesize stimuli-responsive polymers with specific applications in the biomedical and clinical domains. CNFs have emerged as advantageous components for developing self-healing composite hydrogels, which find diverse applications in ophthalmic settings, such as the fabrication of soft contact lenses, flexible corrective lenses, and ocular drug delivery systems. These hydrogels offer ease of maintenance due to their high water content, ensuring lubricity, exceptional oxygen permeability, and resistance to protein deposition. However, the substantial water content of hydrogel-based flexible lenses renders them mechanically vulnerable. Therefore, hydrogel materials intended for ophthalmic use should possess sufficient strength to withstand implantation while retaining a substantial water content and remaining adequately soft and mechanically sturdy. Recent investigations have explored the reinforcement of hydrogels with CNCs and CNFs to address these challenges in ophthalmic applications. Findings from one study demonstrated that CNC-PVA clear hydrogel lenses, containing 90% water content, exhibited rubber-like mechanical properties, flexibility, elasticity, and the ability

**Corresponding author*: debabrata.ghoshdastidar@gnipst.ac.in

Mainak Mukhopadhyay and Debalina Bhattacharya (eds.) Nanocellulose: A Biopolymer for Biomedical Applications, (319–334) © 2024 Scrivener Publishing LLC

to maintain a convex shape. Moreover, these lenses displayed 95% transparency across the visible spectrum and possessed moderate UV absorption capabilities. This chapter aims to elucidate the emerging applications of nanocellulose in ophthalmic settings. Furthermore, it will delineate the existing challenges and propose strategies to overcome them.

Keywords: Nanocellulose, hydrogels, ophthalmic applications

12.1 Introduction

Several nanomaterials have recently been presented as viable replacements for conventional materials in various industries due to their unique features that are affected by their high surface-to-volume ratio. Typically, either physical or chemical processes result in nanoparticles. Both approaches produce nanoparticles with a reduced size, but at a considerable expense and by using hazardous chemicals. Biosynthesis of nanoparticles from biological sources such as enzymes, macromolecules from microorganisms, or phytonutrients from plants is much sought after for various applications, including medical and biological ones, to solve issues connected with toxicity and environmental footprints. Celluloses are acquiring great relevance as possible polymeric nanomaterials amid the wide range of biosynthesized nanomaterials. Cellulose, a natural polymer found in plant cell walls, vegetable fibers, and bacterial cell walls, is a soluble polysaccharide made up of glucose monomer chains. In recent years, nanocellulose has been created from cellulose polymers that were recovered from natural sources using mechanical, chemical, and biological techniques. This polymer, considered an infinite supply of raw ingredients, has the potential to be changed and functionalized for a variety of biological applications [1–3].

Nanocellulose hydrogels are highly hydrated, porous, and mechanically robust soft materials. These hydrogels may be produced from plant or bacterial cellulose. Nanocelluloses may be chemically changed with new functional moieties to improve their characteristics for regulating biological interactions. These hydrogels are used in various biomedical and pharmaceutical applications, such as drug delivery, diagnostics, scaffolds for tissue engineering, etc. [4–6]. Hydrogels have found utility in several ophthalmic contexts, such as in constructing soft contact lenses, developing folding soft corrective lenses, and administering drugs directly to the eye. Due to their high water content, hydrogels are pleasant to wear and offer lubricity, excellent oxygen permeability, and protein deposition resistance [2, 7, 8]. There have been several reports on the development of CNCs and CNF-reinforced hydrogels for ocular applications in recent years. This book chapter will discuss the structure, origin, separation, types, characteristics, and new uses of nanocellulose for ophthalmic applications. We will pay special attention to the existing issues and how to solve them.

12.2 Nanocellulose Overview

12.2.1 Structure, Source, and Isolation of Nanocellulose

The standard formula for carbohydrates is CH_2O, and Gay-Lussac and de Thenard discovered that it contains "45% carbon, 49% oxygen, and 6% hydrogen" in 1811. Anselme Payen, an agricultural scientist, used a systematic acid treatment approach to discover cellulose between 1837 and 1842, a few years after this discovery. He made this finding after seeing that cellulose in young plants had the same chemical make-up. The location of hydroxyl groups in glucose units inside cellulose has been revealed by chemical processes such as "hydrolysis, dimethyl sulfate, esterification, etherification, and metallic sodium reaction in liquid ammonia." In more detail, the scientists have found that the cellulose's glucose ring contains hydroxyl (OH) groups at positions C2, C3, and C6. Two secondary OH groups and one primary OH group are present in these positions, while a secondary OH group is present at the end of the C4 position [1, 9]. After extensive research, disputes in the 20th century were resolved with the aid of contemporary instrumental analysis. Since cellulose is made up of a chain of glucose molecules that have been organized in a ring, it typically takes the appearance of a flat ribbon. The repeating unit, also known as the -1,4-glucosidic bond, is composed of two glucose unit rings joined together by oxygen covalently bonded to C1 of one ring and C4 of the succeeding ring (1-4 linkage). The OH groups and oxygen of nearby ring molecules form intrachain hydrogen bonds to stabilize the connection, resulting in a linear structure for the cellulose chain. In order to make microfibrils, these chains are packed together to form microfibrils. The network of intra- and inter-chain hydrogen bonds in cellulose makes it a stable polymer and gives cellulose fibrils axial stiffness. The majority of cellulose chains $(C_6H_{10}O_5)_n$ [10–13] have a degree of polymerization (n) of between 10,000 and 15,000 units. The degree of polymerization and the length of the polymeric chain have a significant impact on the characteristics of cellulose (Figure 12.1). Native cellulose has both crystalline and amorphous (unordered) domains. The crystallinity may range from 40 to 70% depending on the natural source and extraction technique. Amorphous domains are less

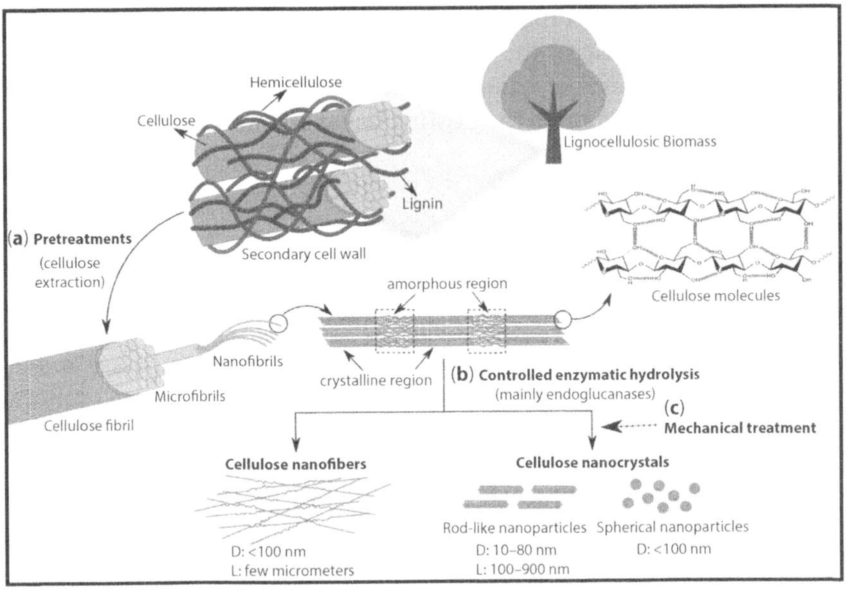

Figure 12.1 Schematic representation of the preparation of nanocellulose through enzymatic hydrolysis followed by mechanical treatment to produce nanocrystals of cellulose with different shapes and sizes. Diagram also represents the basic molecular structure of cellulose. Reprinted with permission from Ref. [13] as distributed by creative common license CC BY license, MDPI publisher.

dense than crystalline sections and are more prone to interact with other chemical groups. Comparatively to its amorphous form, cellulose's crystalline domains are usually more resistant to mechanical, chemical, and enzymatic processes. Numerous polymorphs of cellulose, including cellulose I, II, III, and IV, can exist. These polymorphs are influenced by molecule orientations, van der Waals forces, intra- and intermolecular interactions, and isolation and treatment methods. Chemical or thermal techniques can be used to change one cellulose polymorph into another [12, 14, 15]. Natural crystalline cellulose produced by diverse organisms is known as cellulose I. It can change into cellulose II or III and is thermodynamically stable. In actual application, cellulose II has proven to be the most stable substance. Regeneration (which entails solubilization and re-crystallization) and mercerization (treatments with aqueous sodium hydroxide) are the two methods that can produce it. By heating cellulose I or II with liquid ammonia, cellulose III can be created, and cellulose IV can be created with additional heat treatments. Different processing techniques can be used to produce nanocrystal cellulose (NCC), which can range in size from 100 nm to 3000 nm and is also influenced by the origin of the cellulose. Even smaller than 100 nm is the size of the NCC created from the microcrystals. The size of the cellulose crystals is mostly influenced by their origin and isolation technique [3–16].

A wide range of plant and animal biomass can be used to extract cellulose. A wide range of materials, including "wood pulp (softwood and hardwoods), vegetation, grass, field crops and their wastes, and animal, algal, bacterial, and waste paper," can be used to create cellulose. The many characteristics of cellulose may depend on its natural source, its age and origin, pretreatment, processing steps, and reaction circumstances. Because they are easy to gather, refine, and process, plant and wood pulps are highly sought-after cellulose sources [17, 18]. A complete list of plants from which cellulose can be extracted can be found in Table 12.1. Although wood pulp

Table 12.1 A comprehensive list of plants that may be utilized to extract cellulose.

Types of nanocellulose	Sources	Example
Nanocrystalline cellulose, nanofibril cellulose	Wood	Hardwood (Eucalyptus, Aspen, Balsa, Oak, Elm, Maple, and Birch) and softwood (Pine, Juniper, Spruce, Hemlock, Yew, Larch, and Cedar) pulp
	Non-wood plants	Sisal, Kenaf, Pineapple leaf fibers, Hemp, Flax, Bamboo, Sunflower stalks
	Agroforestry residues	Branches, Cotton stalks, Grain straw, Corn cob, Areca nut husk, Coir fibers, Peanut shells
	Agricultural wastes	Agave bagasse, Lemon seeds, Sugar beet pulp, Banana peels, Pinecone biomass, Potato peel, Garlic straw residues
	Animal	Chordata, Tunicates, *Styela clava*
Bacterial cellulose	Bacteria	*Salmonella*, *Acetobacter*, *Gluconacetobacter*, *Sarcina*, *Pseudomonas*, *Azotobacter*, *Agrobacterium*, *Rhizobium*, *Alcaligenes*, *Aerobacter*, *Rhodobacter*

is the most common source, cellulose can also be obtained from tunicates, algae, and bacteria. In the cell wall of many algae species, cellulose microfibrils are produced. The structure of the cellulose microfibrils in various algae species differs greatly because of changes in the production process. Similar to this, a wide variety of bacterial strains are added to the cellulose harvesting procedure. The mantle of a tunicate is made up of cellulose microfibrils wrapped in a protein matrix, making it the only animal whose cells can synthesize this type of cellulose [19–21].

The procedure entails two procedures to isolate cellulose particles from cellulose. To guarantee that the raw material responds consistently over the following stages, the first stage includes the purification and homogenization pretreatment. Later stages of the treatment for wood and plants involve partial or complete removal of matrix components like hemicellulose and lignin and isolation of individual complete fibers. For cellulose made from algae, the pretreatment procedures usually start with techniques for growing the algae, then purifying steps to get rid of the wall matrix components. These "purified" cellulose components are divided into their microfibrillar and/or crystalline parts in the second step. The three processes are mechanical treatment, acid hydrolysis, and enzyme hydrolysis [3, 22, 23]. Multiples of these techniques were really utilized consecutively or in concert with one another to create the desired particle form. In order to separate cellulose fibrils from WF, PF, MCC, tunicate, algae, and bacterial source materials, mechanical methods, such as "high-pressure homogenizers, grinders, refiners, cryocrushing, high-intensity ultrasonic treatments, and microfluidization," have been used (Figure 12.2). Typically, these reactions cause the longitudinal axis of the cellulose microfibrillar framework to transversely break, which releases long cellulose fibrils, also referred to

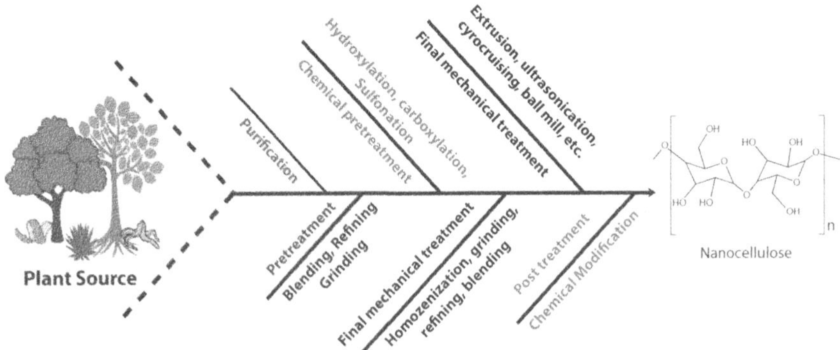

Figure 12.2 Fishbone diagram of isolation and preparation procedure of nanocellulose.

as microfibrillated cellulose. Acid hydrolysis has separated crystalline particles from several cellulose sources in the alternate technique. The initial components are mixed with water and then a certain concentration of acid in this process, for instance, hydrochloric acid, nitric acid, sulfuric acid, etc. The mixers are then washed and cleaned before being dialyzed. The nanoscale crystalline products were then obtained using an ultrasonic treatment [2, 24, 25].

12.2.2 Types and Properties of Nanocellulose

Cellulose can be divided into different categories based on the size and shape of its fibers, and this division determines the characteristics and uses of the material. Cellulose comes in a variety of forms, including cellulose nanoparticles, cellulose nanocrystals, nanofibrils, nanofibers, and bacterial cellulose, among others. Compared to other types, cellulose nanofibrils are employed more frequently. Varied types of nanocellulose have varied qualities that affect its viability and functionality. Nanocrystal cellulose is composed of cellulose crystals with dimensions of 5–70 nm in width and several hundred microns in length. The most common method for producing nanocrystal cellulose involves subjecting an amorphous and pure supply of cellulose to ultrasonic treatment, along with acid hydrolysis and heat-controlled procedures that remove some of the cellulose [26, 27]. As part of the cellulose fiber extract, certain sections of amorphous, hydrolyzed cellulose with improved crystallinity are generated using plant-derived nanocrystal celluloses. The second type of cellulose fibers, known as nanofibrils, are composed of fibrils with densely packed micrometer-length chains as well as crystalline and amorphous cellulose domains. At relatively low concentrations, long particles increase suspensions by increasing the viscosity of the aqueous medium. Several mechanical processes, including grinding, cryocrushing, and milling [28], are used to transform the wood microfibers into nanofibrils. In addition to mono- and polysaccharides, organic acids, and alcohol, the numerous bacterial microorganisms that are employed to make the bacterial celluloses offer a range of other carbon sources as well. Bacterial cellulose nanofibers are typically 20–100 nm in diameter, 3000–9000 times more polymerized, and 80–90% crystalline. It is produced via a process called fermentation that enables *in situ* cellulose shape, also referred to as bioshaping. In nutrient-rich aqueous media, a variety of microorganisms are frequently developed during the course of 1 to 2 weeks of incubation. These germs are mostly *Acetobacter G. xylinus*. Despite having many of the same chemical characteristics as plant cellulose, bacterial celluloses do not need to be pretreated to remove lignin and hemicellulose because they already have free functional sites. *In vivo* tissue scaffolds and pharmaceutical implants are its principal applications [1, 27, 29].

Due to its remarkable properties and distinct structural makeup, nanocellulose is perfect for a range of biomedical applications, including nanocomposite, nano/micro-implants, treatments, drug delivery, functional materials, etc. These properties can largely be divided into mechanical, chemical, and biological categories.

(a) Nanocellulose fibers' mechanical properties can be described using their mechanical strength. These fibrils are composed of crystalline and amorphous components in varying ratios. While the amorphous portion of the material adds plasticity and flexibility, the crystalline cellulose sections give the material elasticity and rigidity. As a result, the amount of these components plays a crucial role in shaping the material's characteristics. Its source and processing methods alone determined this crystallinity. Numerous studies have assessed the mechanical strength of nanocellulose using Young's modulus. It has discovered a range of elastic modulus in the axial direction of 110–220 gigapascal (GPa), which corresponds to around 25% of the strength of a carbon nanotube.

(b) The ability of distinct organic functional moieties to bind is distinguished by their primary chemical properties. Three hydroxyl groups are present on each of the continuous 1,4-D-glucose monomers that make up the polysaccharide chain of cellulose. Within the same sequence ("intramolecular hydrogen bonding") and with a different sequence ("intramolecular hydrogen bonding"), these groups allow the creation of powerful hydrogen bonds with the adjacent glucose unit, which most likely also contributed to the cellulose's reactive surface and provided a special platform for the creation of novel functional nanomaterials by chemical alterations like esterification, polymerization, and oxidation for a range of applications.

(c) Surface charges are yet another crucial chemical property of nanocellulose. The negatively charged sulfate ester (OSO-) groups on the surface of nanocrystalline cellulose are necessary for steady dispersion in an aqueous medium. During the sulfuric acid hydrolysis of cellulose, this unfavorable sulfate ester group is created. It was claimed, in particular, that the sulfated nanocellulose's decreased zeta potential value was caused by the surface's intensely negative charge. The nanocellulose dispersion demonstrated remarkable phase stability as the concentration of the negative sulfonic group rose. However, compared to nanocellulose produced by another acid hydrolysis, nanocellulose containing a sulfonic group is less thermostable.

(d) For biomedical applications, nanocellulose offers good biocompatibility. Due to the formulations' *in vivo* biocompatibility, numerous researchers have done a huge number of studies using nanocellulose-based materials. The biocompatibility of cellulose produced by bacteria is superior to that of all other kinds of cellulose.

(e) Another quality of nanocellulose is its low toxicity. Nanocellulose's minimal toxicity has been demonstrated by researchers through *in vivo* tests, opening up a wide range of biological possibilities.

(f) Another significant obstacle to the use of nanocellulose *in vivo* is its biodegradability. Numerous studies have demonstrated that nanocellulose does not completely break down within animals because of its dependence on the origin of nanocellulose for its crystallinity ratio. However, these non-biodegradable characteristics also offer a chance to unlock new biomedical benefits. This nanocellulose can be used to build scaffolds for implants, tissue regeneration, etc. These implants and scaffolds could offer the body or specific organs long-term support without producing any negative effects. According to certain research, effective nanocellulose modification slows down biodegradation.

12.3 Nanocellulose Compatibility for Ophthalmic Applications

Ocular disorders are a highly common problem in modern society, with numerous known and unknown causes leading to blindness and impaired vision. According to the World Health Organization, 2.2 billion individuals worldwide have nearsightedness or farsightedness, and half of these instances are preventable with the proper treatment. However, people of all ages may suffer from vision loss. The bulk of those who are blind or have vision problems are older than 50. The main causes of vision impairments include uncorrected refractive errors, age-related macular degeneration (AMD), diabetic retinopathy, cataracts, infections, glaucoma, and traumas [30, 31]. Various ocular preparations, including drugs and corrective lenses, may be used to prevent and cure these conditions. Corneal regeneration implants, eyeglasses, and contact lenses are primarily used for corrective vision. These contact lenses, called corneal regeneration implants, are used for corrective, cosmetic, and therapeutic purposes. Hard and soft contact lenses are the two primary categories of contact lenses. The low oxygen permeability of hard contact lenses led to undesirable clinical outcomes such as corneal hypoxia and different forms of swelling.

The development of hydrogel materials with a high water content and a soft texture may improve oxygen permeability and user comfort. Soft contact lenses available on the market can have a water content ranging from 40% to 75%, depending on the type of product. The oxygen permeability and comfort of these thin, soft contact lenses are excellent. Increasing the water content to decrease protein deposition is impossible since hydrogels are mechanically weak, affecting their usage and handling. High water content must be balanced with strength and flexibility. To overcome the water content versus mechanical strength problem, the hydrogel matrix could be reinforced with another mechanically strong component. Various materials might be utilized for strengthening, but the biggest problem is ensuring the finished product's optical characteristics match ophthalmic criteria. The ideal material among them is nanocellulose [8, 32, 33]. Due to the presence of many hydrogen bonds and unique properties in NC, hydrogels are softer, mechanically more robust, water-absorbent, water-retentive, and medically essential. Using NC with other synthetic polymers (e.g., PVA) makes it more beneficial for ophthalmic preparations.

12.4 Nanocellulose Hydrogel

In the domains of drug delivery and regenerative medicine, ophthalmic hydrogels, like other biomaterials, must fulfill particular parameters to guarantee biocompatibility, non-toxicity, non-inflammation, non-immunogenicity, biodegradability, stability, and compliance with natural tissues. The hydrogel development must occur in a safe environment devoid of hazardous compounds, heat, or protons and under physiological temperature and ionic strength parameters. The hydrogel's qualities may be varied and are intended to closely replicate biological matrices via high hydrophilicity, high hydration, and closely matched mechanical properties. These properties help keep biologic pharmaceuticals in their physicochemical condition for longer periods, particularly when contrasted to more rigid, hydrophobic polymer matrices employed for drug encapsulation and release [34–36]. Hydrogels are polymeric substances that may hold substantial water inside their structure. Several variables may influence the physical characteristics of hydrogels, including the kind of polymers employed, the synthesis technique, and the water content contained in the gel. Hydrogels can expand to several times their original volume in reaction to environmental changes such as temperature, pH, or the presence of specific ions. In addition, their elastic and viscous characteristics are similar to those of soft tissues, making them excellent for regenerative medicine and drug administration applications. Often, hydrogels are translucent, which is especially helpful for applications requiring clarity, such as contact lenses and artificial corneas. The adhesion capabilities of hydrogels vary depending on the type of polymeric materials and fabrication methods utilized, with some adhering strongly to tissues and others less so. In addition, the biodegradability of hydrogels may vary depending on the polymers and synthesis processes used, with certain hydrogels degrading in the body over time while others are more durable. Overall, the physical features of hydrogels make them a flexible material that may be used for several applications in drug administration and regenerative medicine [37–39]. Due to their unique features and benefits over typical hydrogels, nanocellulose-based hydrogels can potentially revolutionize ophthalmic applications. Among the uses for nanocellulose-based hydrogels in ophthalmic applications are as follows:

- Nanocellulose-based hydrogels exhibit enhanced mechanical characteristics, including higher stiffness and tensile strength. This may make them more appropriate for applications needing significant mechanical strength, such as corneal restoration.
- Nanocellulose is biodegradable, making nanocellulose-based hydrogels a more sustainable alternative for ophthalmic applications requiring biodegradable materials.
- Nanocellulose is a biocompatible substance that the human body tolerates well, making it suitable for use in ophthalmic applications where biocompatibility is crucial.
- Controlled drug release: The unique structure of nanocellulose-based hydrogels may allow for the controlled release of pharmaceuticals, making them a good candidate for drug delivery applications.
- Reduced inflammation: Nanocellulose-based hydrogels have been proven to reduce inflammation compared to standard hydrogels, making them a suitable choice for applications where lowering inflammation is critical, such as in glaucoma filtration devices.

The potential benefits of nanocellulose-based hydrogels over conventional hydrogels in ophthalmic applications make them desirable for more study and development in this area [34, 40]. Specific requirements must be met during manufacture to assure the safety and effectiveness of ophthalmic nanocellulose-based hydrogels for ocular applications. Among these requirements are as follows:

- Biocompatibility: The hydrogel must be biocompatible, which means that it should not produce an allergic response or inflammation when touching the eye or other ocular tissues.
- Drug delivery efficiency: If the hydrogel is utilized for medication delivery, it must release the drug efficiently over a regulated period.
- Mechanical qualities: To provide enough support and avoid extrusion or movement of the hydrogel inside the eye, the mechanical properties of the hydrogel must match or be near those of the ocular tissues.
- Transparency: The hydrogel must be transparent to enable normal eyesight without altering the cornea's clarity.
- Sterilization: To eradicate potentially hazardous germs, the hydrogel must be sterilized before use.
- Long-term stability: The hydrogel must be stable over time while retaining its mechanical and drug delivery capabilities.
- Non-toxicity: The hydrogel must be non-toxic to ocular tissues and not produce adverse effects.

These parameters must be met to ensure the safety and effectiveness of ophthalmic nanocellulose-based hydrogels in various ocular applications such as medication administration, corneal healing, and glaucoma filtration devices [41–43].

12.5 Ophthalmic Application of Nanocellulose

Contact lenses do play an essential role in modern eye care. Since they are lightweight and virtually invisible, they are popularly used as corrective vision lenses. Contact lenses should have optimum mechanical properties, oxygen permeability, and water content. The hard contact lens has low oxygen permeability and causes corneal hypoxia and edema. Moreover, the life span of such lenses is low due to the deposition of proteinaceous materials. An increase in water content enhances the oxygen permeability and softness of the lens and keeps the lens clean. However, a high water content makes the lens mechanically weak. The conflict between intrinsic water content and mechanical strength can be solved by reinforcing the hydrogel material with mechanically strong components without affecting the lens's optical properties. Cellulose nanofibers derived from 2,2,6,6-tetramethylpiperidine-1-oxyl radical (TEMPO)-mediated oxidation of native cellulose (e.g., microcrystalline cellulose) are most commonly used for this purpose [44–46].

Polyvinyl alcohol (PVA) and PVA/chitin-based composite hydrogels are famous for preparing soft contact lenses and artificial corneas (keratoprosthesis). PVA hydrogel is obtained by quenching a solution of PVA to -20 °C. The molecular weight, degree of deacetylation, concentration of PVA, composition of solution (water and DMSO ratio), and quenching temperature determine the quality of PVA hydrogel [45]. Tummala *et al.* reported the ophthalmic use (contact lens) of TEMPO-oxidized nanocellulose-reinforced PVA hydrogel having >90% water content [46]. The matrix of the lens consisted of an open and macroporous honeycomb structure. The lens had good elasticity. It could be manually pulled without rupture and readily returned to its original form without deformation upon release [46]. Though it had 95% transparency in the visible range, the UV radiation blocking capability was good enough [46]. Later, they varied the aspect ratio of nanocellulose materials (cellulose nanocrystals—CNCs, cellulose nanofibrils—CNFs) and PVA to get the hydrogel with optimum physicochemical properties [45]. Both CNC- and CNF-reinforced hydrogel lenses had a refractive index close to that of water and mechanical strength suitable for easy handling. However, the CNF-reinforced optimized hydrogels had 60% transparency for UV-A (320–400 nm) and 50% transparency for UV-B (290–320 nm). In contrast, the CNC-reinforced hydrogels had 80% transparency for UV-A and 60% transparency for UV-B radiation. Thus, CNF-reinforced gels had better UV radiation scatter properties. Moreover, these nanocellulose-reinforced PVA hydrogels had excellent biocompatibility with human corneal epithelial cells [33].

12.6 Conclusion and Future Aspect

Nanocelluloses such as nanocrystal cellulose (CNC), nanofibril cellulose (CNF), and bacterial nanocellulose have distinctive qualities such as low toxicity, biocompatibility, biodegradability, and improved binding capacity. These qualities make them potentially valuable materials for ophthalmic applications. However, only a few research groups have investigated their applications in ophthalmology. They have been used in composition with PVA and chitosan-based hydrogel for the preparation of contact lenses with optimum mechanical strength. There may be possibilities of using nanocellulose as drug carriers for sustained drug release in cul-de-sac. They may be investigated for the development of surgical sutures and medical devices for the eye. The scientists in the relevant field should explore these opportunities and develop suitable preclinical models and regulatory guidelines.

References

1. Sheikhi, A., Emerging cellulose-based nanomaterials and nanocomposites, in: *Nanomaterials and Polymer Nanocomposites*, Amsterdam, Netherlands, Oxford OX5 1GB, United Kingdom, Cambridge, MA, United States pp. 307–351, Elsevier, 2019.
2. Trache, D. *et al.*, Nanocellulose: From fundamentals to advanced applications. *Front. Chem.*, 8, 392, 2020.
3. Moon, R.J., Martini, A., Nairn, J., Simonsen, J., Youngblood, J., Cellulose nanomaterials review: Structure, properties and nanocomposites. *Chem. Soc. Rev.*, 40, 3941–3994, 2011.
4. Plackett, D., Letchford, K., Jackson, J., Burt, H., A review of nanocellulose as a novel vehicle for drug delivery. *Nord. Pulp Pap. Res. J.*, 29, 105–118, 2014.
5. Tan, T.H., Lee, H.V., Yehya Dabdawb, W.A., Abd Hamid, S.B.B.O.A., A review of nanocellulose in the drug-delivery system. *Mater. Biomed. Eng. Nanomater. Drug Deliv.*, 131–164, 2019, https://doi.org/10.1016/B978-0-12-816913-1.00005-2.
6. Seddiqi, H. *et al.*, Cellulose and its derivatives: Towards biomedical applications. *Cellulose*, 28, 1893–1931, 2021.
7. Shojaeiarani, J., Bajwa, D., Shirzadifar, A., A review on cellulose nanocrystals as promising biocompounds for the synthesis of nanocomposite hydrogels. *Carbohydr. Polym.*, 216, 247–259, 2019.
8. Tummala, G.K. *et al.*, Hyperelastic nanocellulose-reinforced hydrogel of high water content for ophthalmic applications. *ACS Biomater. Sci. Eng.*, 2, 2072–2079, 2016.
9. Hon, D.N.S., Cellulose: A random walk along its historical path. *Cellulose*, 1, 1–25, 1994.
10. Nishiyama, Y., Structure and properties of the cellulose microfibril. *J. Wood Sci.*, 55, 241–249, 2009.
11. Azizi Samir, M.A.S., Alloin, F., Dufresne, A., Review of recent research into cellulosic whiskers, their properties and their application in nanocomposite field. *Biomacromolecules*, 6, 612–626, 2005.
12. Klemm, D., Heublein, B., Fink, H.P., Bohn, A., Cellulose: Fascinating biopolymer and sustainable raw material. *Angew. Chem. Int. Ed.*, 44, 3358–3393, 2005.

13. Michelin, M., Gomes, D.G., Romaní, A., de Polizeli, M. L. T., M., Teixeira, J.A., Nanocellulose production: Exploring the enzymatic route and residues of pulp and paper industry. *Molecules*, 25, 15, 3411, 2020.
14. Thakur, V.K., *Lignocellulosic polymer composites: Processing, characterization, and properties*, vol. 9781118773, Wiley Blackwell, Scrivener Publishing, Beverly, MA, 2014.
15. Thakur, V.K., *Nanocellulose polymer nanocomposites: Fundamentals and applications*, vol. 9781118871, Wiley Blackwell, Scrivener Publishing, Beverly, MA, 2014.
16. Ishikawa, A., Okano, T., Sugiyama, J., Fine structure and tensile properties of ramie fibres in the crystalline form of cellulose I, II, IIII and IVI. *Polym. (Guildf)*, 38, 463–468, 1997.
17. Klemm, D. *et al.*, Nanocellulose as a natural source for groundbreaking applications in materials science: Today's state. *Mater. Today*, 21, 720–748, 2018.
18. Van De Ven, T.G.M. and Sheikhi, A., Hairy cellulose nanocrystalloids: A novel class of nanocellulose. *Nanoscale*, 8, 15101–15114, 2016.
19. Kim, N.H., Herth, W., Vuong, R., Chanzy, H., The cellulose system in the cell wall of micrasterias. *J. Struct. Biol.*, 117, 195–203, 1996.
20. T.I., and Sugiyama, J., Nanodomains of Iα and Iβ cellulose in algal microfibrils. *Macromolecules*, 31, 6275–6279, 1998.
21. Huang, Y. *et al.*, Recent advances in bacterial cellulose. *Cellulose*, 21, 1–30, 2014.
22. Trache, D., Hussin, M.H., Haafiz, M.K.M., Thakur, V.K., Recent progress in cellulose nanocrystals: Sources and production. *Nanoscale*, 9, 1763–1786, 2017.
23. Nechyporchuk, O., Belgacem, M.N., Bras, J., Production of cellulose nanofibrils: A review of recent advances. *Ind. Crops Prod.*, 93, 2–25, 2016.
24. Park, N.M., Choi, S., Oh, J.E., Hwang, D.Y., Facile extraction of cellulose nanocrystals. *Carbohydr. Polym.*, 223, 115114, 2019.
25. Agbor, V.B., Cicek, N., Sparling, R., Berlin, A., Levin, D.B., Biomass pretreatment: Fundamentals toward application. *Biotechnol. Adv.*, 29, 675–685, 2011.
26. Jeevanandam, J. *et al.*, Nanocelluloses: Sources, types, unique properties, market, and regulations, in: *Handbook of Nanocelluloses*, pp. 3–34, Springer, Cham, 2022.
27. Gatenholm, P. and Klemm, D., Bacterial nanocellulose as a renewable material for biomedical applications. *MRS Bull.*, 35, 208–213, 2010.
28. Abitbol, T. *et al.*, Nanocellulose, a tiny fiber with huge applications. *Curr. Opin. Biotechnol.*, 39, 76–88, 2016.
29. Reiniati, I., Hrymak, A.N., Margaritis, A., Recent developments in the production and applications of bacterial cellulose fibers and nanocrystals. *Crit. Rev. Biotechnol.*, 37, 510–524, 2017.
30. Bourne, R.R.A. *et al.*, Causes of blindness and vision impairment in 2020 and trends over 30 years, and prevalence of avoidable blindness in relation to VISION 2020: The right to sight: An analysis for the global burden of disease study. *Lancet Glob. Health*, 9, e144–e160, 2021.
31. Bourne, R.R.A. *et al.*, Causes of blindness and vision impairment in 2020 and trends over 30 years, and prevalence of avoidable blindness in relation to

VISION 2020: The right to sight: An analysis for the global burden of disease study. *Lancet Glob. Health*, 9, e144–e160, 2021.
32. Li, W. *et al.*, Synthesis and characterization of pH-responsive PEG-poly(β-amino ester) block copolymer micelles as drug carriers to eliminate cancer stem cells. *Pharmaceutics*, 12, 111, 2020.
33. Tummala, G.K., Lopes, V.R., Mihranyan, A., Ferraz, N., Biocompatibility of nanocellulose-reinforced PVA hydrogel with human corneal epithelial cells for ophthalmic applications. *J. Funct. Biomater.*, 10, 35, 2019.
34. Chang, D., Park, K., Famili, A., Hydrogels for sustained delivery of biologics to the back of the eye. *Drug Discovery Today*, 24, 1470–1482, 2019.
35. Kirchhof, S., Goepferich, A.M., Brandl, F.P., Hydrogels in ophthalmic applications. *Eur. J. Pharm. Biopharm.*, 95, 227–238, 2015.
36. Aghamirsalim, M. *et al.*, 3D printed hydrogels for ocular wound healing. *Biomedicines*, 10, 1562, 2022.
37. Chen, G. *et al.*, Applications of hydrogels with special physical properties in biomedicine. *Polymers*, 11, 1420, 2019.
38. Kaczmarek, B., Nadolna, K., Owczarek, A., The physical and chemical properties of hydrogels based on natural polymers, in: *Hydrogels Based on Natural Polymers*, pp. 151–172, Elsevier, Radarweg, AE Amsterdam, Netherlands, The Boulevard, Langford Lane, Kidlington, Oxford OX5 1GB, United Kingdom, Cambridge, MA, United States, 2019.
39. Lynch, C.R. *et al.*, Hydrogel biomaterials for application in ocular drug delivery. *Front. Bioeng. Biotechnol.*, 8, 228, 2020.
40. He, P. *et al.*, Nanocellulose-based hydrogels as versatile drug delivery vehicles: A review. *Int. J. Biol. Macromol.*, 222, 830–843, 2022.
41. Luo, Y., Li, G., Chen, L., Hong, F.F., Preparation and evaluation of bacterial nanocellulose/hyaluronic acid composite artificial cornea for application of corneal transplantation. *Biomacromolecules*, 24, 201–212, 2023.
42. Anton-Sales, I. *et al.*, Bacterial nanocellulose as a corneal bandage material: A comparison with amniotic membrane. *Biomater. Sci.*, 8, 2921–2930, 2020.
43. Anton-Sales, I., Koivusalo, L., Skottman, H., Laromaine, A., Roig, A., Limbal stem cells on bacterial nanocellulose carriers for ocular surface regeneration. *Small*, 17, 2003937, 2021.
44. Saito, T., Kimura, S., Nishiyama, Y., Isogai, A., Cellulose nanofibers prepared by TEMPO-mediated oxidation of native cellulose. *Biomacromolecules*, 8, 2485–2491, 2007.
45. Tummala, G.K., Rojas, R., Mihranyan, A., Poly(vinyl alcohol) hydrogels reinforced with nanocellulose for ophthalmic applications: General characteristics and optical properties. *J. Phys. Chem. B*, 120, 13094–13101, 2016.
46. Tummala, G.K. *et al.*, Hyperelastic nanocellulose-reinforced hydrogel of high water content for ophthalmic applications. *ACS Biomater. Sci. Eng.*, 2, 2072–2079, 2016.

13

Application of Nanocellulose as Implant and Grafting Materials

Shubhangi Paliwal* **and Anushka Lather**

Department of Bio Sciences and Biotechnology, Banasthali Vidyapith, Niwai, Tonk, Rajasthan, India

Abstract

Nanocellulose, a novel and advantageous natural substance, is derived from cellulose. Nanocellulose has gained considerable popularity for application as a very important material in the field of biomedicine due to its extraordinary properties, namely, its unique surface chemistry, physical features, and good biological capabilities (mainly biocompatibility, low toxicity, and biodegradability). The research and study of nanocellulose, when done in the last few years, has consequently resulted in the successful creation of sophisticated nanofibers from cellulose, which have immense and wonderful properties and various functional as well as morphogenetic characteristics. Surface alterations, photonics, films and foams, nanocomposites, and devices used in medicine are the chief fields of nanocellulose research because numerous applications, which include flexible optoelectronics and tissue regeneration scaffolds, are possible for these small nanocellulose fibers. Owing to the increase in interest in tissue-engineered products for both wound care and the regeneration of injured or ill organs, nanocellulose has lately attracted a lot of attention. Cellulose nanocrystals (CNCs), cellulose nanofibrils (CNFs), and bacterial cellulose (BC) are the three different kinds of nanocellulose derived from native cellulose, a polymer that is intriguing for use in tissue engineering as implants and scaffolds. From plant cell walls, cellulose nanocrystals and nanofibrils can both be obtained. Bacteria like *Acetobacter xylinum* generally synthesize bacterial cellulose (BC) or microbial cellulose (MC) in pure form. Due to its superior mechanical qualities and biocompatibility, nanocellulose shows potential as substitutes or medical biomaterials such as soft tissue—ligament, meniscus, and cartilage replacements; nucleus pulposus replacement; blood vessel

**Corresponding author*: shubhangipaliwal0910@gmail.com

Mainak Mukhopadhyay and Debalina Bhattacharya (eds.) Nanocellulose: A Biopolymer for Biomedical Applications, (335–358) © 2024 Scrivener Publishing LLC

replacement; and antimicrobial nanomaterials, but most important of all is its vital role as advanced nanomaterials for tissue repair, regeneration, and healing, which includes skin tissue repair and wound healing, and bone tissue regeneration and healing, which will be discussed further in this chapter in detail.

Keywords: Nanocellulose, biocompatibility, biodegradability, biomedical applications, nanomaterials, regeneration

13.1 Introduction

Materials made from naturally renewable resources, like cellulose, alginate, gelatin, chitosan, starch, elastin, and collagen, have gained significant popularity in the biomedicine field recently for use in a variety of applications, such as biological detections, scaffolds for the engineering of tissue, vascular grafts, implants, cancer targeting, vascular grafts, drug delivery, and cornea replacement [1–4]. The biosphere's most prevalent organic compound is cellulose, an intriguing biopolymer. Numerous living things, including bacteria, higher plants, fungi, and sea creatures, are able to synthesize it biologically. Being regarded as a source of endless feedstock, it is of utmost use as a product that is environmentally friendly as well as biocompatible. Natural regeneration occurs often and in a short amount of time; bioproduction is predicted to reach around 7.5 × 1010 metric tons every year [4, 5]. This material is an alluring one that is renewable and is used for its applications in cutting-edge industries due to the rising demand for materials that have a high performance with specialized physical qualities and mechanical properties [6]. These natural polymers have been researched for use in everything from items that cannot be implanted, like wound dressings, to items like vascular grafts, which are implantable. Biomedical materials made out of cellulose, which is the most abundant as well as the most renewable resource of polymers, are currently available and have remarkable mechanical characteristics, biodegradability, and low cytotoxicity [7]. "Nanocellulose" refers to processed materials or cellulosic extracts having defined structural dimensions on the nanoscale, and it is the end result of biosynthesis obtained from microbes, plants, and animals [6]. A homopolymer of carbohydrates called cellulose includes in it the glucopyranose units joined by 1,4-glycosidic linkages [8]. Although very insoluble and inelastic, cellulose fibrils are able to provide mechanical support to the tissues at the location where they are found [9]. Three different kinds of nanocellulose exist: cellulose nanofibrils (CNFs), also known by other names as nanofibrillated cellulose (NFC), cellulose nanocrystals (CNCs), nanocrystalline cellulose (NCC), cellulose nanowhiskers (CNWs), and bacterial cellulose (III) (BC). When nanoparticles are extracted using various techniques, the resultant particles have different surface chemistries, crystallinities, and mechanical properties [6]. Advantageous traits of cellulose have been used recently in research and development, including areas of environment, biomedicine, water, and energy-related domains. The presented chapter focuses on the utilization of cellulose in the sector of biomedicine, keeping in mind specifically the developments that are most current with the help of references related to the implant and grafting applications of nanocellulose.

13.2 Chemical as Well as the Structural Makeup of Cellulose Fibers

Being a homopolymer, which is natural with a high molecular weight, cellulose is made up of anhydro-D-glucose units linked by a 1, 4, or 5 linkage which can be seen from the Figure 13.1. By joining the H and OH groups and getting rid of water, a connection occurs between the units of sugar as descripted with the help of the Figure 13.2. A disaccharide, cellobiose, is created by joining two of these sugars [10]. The molecules of glucose are organized into pyranoses, or 6-membered rings. In one of the pyranose rings, a C1-acetal connection is found by using a single oxygen atom to the C-4 of the other ring. The cellulose polymer containing glucose units is known as anhydroglucose units because the water molecule is eliminated during this reaction, which is also participated in by an alcohol and a hemiacetal, resulting in an acetal [11]. Three distinct anhydroglucose units are present in the molecule of cellulose: rings in the interior are joined at C1 and C4, and the reducing end has a free hemiacetal (or aldehyde) group at C-1 [12]. The glucose molecule's carbons 2, 3, 4, and 5 all have fixed stereochemistry. This is not the case in the pyranose structure, as it has two alternative stereochemistry because the hydroxyl, which is present at carbon number 4, approaches and forms a bond with the carbonyl at carbon number 1 from both sides. Hence, the hydroxyl group of carbon at position 1 is said to be in the configuration if it comes from sides of the ring similar to the ring of carbon at position 6. The one numbered carbon, or C-1, in cellulose has an orientation that is opposite in nature and thus gives rise to the configuration that is referred to as the β configuration [13]. The molecular chain of cellulose is aligned in a line that is almost straight because the functional groups are at equatorial positions which is represented in the Figure 13.3 [11].

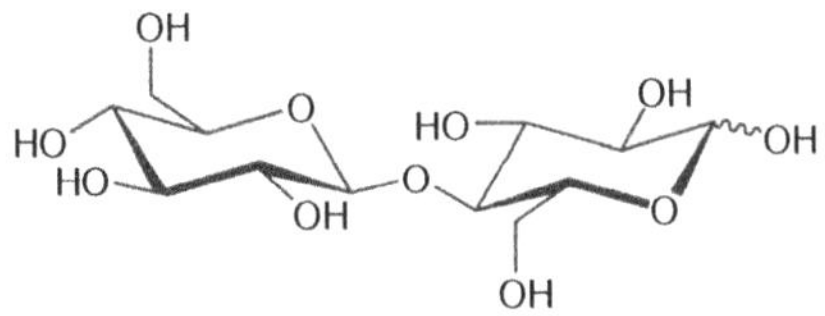

Figure 13.1 Cellulose unit [10].

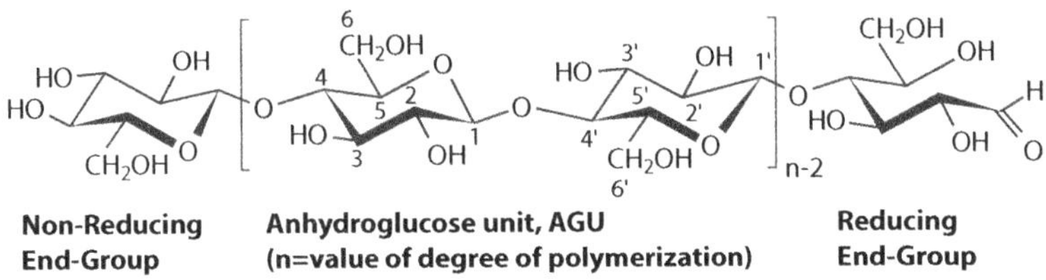

Figure 13.2 Molecular makeup of cellulose [10].

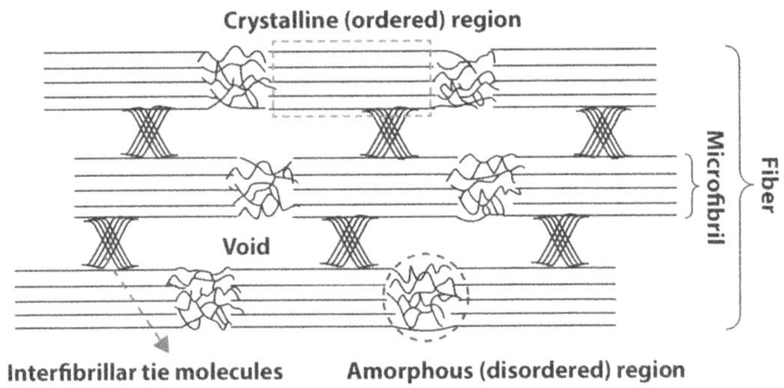

Figure 13.3 Schematics of an idealized cellulose fiber in Figure 13.3 depict one possible arrangement of the crystalline and amorphous areas [7].

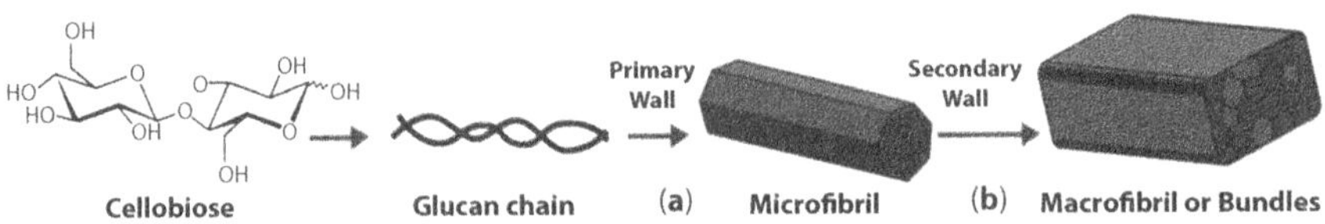

Figure 13.4 Individual chains of glucose residues are polymerized into (a) microfibrils and then (b) bundles or commonly called macrofibrils (redrawn from [10]).

The above Figure 13.4 shows the cellulose, in nature, does not exist as an isolated single molecule. The molecules are spun in a hierarchical order at the site of biosynthesis and hence finally produce cellulose, which is now formed as an individual molecule. Plants' main cell walls are made of individual chains of polymerized glucose residues. A microfibril is created later by joining 36 chains. Microfibrils frequently further associate and turn into the structures known as bundles at the time when the plant's secondary wall is formed. The chain length can vary, though, and can be anywhere between 2000 and 20,000 glucose residues [10]. An intra- and intermolecular hydrogen-bond network that is both powerful and extremely complex stabilizes the crystallites of cellulose that are firmly packed together in the ordered zones. It is possible for cellulose to exist in a variety of polymorphs or allomorphs depending on the bonding of hydrogen, which then forms the network and finally the molecular orientation. They are I, II, IIII, IIIII, IVI, and IVII, which are cellulose's six interconvertible polymorphs. While this is happening, the amorphous regions are dispersed along the elementary fibril as chain dislocations, where the microfibers are twisted and tilted due to internal fiber strain [14]. It has been assumed that native cellulose possesses the cellulose number I form.

13.3 Cellulose—Origin, Extraction, and Sources

Numerous sources are used to obtain cellulose. The traditional and main source is derived from higher-plant wood. The agricultural waste products, for example, empty fruit bunches, corncobs, wheat rice, wheat straw, corn stover, sugarcane bagasse, and coconut husks, are further sources of cellulose [15]. Aside from being derived from plants, the fibers of cellulose are also obtained from the extracellular surfaces of some of the bacteria from the genera *Sarcina*, *Acetobacter*, *Rhizobium*, *Agrobacterium*, *Pseudomonas*, and *Alcaligenes* [16]. The matrix formed by the hemicellulose, which is a non-cellulosic amorphous polysaccharide along with the lignin, has microfibrils of the cellulose embedded in it, where wood contains plant cellulose in the tissues of the xylem [17]. Lignin acts as a stiffener, preserving water in fibers, which in turn helps make the stem of the plant resistant to the stresses of gravity and the wind and offers defense against biological assault. It is thought that cellulose and lignin are more compatible when hemicellulose is present [18]. Hemicellulose and lignin are found in PC, making it less desirable for use in biological applications. In fact, lignin would not biodegrade in the body's environment, and it would be difficult to predict what would happen if it stayed there for a period of time that was too long. Furthermore, the leftover compounds, along with their metabolites, can also have negative effects on the body because of the very strong chemicals used in the removal of hemicellulose and lignin compounds.

Besides plants, microorganisms also produce cellulose, mainly bacterial cellulose, or BC. Although bacterial cellulose and plant cellulose are chemically similar, their macromolecular structure and arrangement vary, and as a result, their characteristics are different [19]. The FDA has declared BC to be "generally recognized as safe". Because BC lacks lignin, hemicellulose structures, and sulfur and requires no chemical manufacture or processing, it is harmless. This is because it is free of these substances. Additionally, purifying techniques enable the production of BC with an endotoxin content that is within the FDA-recommended range for biomedical uses [20]. The smallest naturally occurring fibers have measurements of about 1.5 nm in width after BC aggregates after their production, After that, BC subfibrils solidify into bundles and finally into ribbons [21]. The dimensions of BC ribbons have been measured and reported by a number of researchers. The research has stated that the ribbons had dimensions of 3.2–133 nm and 4.1–117 nm, respectively [22–24]. According to some researchers, these ribbons (also known as fibrils) can reach dimensions of between 10 and 100 nm in width and almost 10 m in length, which depends totally on the environment as well as the type of culture medium used in the process [9, 25–27]. These ribbons obtained from the bacterial cellulose form a structure that is dense as well as reticulated and is stabilized with the help of strong bonding, which is generally hydrogen bonding.

Additionally, plant cellulose is generally less crystalline than bacterial cellulose [21], which is another difference. Additionally, unlike plant cellulose, which requires chemical treatments to remove lignin and hemicelluloses, cellulose obtained from bacteria is hydrated to a higher extent and has a relatively pure form of cellulose in the membrane [28, 29]. The most significant enzyme in bacterial cellulose synthesis is cellulose synthase, and cyclic diguanylate monophosphate, also called c-di-GMP, serves it by activating it. A biological process revealed that glucose could be used to produce cellulose. Glucose and sucrose are primarily involved in the bacterial cellulose manufacturing process as the two major sources of carbon. However, a variety of different carbon substrates, including fructose, arabitol, mannitol, and glycerol, can be used to produce cellulose. In a recent study, it was found that after 2–4 days of culture, glucose with a generally lower concentration (less than 15 g/L) might yield cellulose, which weighs about 0.6 g when 1 g of glucose is used [30–34]. In addition to the carbon sources mentioned before, *Acetobacter* uses four carbon sources, which are gluconate, lactate, citrate, and gluconolactone [21].

13.4 Isolation, Chemical Transformations, and Purification of Cellulose

13.4.1 Nanofibrillar Cellulose (NFC)

As was previously noted, fibers that are long, thin filaments and are mechanically strong are seen in the case of native cellulose. These alternate between domains that are crystalline and non-crystalline, and they are generally called microfibrils. Nanofiber cellulose, or NFC, which is employed in the applications of biomedicine, can be produced from microfibrils. To obtain NFC, a variety of techniques involve mechanical shearing, and along with this, a mix of techniques might be used that involve mechanical as well as chemical phenomena. For instance, research and studies by Surip *et al.* [35] created nanofiber cellulose, or NFC, from kenaf fiber that was treated first by the alkali and then followed by the treatment with an acid that is mainly hydrochloric acid, prior to being processed using pulverization mechanically along with the cryo-crushing. These NFCs ranged in diameter from 40 to 200 nm. Generally speaking, nanofiber cellulose is made from cellulose with a high aspect ratio (that is, the ratio of length to width). The width of the fibril ranges from 20 to 40 nm, with several micrometers in length [36]. Bacterial cellulose is a form of cellulose that is very pure; unlike plant cellulose, it does not require chemical processing to get rid of lignin and hemicelluloses. Because of these factors, BC is frequently preferred over cellulose derived from other sources. However, there are indications that BC may have an irregular structural shape and may be challenging to encapsulate compounds, making plant-derived cellulose preferable [37].

13.4.2 Cellulose Nanocrystal or CNC

Hydrolysis with the help of an acid that is strong causes cellulose, be it macro or microscopic in nature, to transversely cleave along the regions that are amorphous, producing cellulose nanocrystal, or CNC, a rod-like substance [38]. CNCs typically have a diameter of 2 to 20 nm and a length widely distributed among the ranges of 100 to 600 nm as can be seen from the Figures 13.5 and 13.6 [39]. Due to its special characteristics, such as its high surface area (it has a surface area of several hundred meters per gram) [40], low density (which is almost 1.61 g/cm^3) [21], availability, and biodegradability, along with the fact that it is renewable, CNC has gained not only popularity but also a remarkable amount of interest in the research and development of nanocomposites. CNC is an attractive and highly promising choice that can be utilized as a green reinforcing material for

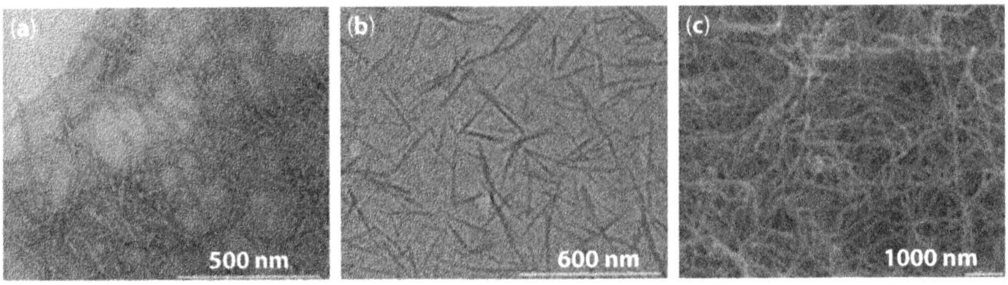

Figure 13.5 Nanocellulose typically exists as the cellulose nanofibrils, CNC, or BC showing cellulose nanofibrils of length (a) 500nm (b) 600 nm (c) 1000nm [73].

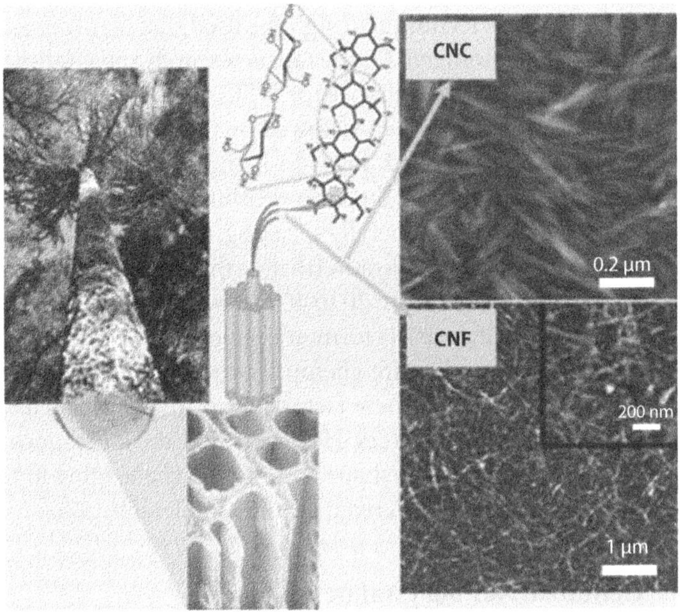

Figure 13.6 Schematic representations of cellulose nanocrystals and nanofibrils [77].

composites due to its outstanding mechanical strength and also having a Young's modulus of about 150 GPa [41–43]. For the preparation of cellulose nanocrystals, mainly acids like HCl or hydrochloric acids as well as sulfuric acids have been widely utilized. Utilizing two distinct acids results in CNC with various characteristics. CNC has a limited ability to diffuse in aqueous media and has a propensity to flocculate if it is created by hydrolysis in hydrochloric acid [44]. Sulfuric acid, on the other hand, encourages the mixing or interaction of cellulose nanocrystals in water, as a result of an interaction between the OH groups present on the cellulose surface and

Table 13.1 Types, common names, sources, size and formation of nanocellulose materials.

Types of nanocellulose	Similarly called as-	Common sources	Size and formation
Cellulose Nanocrystals (CNC)	• Rodlike cellulose microcrystals. • Whiskers. • Crystallites. • Nanocrystalline cellulose.	• Flax • Hemp • Wood • Cotton • Mulberry • Wheat Straw • Ramie • Bark • Tunicin • Avicel • Cellulose from bacterial and algae.	• Acid hydrolysis of cellulose. • Diam-5-70nm. • Length-100-250nm.
Cellulose Nanofibrils (CNF)	• Nanofibrillated cellulose. • Microfibrils.	• Potato tuber • Sugar beet • Wood • Flex • Hemp	• Delamination of wood pulp by mechanical pressure. • Diam-5-60nm. • Length-several micrometers.
Bacterial Cellulose (BC)	• Biocellulose. • Microbial cellulose. • Bacterial Cellulose.	Sugars and alcohols of low molecular weight.	• Bacterial synthesis. • Diam-20-100nm. • Different types of nanofibre networks.

the sulfate esters, which have charged surfaces. The thermostability of the nanocrystals is jeopardized by the addition of charged sulfate groups [45]. The rheological behavior of the two acids is another difference that has been noted. The suspensions produced by the hydrolysis of sulfuric acid do not exhibit viscosity, which is entirely dependent on time. Contrarily, the suspension in the hydrochloric acid displayed anti-thixotropic behavior below 0.3% of concentration and thixotropic behavior over 0.5% (w/v) of concentration [44]. Table 13.1 summarizes how changes in temperature, hydrolysis duration, and acid concentration result in CNC with a variety of morphologies and properties. Cellulose nanocrystals have the ability to be employed in industries that use structural composites to create reinforcement materials. Therefore, numerous attempts have been made to include CNC in the field of biomedicine in order to enhance the mechanical characteristics of hydrogel utilized for drug administration [46, 47].

13.4.3 Microcrystalline Cellulose (MCC)

During spray-drying methods, the hydrogen bonds that are strongly formed between cellulose nanocrystals, or CNC, encourage aggregation again [48]. As a result of this process, microcrystalline cellulose (MCC), another type of cellulose structure, is created. MCC typically has a length dimension of more than 1 m. MCC has appealing binding characteristics and is stable, chemically inert, and physiologically inert [12]. The following applications for commercially available MCC are listed: (1) rheology control agent; (2) pharmaceutical industry binder; (3) tablet binder; (4) food applications; and (5) texturizing agent and fat replacer [12].

13.4.4 Nanocellulose—Chemical Alterations

Polymer grafting, esterification, carboxylation, cationization, and silylation are a few of the chemical processes that can modify nanocellulose to yield derivatives with a wider range of uses [49]. With the intention of (1) increasing the amount of stable charges, be they negative charges or positive charges, for better dispersion, and (2) aiding in the formation of matrices in hydrophobic and non-polar nanocomposites by improving the compatibility factor [12], certain functional groups are introduced during modification. Acetylation and carboxymethylation are two of the most prevalent chemical alterations, as will be explained below. Acetic acid has acetic anhydride in it, which is used in creating cellulose acetate (CA) [50–52]. The carboxymethyl group, also known as the carboxymethyl group, replaces the hydroxyl, or OH group, to produce carboxymethyl cellulose [53]. CMC has numerous uses in industries like pharmacy as well as beauty and care, the food industry, and many other industries [54]. While the aforementioned chemical alterations can enhance the qualities of nanocellulose in connection with the particular application, there are a few things to keep in mind while using them for biological purposes. First, derivatization has the potential to raise the cost of the final product. Second, cellulose can undergo chemical changes that make it "less natural" and less likely to receive regulatory approval. Regarding this final consideration, it must be noted that there is a dearth of information regarding the hazardous potential of such adjustments. A recent study skillfully investigated the possible toxicity of cellulose alterations in the animal model of embryonic zebrafish [55]. This research investigates the impacts influencing and affecting a sophisticated organism, along with the instructive area of developing and growing vertebrates, in contrast to earlier studies that were limited to the investigation of the effects on the levels of cells only. The consequences of adding ethoxyethanol, sulfur, hexamethylenediamine, carboxyl, and ethylenediamine groups to cellulose were among the things the writers assessed. In terms of acute exposure, the data appear to corroborate the absence of any major toxicity in the animal model under consideration. Obviously, further research is required to fully understand the effects of prolonged exposure [56].

13.5 Application of Nanocellulose as Implant and Grafting Materials

In particular, most of the areas of biomedicine, which include either the delivery of drugs or the replacement of the cartilages, or be it engineering of the tissues, the dressing of wounds, or commonly known wound healing, or certain applications in the cardiovascular system, or any other type of implantation in the medicinal fields, all make use of nanocellulose, a biomaterial that is extremely suitable for the biomedical industry. The most interesting and vast field of engineering tissues uses technologies and methods of life science to form natural replacements for the biological systems, which helps them to be able to maintain, improve, and restore tissue functions that have been damaged [57]. The cells are typically grown on biodegradable scaffolds in the process of engineering tissues that are biodegradable, and they match the extracellular matrix, which ranges in size from nanometers to micrometers [58].

To qualify as regenerative medicine, biomaterials employed in tissue engineering must adhere to a set of strict physical and chemical specifications. Tissue regeneration material must have strong physical characteristics (e.g., artificial blood vessels as well as other soft tissue repairs require tensile strength, and engineering of bone tissue needs compressive strength). Some factors of chemistry, which are mainly the surface chemistry of the materials, are important, along with the needs of materials to be made keeping in mind a particular application. For repair of the general tissue, materials should have a surface that must enable protein binding and provide support to promote cell adhesion. Not only this, but the materials should have a surface that prevents the accumulation of blood cells, which would thus result in the thrombosis needed for the repair of the blood vessels. The water-holding capacity should be high in order to use it as a material best suited for the purpose of dressing wounds [59]. Nanocrystalline cellulose holds promise as a cell attachment and proliferation medium due to its superior mechanical qualities and biocompatibility. In addition, many biological species have been cultivated using nanocellulose biomaterials such as sponges, composites, membranes, hydrogels, and electrospun nanofibers. Due to its biodegradability, low toxicity, and high porosity, bacterial nanocellulose is thought to be the most widely used source of nanocellulose for cell culture media. The typical crucial consideration of this biomaterial is the replacement of healthy tissue in place of damaged and destroyed tissue, along with its proper functioning. It should coincide with the time of tissue creation. During the wound dressing and healing process of the skin, the most crucial and important quality is to absorb the exudate during the above process. At the surface of wounds, due to the extreme permeability and subsequent strong adherence, the conventional materials for healing the tissues of the skin appear unsuitable for usage and instead cause fresh trauma when removed. Due to its appealing qualities, which include a high holding capacity of water at a higher rate, a degree of flexibility at higher rates, and a degree of strong mechanical strength, NCC, particularly BC, is the best choice as a replacement for the orthodox methods used in olden times for the repair of tissues of the skin in order to heal wounds [60, 61]. Cellulose nanocrystals are combined with films of collagen that have improved thermomechanical properties and biocompatibility, and as a result, Li *et al.* have shown that it is suggested for application in the engineering of tissues in the case of wound dressings [62]. Another technique for engineering tissues is bone regeneration, which replaces lost bone due to conditions like periodontal disease, neoplasia, congenital malformations, trauma, or trauma-related deformities. A study on bone regeneration employing a hydrohyapatite/cellulose nanocrystals/silk fibroin (HA/CNC/SF) scaffold was done by Chen *et al.* in a rat calvarial defect model. The collected data demonstrate that, following 12 weeks of implantation, the rat calvarial lesion was repaired with only a few newly produced bones. Additionally, it was discovered that the scaffold's rate of disintegration and bone regeneration matched well [63]. In circumstances like burns, diabetic ulcers, and the healing process, BC has been demonstrated to be the most efficient method used for the purpose of dressing wounds. Replacements of the meniscus, wound dressings for skin healing, and blood vessels made artificially are just a few of the biological uses of BC [64–66] (Figure 13.7). Figure 13.8 shows the bacterial cellulose graft with measurements of 6 mm that was extracted after 4 weeks of using it as the bypass in the infrarenal aorta. The graft performed mechanically without any issues. The tube made up of the bacterial cellulose has appropriate qualities in terms of its mechanical strength as a vascular transplant because there were no issues like leakages or ruptures, proving it best in terms of mechanical firmness. Also, thrombosis was not observed after the opening of the tube along its longitudinal axis. As seen in the histology section's Figure 13.8b, the inside surface of the BC's lumen appeared to be very soft and easy and included good cells in a single layer. Therefore, it may be concluded that the bacterial cellulose inner layer, which was soft and smooth and created in the bioreactor that was vertical as reported in the research, performs better as a vascular graft.

A few products, such as Nanoskin®, Biofill®, Bioprocess®, and XCell®, are now on the market for wound dressings made from nanocellulose, specifically bacterial cellulose [70, 71]. The first two products are used as skin that is temporary but also artificial in the treatment of several skin issues like ulcers, wounds on the skin, and skin burns. Another medicine on the market, Gengiflex®, is used to treat periodontal disorders. These products are different, having certain variations due to the differences in the processes

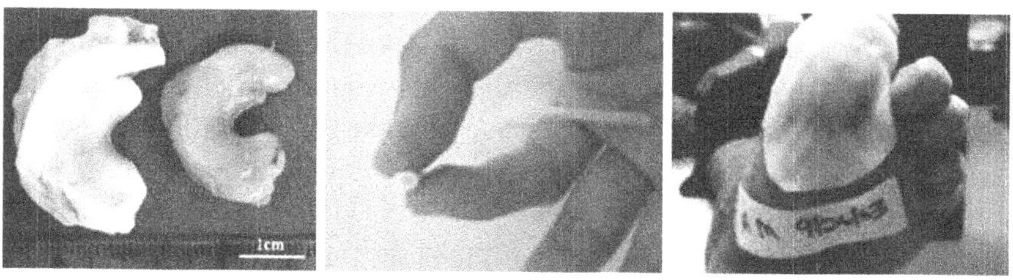

Figure 13.7 Replacements of meniscus (the given figure shows the meniscus of pig on the left, and the image on the right is showing the bacterial cellulose meniscus), dressing of wounds for the healing of skin, and engineering of tissues along with the blood vessels made artificially are a few examples of BC's biological applications [67–69].

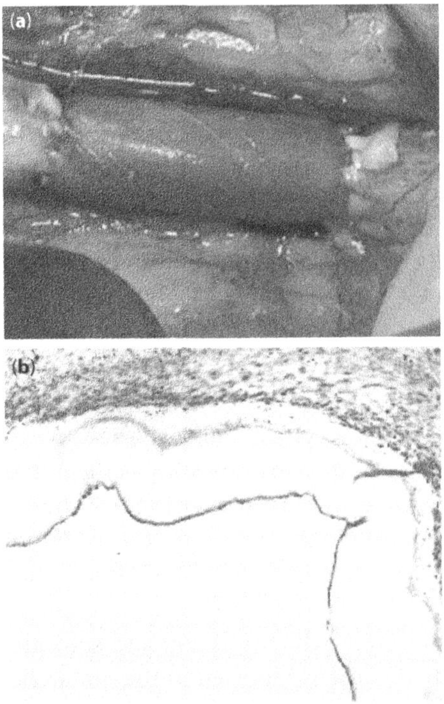

Figure 13.8 For vascular grafts of small diameter as a replacement material, the nanocellulose commonly used is bacterial cellulose obtained mainly by *Acetobacter xylinum* and is utilized primarily, demonstrating encouraging results. These BC tubes have undergone infrarenal aortic bypass testing in pigs, and the results are encouraging for their potential application as grafts of the vascular systems in the coming years. (a) As an infrarenal aortic bypass, a 6-mm BC graft. (b) After 1 month of implantation in the pig aorta, histology studies done at a routine level have shown a beautiful cell layer that was single inside of the surface of the bacterial cellulose [69].

by which they are manufactured and synthesized, such as differences in the concentrations taken initially from those of the sources of carbon; the ratio of surface to volume may be dissimilar; and fermentation periods may also vary from one to the other. We can take the case of Biofill®, which only needs 2 days for fermentation, whereas Gengiflex® needs 8 days [72]. The series XCell® for the treatment and care of skin wounds was created by Xylos Corporation. These solutions support the proper balance of moisture needed when there is a need for healing by allowing appropriate hydration and absorption at much faster rates [73]. In 2007, according to Czaja *et al.*, the characteristics of the membranes made up by the bacterial cellulose had similarities to those of the perfect material for a wound dressing.

A good wound dressing material, for instance, comes in a variety of different structures, which include numerous shapes and sizes; also, the bacterial cellulose membranes can be shaped on the spot. Similarly, membranes obtained from bacterial cellulose are made of a structure with nanopores that prevent germs from penetrating at the site of the wound, and wound dressing materials offer a physical barrier against bacterial infections. Figure 13.9 illustrates the use of Nanoskin®, created by Innovatec's Biotechnology Research and Development, as the best choice for healing wounds by dressing and being successful in treating a diabetic ulcer [70].

Tubes that are made artificially could perhaps replace vascular grafts, be it at a small size of about 4 mm or at a bigger size of about 6 mm, which is another usage for bacterial cellulose. An example is Bacterial Synthesized Cellulose (BASYC®). Figure 13.10 displays BASYC® tubes with various wall thicknesses, inner diameters, and lengths. The arrow-denoted BASYC® tube is adequate for experimental microsurgical needs. This tube has measurements of about 5 mm in length and a diameter of about 1 mm as

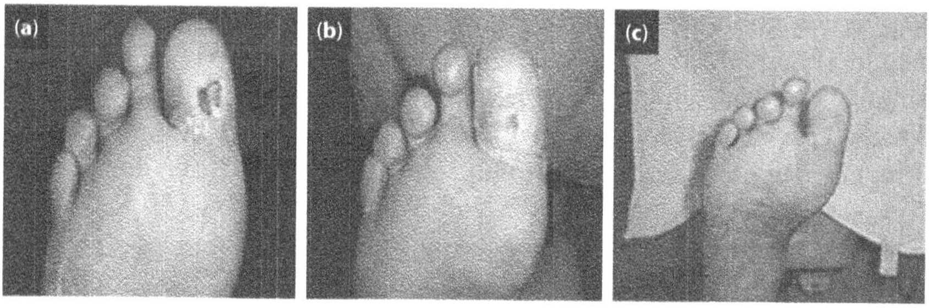

Figure 13.9 Using Nanoskin®, a diabetic ulcer was totally cured after 7 days [70].

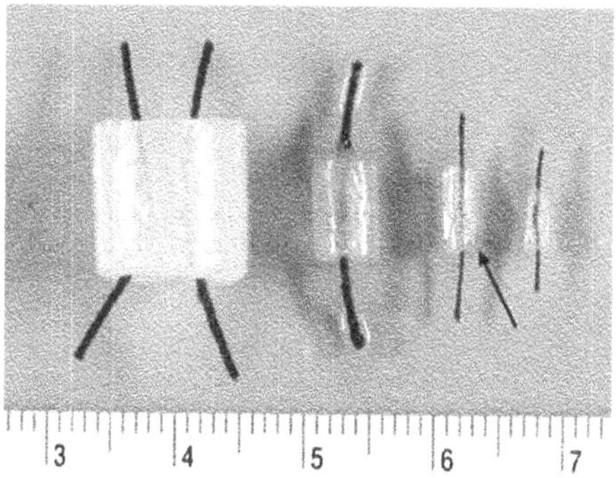

Figure 13.10 BASYC tubes [74].

the inner diameter, and the thickness of the wall measures about 7 mm [74]. The use of BASYC® has been demonstrated successfully as a blood vessel that was produced artificially in the case of organisms like pigs as well as the rats used for the minor surgery, which can also be said to be a microsurgery because of its great mechanical strength. These tubes formed with bacterial cellulose, as opposed to traditional synthetic vascular graft materials, can be used for vascular conduits with small diameters, and they can limit the stenosis and thrombus induction phenomena. Figure 13.11 depicts a rat's static nerve healing 10 weeks following BASYC® implantation. The BASYC® tube is depicted in Figure 13.11b as having small vessels within and being wrapped in connective tissue [74]. In an effort to simulate human coronary arteries, in 2013, research and a study by Azevedo *et al.* created tubes that not only were hollow but also had small diameters made of mixes of cellulose and chitosan [75]. Due to their biocompatibility and lack of toxicity, these materials have great and immense advantages in the research and developments in the field of biomedicine. Firstly, it was summarized that the cellulose-to-chitosan ratio control can enable, enhance, or change the mechanical properties. The ratio of cellulose and chitosan was kept at 5:5 in order to create a replica of the coronary arteries of human beings (Figure 13.12). This study demonstrated that hollow tubes are a suitable material candidate as materials to be used in the bypass or graft for the coronary artery, along with describing the promising mechanical strength, support, and compatibility qualities with the cells for further future research, development, and investigation.

In 2013, Nimeskern *et al.* tried to replace cartilages of the ear in order to investigate bacterial cellulose and its properties on the grounds of mechanical strength [76]. To comprehend the impact on the mechanical properties of the BC samples, various BC pellets with various percentages of cellulose were investigated. Results were better when samples included

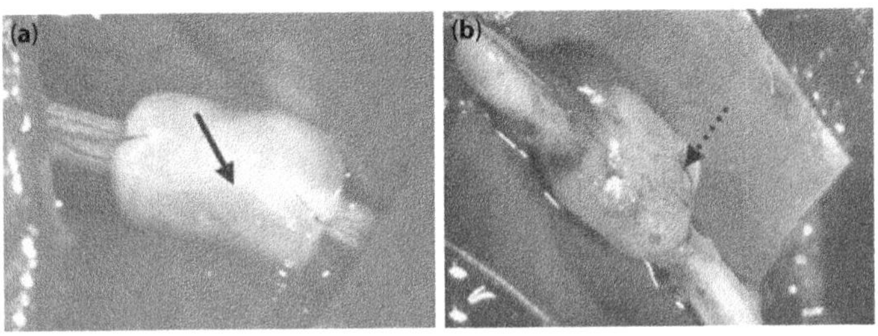

Figure 13.11 Seven BASYC® tubes were used during micronerve surgery on a rat: one was used as a cover protecting the sciatic nerve following the procedure, and the other was used 10 weeks later [74].

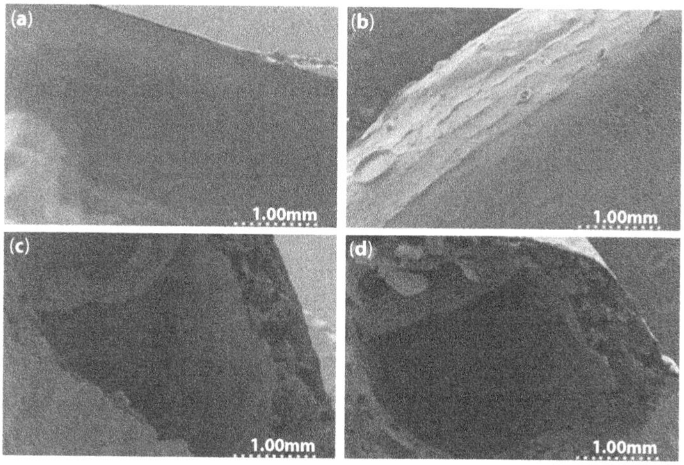

Figure 13.12 SEM photos of the hollow tubes' inner and outer areas shown in that of the images (a) and also in (b), and their cross sections are shown clearly in the figures named (c) and (d) [75].

a higher percentage of cellulose. They came to the conclusion that BC can achieve mechanical qualities that are important for replacing ear cartilage. Additionally, it was determined that it is feasible to manufacture patient-specific ear shapes (Figure 13.13).

The World Health Organization (WHO) estimates that cardiovascular disease-related deaths account for 30% of all fatalities worldwide, and thousands of patients undergo heart bypass surgery each year. In this procedure,

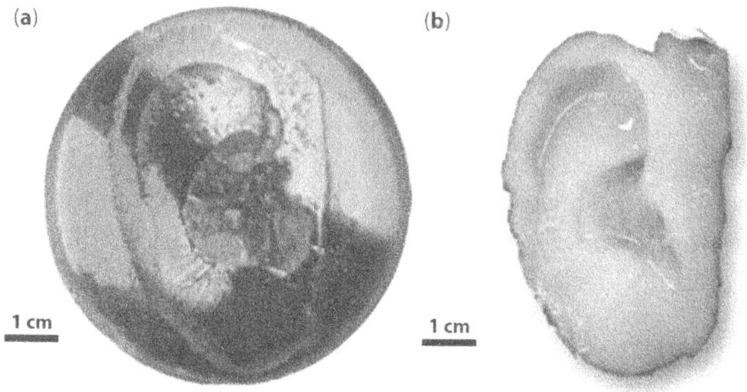

Figure 13.13 (a) Use of a negative silicone mold that is negative, helping direct the bacteria present at the time of the bacterial culture, results in the replication of the major and large-scale features of the outer ear [76]. (b) 3D BC implant prototype.

vessels taken from the patient's thorax or leg are used to replace missing arteries. Nanocellulose is an ideal material to form a matrix that is biocompatible and totally apt to create a fibrin-like matrix, resulting in a very unique biomaterial. Because of this, it is used as a replacement for small-diameter vascular grafts. Not only this, but the use of nanocrystalline cellulose is also used in cases of diseases related to the cardiovascular system in living organisms. This biomaterial, which has both the properties of nano dimensions and the nanocrystalline strength of nanocellulose, can be utilized to replace an entire segment of a blood vessel or to repair damaged and diseased blood vessel tissue. Several synthetic fabric grafts, including Gore-Tex, polyurethane, and Dacron, are utilized to substitute large-diameter (>6 mm) blood vessels. However, these blood channels have a low long-term success rate when used in small-diameter vessels like coronary bypass or peripheral vascular blood vessels because they are prone to blood clots [78]. Based on recent research by Brown *et al.*, replacement vascular applications of nanocrystalline cellulose (NCC)-reinforced fibrin nanocomposites were explored. Additionally, the structural abilities of nanocellulose, that is, its porous as well as permeable nature, help provide a comfortable environment for the invading cells, promoting new tissue growth and development. The particles of nanocrystalline cellulose are therefore promoted for use in the sector of biomedicine, for example, in implantations and wound healing, vascular, artificial liver, bones, cartilage, etc. [79].

13.6 Conclusion

For biologists and material scientists, the development and growth in the field of innovative materials widely used in the areas of biomedicine coming out of polymers, which are mainly natural and are used in practical aspects as well as applications, including clinical laboratories, is constantly a matter of great concern. The word "biocellulose" is sometimes used in research to refer to information on nanocellulose and its use. This is because of the material's special qualities and potential for use in the investigation of a variety of biomedical materials. This chapter's goal was to explain recent developments in the field of biomedicine that have used nanocellulose, using specific examples from the most recent research. Diverse investigations and new horizons, along with future plans on nanocellulose-based biomedical materials, are highlighted and noted, from the molecular level (cellular culturing) to macroscopic biomaterials (tissue healing, regeneration, grafting, implantation, etc.). To create a cell-friendly environment, nanocellulose-based biomaterials are widely used, resulting in the attachment of cells and their proliferation as a very different tissue scaffold with advantages like biocompatibility and appropriate mechanical properties comparable to natural tissue. On many living beings, nanocellulose-based biomaterials have been observed and cultivated, among which forms like hydrogels, composites, electrospun nanofibers, sponges, or membranes were mainly noticeable. One of the most considered options for implants is BC, which serves as the best cell culture medium, and the credit goes to its properties like maximum porosity and minimum cytotoxicity. Not only this, but development and growth in the fields related to biomaterials like nanocellulose as a replacement or medical biomaterial for soft tissue, the nucleus pulposus, blood vessels (vascular grafts), and other body parts are aided by its very good and effective properties like mechanical strength and biocompatibility, which are not only long-lasting but also strong. According to the results of several experiments on living organisms, mainly animals, conducted prior to research in clinical laboratories, nanocellulose has been widely referred to as the replacement of the blood vessels and is highly appealing and advantageous. The majority of works on research on nanocellulose as soft tissue and nucleus pulposus are still in the early stages and mostly concentrate on comparing the various features of nanocellulose-based materials with actual organs. The materials used in the repair of skin, commonly called the dressing of the wounds, have the ability to absorb exudate during the process of wound dressing, resulting in its removal from the place or a wound surface when the healing process is complete, which marks its quality, which is quite crucial. Traditional materials used in the repair of skin tissue, which include gauze, have a permeability that is too strong and causes adhesion that is much more tightly placed on the parched wound surface and might result in fresh trauma when removed. Given its important biological features, nanocellulose (particularly BC) has gradually gained growing interest for innovative wound treatment and is emerging as the most favorable grafting and implant material.

References

1. Dumanli, A.G., Nanocellulose and its composites for biomedical applications. *Curr. Med. Chem.*, 24, 512–528, 2017.
2. Jofri, M. and Foster, E.J., Recent advances in nanocellulose for biomedical applications. *J. Appl. Polym. Sci.*, 132, 41719–41737, 2015.
3. Babu, R.P., O'connor, K., Seeram, R., Current progress on bio-based polymers and their future trends. *Prog. Biomater.*, 2, 8–23, 2013.

4. Ivanova, E.P., Bazaka, K., Crawford, R.J., New functional biomaterials for medicine and healthcare, in: *New Functional Biomaterials for Medicine and Healthcare*, pp. 32–70, Woodhead Publishing Limited, UK, 2014.
5. Trache, D., Khimeche, K., Mezroua, A. *et al.*, Physicochemical properties of microcrystalline nitrocellulose from Alfa grass fibres and its thermal stability. *J. Therm. Anal. Calorim.*, 124, 1485–1496, 2016.
6. Nanomaterials, G., Klemm, D., Klemm, D., Kramer, F., Moritz, S., Lindstroöm, T., Ankerfors, M., Gray, D., Dorris, A., Reviews nanocelluloses: A new family of nature-based materials. *Angew. Chem. Int. Ed.*, 50, 5438–5466, 2011.
7. Lin, N. and Dufresne, A., Nanocellulose in biomedicine: Current status and future prospect. *Eur. Polym. J.*, 59, 302–325, 2014.
8. Qin, C., Soykeabkaew, N., Xiuyuan, N., Peijs, T., The effect of fibre volume fraction and mercerization on the properties of all-cellulose composites. *Carbohydr. Polym.*, 71, 458–467, 2008.
9. Yamanaka, S., Watanabe, K., Kitamura, N., Iguchi, M., Mitsuhashi, S., Nishi, Y., Uryu, M., The structure and mechanical properties of sheets prepared from bacterial cellulose. *J. Mater. Sci.*, 24, 3141–3145, 1989.
10. Delmer, D. and Amor, Y., Cellulose biosynthesis. *Plant Cell*, 7, 987–1000, 1995.
11. Kalia, S., Dufresne, A., Cherian, B., Kaith, B., Avérous, L., Njuguna, J., Nassiopoulos, E., Cellulose-based bioand nanocomposites: A review. *Int. J. Polym. Sci.*, 2011, 837875, 35, 1–35, 2011.
12. Habibi, Y., Lucia, L., Rojas, O., Cellulose nanocrystals: Chemistry, self-assembly, and applications. *Chem. Rev.*, 110, 3479–3500, 2010.
13. Brown, R.M.J., The biosynthesis of cellulose. *Pure Appl. Chem.*, 33, 1345–1373, 1996.
14. Rowland, S. and Roberts, E.J., The nature of accessible surfaces in the microstructure of cotton cellulose. *J. Polym. Sci. Part A Polym. Chem.*, 10, 2447–2461, 1972.
15. Mood, S., Golfeshan, A., Tabatabaei, M., Jouzani, G., Najafi, G., Gholami, M., Ardjmand, M., Lignocellulosic biomass to bioethanol, a comprehensive review with a focus on pretreatment. *Renew. Sustain. Energy Rev.*, 27, 77–93, 2013.
16. El-Saied, H., Basta, A., Gobran, R., Research progress in friendly environmental technology for the production of cellulose products (bacterial cellulose and its application). *Polym.-Plast. Technol. Eng.*, 43, 797–820, 2004.
17. Dammström, S., Salmén, L., Gatenholm, P., The effect of moisture on the dynamic properties of bacterial cellulose/glucuronoxylan nanocomposites. *Polymer*, 46, 10364–10371, 2005.
18. Kalia, S., Kaith, B., Kaur, I., Pre-treatments of natural fibers and their application as reinforcing material in polymer composites-a review. *Polym. Eng. Sci.*, 49, 1253–1272, 2009.
19. Iguchi, M., Yamanaka, S., Budhiono, A., Review bacterial cellulose—A masterpiece of nature's arts. *J. Mater. Sci.*, 35, 261–270, 2000.

20. Food and Drug Administration, Pyrogens and endotoxins testing: Questions and answers, in: *Guidance for Industry*, FDA, Silver Spring, MD, USA, 2012.
21. Jonas, R. and Farah, L., Production and application of microbial cellulose. *Polym. Degrad. Stab.*, 59, 101–106, 1998.
22. Zaar, K., The biogenesis of cellulose by Acetobacter xylinum. *Cytobiologie*, 16, 1–15, 1977.
23. Brown, R.M.J., Willison, J., Richardson, C., Cellulose biosynthesis in Acetobacter xylinum: Visualization of the site of synthesis and direct measurement of the *in vivo* process. *Proc. Natl. Acad. Sci. U.S.A.*, 73, 4565–4569, 1976.
24. Yamanaka, S., Ishihara, M., Sugiyama, J., Structural modification of bacterial cellulose. *Cellulose*, 7, 213–225, 2000.
25. Ben-Hayyim, G. and Ohad, I., Synthesis of cellulose by Acetobacter xylinum. VIII. On the formation and orientation of bacterial cellulose fibrils in the presence of acidic polysaccharides. *J. Cell Biol.*, 25, 191–207, 1965.
26. Czaja, W., Krystynowicz, A., Bielecki, S., Brown, R.M., Jr., Microbial cellulose-the natural power to heal wounds. *Biomaterials*, 27, 145–151, 2006.
27. Hult, E., Yamanaka, S., Ishihara, M., Sugiyama, J., Aggregation of ribbons in bacterial cellulose induced by high pressure incubation. *Carbohydr. Polym.*, 53, 9–14, 2003.
28. Wan, W., Hutter, J., Millon, L., Guhados, G., Bacterial cellulose and its nanocomposites for biomedical applications, in: *Cellulose Nanocomposites. Processing Characterization, and Properties*, K. Oksman, and M. Sain, (Eds.), American Chemical Society, Washington, DC, USA, 2006.
29. Barud, H., Barrios, C., Regiani, T., Marques, R., Verelst, M., Dexpert-Ghys, J., Messaddeq, Y., Ribeiro, S., Selfsupported silver nanoparticles containing bacterial cellulose membranes. *Mater. Sci. Eng. C*, 28, 515–518, 2008.
30. De Wulf, P., Joris, K., Vandamme, E.J., Improved cellulose fromation by an Acetobacer xylinum mutant limited in (keto)gluconate synthesis. *J. Chem. Technol. Biotechnol.*, 67, 376–380, 1996.
31. Oikawa, T., Morino, T., Ameyama, M., Production of cellulose from D arabitol by Acetobacter xylinum KU-1. *Biosci. Biotechnol. Biochem.*, 59, 1564–1565, 1995.
32. Oikawa, T., Ohtori, T., Ameyama, M., Production of cellulose from D mannitol by Acetobacter xylinum KU-1. *Biosci. Biotechnol. Biochem.*, 59, 331–332, 1995.
33. Masaoka, S., Ohe, T., Sakota, N., Production of cellulose from glucose by Acetobacter xylinum. *J. Ferment. Bioeng.*, 75, 18–22, 1993.
34. Toyosaki, H., Naritomi, T., Seto, A., Matsuoka, M., Tsuchida, T., Yoshinaga, F., Screening of bacterial cellulose-producing Acetobacter strains suitable for agitated culture. *Biosci. Biotechnol. Biochem.*, 59, 1498–1502, 1995.
35. Surip, S., Wan Jaafar, W., Azmi, N., Anwar, U., Microscopy observation on nanocellulose from kenaf fibre. *Adv. Mater. Res.*, 488, 72–75, 2012.

36. Siró, I. and Plackett, D., Microfibrillated cellulose and new nanocomposite materials: A review. *Cellulose*, 17, 459–494, 2010.
37. Shin, E.J., Choi, S.M., Singh, D., Zo, S.M., Lee, Y.H., Kim, J.H., Han, S.S., Fabrication of cellulose-based scaffold with microarchitecture using a leaching technique for biomedical applications. *Cellulose*, 21, 3515–3525, 2014.
38. Randy, B., The cellulose micelles. *Tappi*, 35, 53–58, 1952.
39. Hubbe, M., Rojas, O., Lucia, L., Sain, M., Cellulosic nanocomposites: A review. *BioResources*, 3, 929–980, 2008.
40. Šturcova, A., Davies, G., Eichhorn, S., Elastic modulus and stress-transfer properties of tunicate cellulose whiskers. *Biomacromolecules*, 6, 1055–1061, 2005.
41. Samir, M., Alloin, F., Dufresne, A., Review of recent research into cellulosic whiskers, their properties and their application in nanocomposite field. *Biomacromolecules*, 6, 612–626, 2005.
42. Dufresne, A., Comparing the mechanical properties of high performances polymer nanocomposites from biological sources. *J. Nanosci. Nanotechnol.*, 6, 322–330, 2006.
43. Dufresne, A., Polysaccharide nanocrystals reinforced nanocomposites. *J. Chem.*, 86, 484–494, 2008.
44. Araki, J., Wada, M., Kuga, S., Okano, T., Flow properties of microcrystalline cellulose suspension prepared by acid treatment of native cellulose. *Colloids Surf. A*, 142, 75–82, 1998.
45. Roman, M. and Winter, W., Effect of sulfate groups from sulfuric acid hydrolysis on the thermal degradation behavior of bacterial cellulose. *Biomacromolecules*, 5, 1671–1677, 2004.
46. Johari, N., Ahmad, I., Halib, N., Comparison study of hydrogels properties synthesized with micro- and nano- size bacterial cellulose particles extracted from nata de coco. *Chem. Biochem. Eng. Q.*, 26, 399–404, 2012.
47. Yin, O., Ahmad, I., Amin, M., Effect of cellulose nanocrystals content and pH on swelling behaviour of gelatin based hydrogel. *Sains Malays.*, 44, 793–799, 2015.
48. Levis, S. and Deasy, P., Production and evaluation of size reduced grades of microcrystalline cellulose. *Int. J. Pharm.*, 213, 13–24, 2001.
49. Mondal, S., Preparation, properties and applications of nanocellulosic materials. *Carbohydr. Polym.*, 163, 301–316, 2017.
50. Sassi, J. and Chanzy, H., Ultrastructural aspects of the acetylation of cellulose. *Cellulose*, 2, 111–127, 1995.
51. Li, J., Zhang, L., Peng, F., Bian, J., Yuan, T., Xu, F., Sun, R., Microwave-assisted solvent-free acetylation of cellulose with acetic anhydride in the presence of iodine as a catalyst. *Molecules*, 14, 3551–3566, 2009.
52. Wu, J., Zhang, J., Zhang, H., He, J., Ren, Q., Guo, M., Homogeneous acetylation of cellulose in a new ionic liquid. *Biomacromolecules*, 5, 266–268, 2004.

53. Charpentier, D., Mocanu, G., Carpov, A., Chapelle, S., Merle, L., Muller, G., New hydrophobically modified carboxymethyl cellulose derivatives. *Carbohydr. Polym.*, 33, 177–186, 1997.
54. Toðrul, H. and Arslan, N., Production of carboxymethyl cellulose from sugar beet pulp cellulose and rheological behaviour of carboxymethyl cellulose. *Carbohydr. Polym.*, 54, 73–82, 2003.
55. Harper, B.J., Clendaniel, A., Sinche, F., Way, D., Hughes, M., Schardt, J., Simonsen, J., Stefaniak, A.B., Harper, S.L., Impacts of chemical modification on the toxicity of diverse nanocellulose materials to developing zebrafish. *Cellulose*, 23, 1763–1775, 2016.
56. Endes, C., Camarero-Espinosa, S., Mueller, S., Foster, E.J., Petri-Fink, A., Rothen-Rutishauser, B., Weder, C., Clift, M.J., A critical review of the current knowledge regarding the biological impact of nanocellulose. *J. Nanobiotechnol.*, 14, 78, 2016.
57. Dhandayuthapani, B., Yoshida, Y., Maekawa, T., Kumar, D.S., Polymeric scaffolds in tissue engineering application: A review. *Int. J. Polym. Sci.*, 2011, 9–11, 2011.
58. Mohan, T., Hribernik, S., Kargl, R., Stana-Kleinschek, K., Nanocellulosic materials in tissue engineering applications, in: *Cellulose—Fundamental Aspects and Current Trends*, InTech, Rijeka, 2015, http://dx.doi.org/10.5772/59889.
59. Wang, J., Zhu, Y., Du, J., Bacterial cellulose: A natural nanomaterial for biomedical applications. *J. Mech. Med. Biol.*, 11, 02, 285–306, 2011.
60. Czaja, W.K., Young, D.J., Kawecki, M., Brown, R.M., The future prospects of microbial cellulose in biomedical applications. *Biomacromolecules*, 8, 1, 1–12, 2007.
61. Vatankhah, E., Prabhakaran, M.P., Jin, G., Mobarakeh, L.G., Ramakrishna, S., Development of nanofibrous cellulose acetate/gelatin skin substitutes for variety wound treatment applications. *J. Biomater. Appl.*, 28, 6, 909–21, 2014.
62. Li, W., Guo, R., Lan, Y., Zhang, Y., Xue, W., Zhang, Y., Preparation and properties of cellulose nanocrystals reinforced collagen composite films. *J. Biomed. Mater. Res. Part A*, 102, 4, 1131–9, 2014.
63. Chen, X., Zhou, R., Chen, B., Chen, J., Nanohydroxyapatite/cellulose nanocrystals/silk fibroin ternary scaffolds for rat calvarial defect regeneration. *RSC Adv.*, 6, 42, 35684–91, 2016.
64. Zugenmaier, P., Conformation and packing of various crystalline cellulose fibers. *Prog. Polym. Sci.*, 26, 9, 1341–1417, 2001.
65. Keshk, S.M.A.S. and Sameshima, K., Evaluation of different carbon sources for bacterial cellulose production. *Afr. J. Biotechnol.*, 4, 6, 478–482, 2005.
66. Jung, J.Y., Park, J.K., Chang, H.N., Bacterial cellulose production by *Gluconoacetobacter hansenii* in an agitated culture without living non-cellulose producing cells. *Enzyme Microb. Technol.*, 37, 3, 347–354, 2005.
67. Svensson, A., Nicklasson, E., Harrah, T., Panilaitis, B., Kaplan, D.L., Brittberg, M., Gatenholm, P., Bacterial cellulose as a potential scaffold for tissue engineering of cartilage. *Biomaterials*, 26, 4, 419–431, 2005.

68. Cockbill, Evaluation *in vivo* and *in vitro* of the performance of interactive dressings in the management of animal soft tissue injuries. *Vet. Dermatol.*, 9, 2, 87–98, 1998.
69. Goodrich, L.R., Moll, H.D., Crisman, M.V., Lessard, P., Bigbie, R.B., Comparison of equine amnion and a nonadherent wound dressing material for bandaging pinch-grafted wounds in ponies. *Am. J. Vet. Res.*, 61, 3, 326–329, 2000.
70. De Olyveira, G.M., Manzine Costa, L.M., Basmaji, P., Xavier Filho, L., Bacterial nanocellulose for medicine regenerative. *J. Nanotechnol. Eng. Med.*, 2, 034001, 2011.
71. Petersen, N. and Gatenholm, P., Bacterial cellulose-based materials and medical devices: Current state and perspectives. *Appl. Microbiol. Biotechnol.*, 91, 1277–1286, 2011.
72. Brown, R., Czaja, W., Jeschke, M., Young, D.J., Multiribbon nanocellulose as a matrix for wound healing, WO 2007/027849 A2, US, 2–6, 2007.
73. Thakur, V.K., *Nanocellulose polymer nanocomposites: Fundamentals and applications*, Wiley, Beverly, MA, 01915-6106, 2014.
74. Klemm, D., Schumann, D., Udhardt, U., Marsch, S., Bacterial synthesized cellulose - artificial blood vessels for microsurgery. *Prog. Polym. Sci.*, 26, 1561–1603, 2001.
75. Azevedo, E.P., Retarekar, R., Raghavan, M.L., Kumar, V., Mechanical properties of cellulose: Chitosan blends for potential use as a coronary artery bypass graft. *J. Biomater. Sci. Polym. Ed.*, 24, 239–52, 2013.
76. Nimeskern, L., Martínez Ávila, H., Sundberg, J. *et al.*, Mechanical evaluation of bacterial nanocellulose as an implant material for ear cartilage replacement. *J. Mech. Behav. Biomed. Mater.*, 22, 12–21, 2013.
77. Zhang, Y., Nypelö, T., Salas, C. *et al.*, Cellulose nanofibrils: From strong materials to bioactive surfaces. *J. Renew. Mater.*, 1, 195–211, 2013.
78. Kannan, R.Y., Salacinski, H.J., Butler, P.E., Hamilton, G., Seifalian, A.M., Current status of prosthetic bypass grafts: A review. *J. Biomed. Mater. Res. Part B: Appl. Biomater.*, 74, 1, 570–81, 2005.
79. Brown, E.E., Hu, D., Abu Lail, N., Zhang, X., Potential of nanocrystalline cellulose–fibrin nanocomposites for artificial vascular graft applications. *Biomacromolecules*, 14, 4, 1063–71, 2013.

14

Use of Nanocellulose for Dental Applications

Rupankar Ghosh[1], Amlan Roychowdhury[2*] and Ditipriya Hazra[3†]

[1]Department of Biosciences, JIS University, Kolkata, India
[2]Center for Healthcare Science and Technology Indian Institute of Engineering Science and Technology, Shibpur, India
[3]Department of Biotechnology, St. Xavier's College, Kolkata, India

Abstract
Nanocellulose is a unique ubiquitously available biomolecule from phytogenic or bacterial sources. The molecular structure is made up of multiple glucose monomers, connected by beta-1,4 glycosidic bonds. The nanocrystalline form of cellulose has been found to be affiliated with excellent biocompatible properties in the medical field, from wound healing to drug delivery agent to even be used as implants. This chapter focuses on the uses of nanocellulose in several areas of dental science.

Keywords: Guided tissue regeneration, soft tissue reinforcement, canker sore, dental root canal treatment, class II furcation treatment, Pickering emulsions

14.1 Introduction

Cellulose is a biopolymer composed of polysaccharides that is abundantly distributed in nature. It is made up of hundreds to thousands of glucose molecules connected by covalent bonds. Cellulose is relatively hydrophilic if compared with other naturally available biopolymers and could be purified in crystalline form. In cellulose, the monomeric glucose units are linked to each other by beta-1,4 glycosidic bonds and the chain length is up to 4 μm. Cellulose is mainly obtained from plants, algae, and microbes.

**Corresponding author*: amlan.iitkgp@gmail.com
†*Corresponding author*: dityрriyahazra@sxccal.edu

Mainak Mukhopadhyay and Debalina Bhattacharya (eds.) Nanocellulose: A Biopolymer for Biomedical Applications, (359–372) © 2024 Scrivener Publishing LLC

Based on its origin, cellulose could be divided into two groups: plant cell wall-based cellulose and bacterial membrane-based cellulose. The cellulose obtained from the primary or secondary cell walls of plant cells is generally contaminated with lignin, hemicelluloses, and other non-cellulosic compounds. In comparison, bacterial membrane-based cellulose offers high levels of purity, high tensile strength, high mechanical properties, and porosity. These properties of bacterial nanocellulose are widely applicable in biomedical fields like tissue engineering, drug delivery, cartilage replacements, cardiovascular applications, wound dressings, and medical implantation.

Cellulose is treated with chemical, mechanical, or enzymatic means to obtain crystalline nanocellulose (CNC) or the nanofibers of cellulose, which have a nanoscale range between 5 and 20 nm [1]. Cellulose-based nanocellulose fibers are even attempted to be used in the field of dentistry because of their properties of imitating the 3D architecture of the extracellular matrix (ECM) of native tissues. It is widely applicable in different areas of dental science, including restoring periodontal and pulp tissues, surgical dressings, and treating intraoral wounds [2].

14.1.1 Nanocellulose versus Cellulose

The nanofibers of cellulose carry biological features like biocompatibility, biodegradability, and low cytotoxicity and also offer a unique platform for addition or alteration to generate additive conjugate compounds. Bacterial nanocellulose is conceived to be a promising and budget-friendly nanomaterial in the biomedical field [1]. The microfibrillar structure of nanocellulose is more homogeneous and is coupled with high tensile strength, a high degree of polymerization, and crystallinity. The crystallinity index (CI), which refers to the quantitative degree of crystallinity in a compound, can touch up to 84–90%. This CI value depends on the C4 carbon atoms present in cellulose; CI values can be determined using NMR spectroscopy [2]. The microfibrillar structure of nanocellulose is more homogeneous and is coupled with high tensile strength, high degree of polymerization, and crystallinity. The crystallinity index (CI), which refers to the quantitative degree of crystallinity in a compound, can touch up to 84–90%. This CI value depends on the C4 carbon atoms present in cellulose; CI values can be determined using NMR spectroscopy [3]. The microfibrillar nanocrystalline structure has good formability and is better at dealing with additive-based modifications than cellulose. The macromolecular cellulose is structurally nonhomogeneous and is less ordered compared to nanocellulose [4].

Bacterial nanocellulose (BNC) falls under the category of GRAS (Generally Regarded as Safe) materials, allowing it to be suitable for a wide range of applicability [5].

14.1.2 Bacterial Nanocellulose Production

Excavation of cellulose (exopolysaccharide) from a bacterial source was first performed by A.J. Brown from *Gluconaceto bacterxylinum* (now known as *Komagataei bacterxylinus*) in 1886 [6]. Current practices include *Agrobacterium* sp., *Pseudomonas* sp., *Rhizobium* sp., and *Escherichia* sp. *In vitro*, if these strains are cultured in media with glucose or other saccharides as carbon sources under aerobic conditions, they secrete a 3D web of cellulose fibrils.

According to the type of culture used for cellulose production, bacterial cellulose (BC) can be classified as follows:

I. Static culture-based membranous form
II. Shaker culture-based microform

Nanofiber cellulose (NFC) is further obtained by processing the cellulose fibers through physical, chemical, or enzymatic treatments [7]. Static-culture-based nanocellulose membranes are more uniform and stable than those in agitated conditions and are thus preferred in different fields. Bioreactor technologies and static intermittent, semi-continuous-fed batch technology methods are more reliable to meet the industry-scale demands of BNC [7].

14.1.3 Plant-Based Nanocellulose Preparation

Various fibrous plants or parts can be used for cellulose extraction. Notably, wood pulps, banana fibers, pineapple leaves, sugar beets, etc. are used as resources. Amim Cl (1-allyl-3-methyl imidazolium chloride), an ionic liquid, is used to extract cellulose from woody plants like pine, poplar, Chinese parasol, and catalpa wood chips [7]. The plant cell wall is a rich source of cellulose; it can be extracted using the following steps: first, the chosen raw material is taken, any non-cellulose impurities are thoroughly checked and eliminated using homogenizers, and high-pressure mechanical defibrillation is achieved. Before mechanical defibrillation, strong acid [H_2SO_4]-based hydrolysis pretreatment is done. This method delaminates cell walls and separates nano-sized particles. It is regarded as sulfite pulping or alkaline Kraft pulping. The concentration of H_2SO_4 can be increased to boost the hydrolysis reaction. Sulfur trioxide group incorporation in the cellulose nanofibers accelerates the thermal degradation process. Performing these methods yields completely separated, surface charged, and highly entangled cellulose nanofibers [8].

14.2 Application of Nanocellulose in Dental Science

The nano-sized fibrils are pseudo-plastic; the crystalline form is transparent, electrically conductive, non-toxic, and stiffer than steel (around 20 GPa). In addition to having a low processing fee, it is an interesting biopolymer in the biomedical field. Major areas that consider the use of bacterial nanocellulose include delivery systems, healing materials, and tissue engineering for regenerative medicine [9].

Compounds like bacterial cellulose-based collagen, gelatin, and fibroin chitosan are of keen interest and are readily employed as suitable materials for artificial blood vessels, micro vessels, and artificial skin production. In the field of dentistry, the effective applications of bacterial cellulose have a long road ahead of them.

As the record goes, BNC has mainly been assigned to the promotion of regeneration of periodontal tissues and surgical dressings, and for treating intraoral wounds and regeneration of pulp tissue.

Oral and dental dispensation of bacterial cellulose comprises the design of implants, scaffolds, and wound dressing materials and is extended up to carriers for drug delivery [2].

The biocompatible nature of nanocellulose, i.e., biodegradability and non-toxicity, further complements its ardent demand in the dental industry as per the recent trends of relying on biocompatible materials and devices. Besides biocompatibility, nanocellulose serves as a constructive and subjective substitute for native tissues. The biocompatible nature of bacterial cellulose allows it to be directly recruited alongside biological tissues.

Practical involvement of bacterial cellulose in dentistry includes guided tissue regeneration technique for periodontal disease treatment, Class II furcation lesion treatment, oral mucosa-based surgical wound treatment, dental pulp tissue treatment, and substitute biomaterials in minimizing free flap donor site morbidity [10].

14.2.1 BNC-Based Membrane Barrier in Guided Tissue Regeneration

BNC has found indirect use in the high-level guided tissue regeneration (GTR) technique as the raw material for membrane barriers used in periodontal surgeries. The GTR technique is a sophisticated approach for the treatment of periodontal diseases like periodontitis, a severe clinical condition due to continuous unhygienic practices. GTR surgery is performed to separate the gum tissues from the tooth area, allowing the surface to be efficiently cleaned and potent infected areas to be removed, followed by regeneration of the infected area.

In the case of regeneration, gum tissues and dental bones grow at varying rates, so it is necessary to distinguish the gum tissues from the regions of bone growth. The membrane barrier is used to promote selective proliferation of tissues; thus, the bone and dental tissues can proliferate orderly without migrating into each other's domains [11, 12].

14.2.2 Periodontal Disease Treatment

Advanced periodontal disease is a clinical condition where the gum tissues are pulled away from teeth, generating pockets for microbes to fill in, which may result in periodontal degradation through infections. The GTR technique aims to promote periodontal regeneration. Instead of the traditional approaches, where gum tissues are removed and exposed bone and roots are re-contoured, the ultimate result is instability and weakness of the teeth. The GTR technique helps teeth gain stability, gum tissue regrowth, and restoration of any lost bone successfully. Additionally, the gaps generated in the gum tissues, which would have accumulated infectious pathogens and microbes, are also cleaned during the process.

In this whole procedure, there is an important role for the barrier that separates the gum tissues from the tooth area. The BNC is known to serve as an ideal physical barrier, bio-absorbable in nature, against microbial infections. It is non-allergic and non toxic in nature and is reported to reduce pain to a significant level when applied as a wound dressing agent. Notably, the commercial membrane Gingiflex is applied alongside bone integrated implants [13].

14.2.3 Class II Furcation Lesion Treatment

A furcation lesion refers to a type of skeletal defect, i.e., dental bone loss. The furcation lesion falls under periodontal diseases, which affects the bottom of the root trunk of a tooth junction, where two or more roots meet. Furcation is associated with only multi-rooted teeth; the upper first premolar, maxillary molar, and mandibular molars are the only ones.

Class II furcation is a severe condition that includes multiple areas of furcal bone loss. A Class II furcation anomaly is at least 3 mm deep, surpassing the buccolingual thickness of the tooth partially, with a limited amount of interradicular bone adhered to the angle of the furcation at one end. In the Class I furcation, the presence of the blind pouch containing plaque is also observed, i.e., the cul-de-sac condition [14]. In Class II furcation, the probability of tooth loss is doubled in comparison to Class I furcation.

Furcation plasty, open debridement, or bur odontoplasty and guided tissue regeneration techniques are utilized to treat the affected bone and make it more cleanable for the suffering patient. The treatment aims at establishing new clinical attachments. BNC serves as a scaffold for treating osteological clinical conditions [15].

14.2.4 Oral Mucosa-Based Surgical Wound Treatment

Surgical wounds were generated due to various surgical procedures like the GTR technique, free flap donor site morbidity, dental bone graft, dental implants, periodontal surgery, tooth extraction, and mucogingival surgery. The BNC is known to serve as a standard biopolymer for wound dressing [13, 14]. The nano-fibrous structure is mechanically stable and flexible in nature, with a minimum of 95% water content in the structure, which helps to prolong dermal hydration in cases of burn-related wounds. The biopolymer produces a moist environment that prevents the loss of excessive fluid through the wound. Evaporation of the water content generates a cooling effect around the wound, offering temporary relief.

The bacterial nanocellulose can be readily drenched with antiseptic solutions like polyhexanide, octenidine, and povidone iodine. The antiseptic-soaked BNC can also be used in treating biofilms on oral wounds, thus promoting better hygiene around the wound. Mouth sores like painful lesions formed in oral soft tissues can also be dressed using BNC.

Mouth sores are a common type of oral lesion that may occur at different times throughout the human life cycle. Cold sores and canker sores are very common types of oral lesions. How BNC-based nano-structured patches loaded with therapeutics are designed for the treatment of aphthous stomatitis ulcers or canker sores is detailed in the following section [16].

14.2.5 Treating Canker Sore

Canker sore or aphthous ulcer is a common clinical condition, under which small lesions occur on intraoral soft tissues inside the mouth and gum tissues, which makes it hard for the patient to consume food or have verbal conversations. BNC is used as the raw material for the production of nanostructured patches or composite drug carriers. The patches consist of a conjugate structure, and diffusion of aqueous solutions of hyaluronic acid and diclofenac in varying concentrations is compiled into wet bacterial nanocellulose, with porous 3D make-up [17].

Following a diffusion and swelling controlled method, the drug is secreted into the oral microenvironment. This method aims at relieving pain and reducing inflammation [20].

14.2.6 Dental Pulp Tissue Treatment and Soft Tissue Reinforcement

The pulp tissue is a group of connective tissue that is located underneath the dentin layer of teeth and acts as a nutritional reserve for the tooth. The sensitive pulp tissue can be exposed to the external microenvironment due to the destruction of the outer layers of the tooth. Such conditions allow external pathogens to bypass directly into the pulp layer. Nerve and blood vessel-rich pulp is easily susceptible to microbes; inflammation in the pulp tissue results in a severe clinical condition called pulpitis [18].

Irreversible pulpitis may lead to pulp necrosis or pulp death, where pain or irritability cannot be perceived as the pulp tissue-bound nerve dies. Pulpectomy is a surgical procedure that is used to eliminate the dead pulp tissue, and as a replacement, the vacuum is filled with soft tissue reinforcement material that can be later reabsorbed by the body. The anti-adhesive nature and mechanical properties of BNC make it a suitable soft tissue reinforcement material.

Bacterial nanocellulose–polypropylene conjugates with anti-adhesive properties serve as a commercial mesh for soft tissue reinforcement. Polypropylene meshes have a tendency to trigger side effects at the implant site, like bowel perforation and chronic pain. The side effects can be overcome by the application of conjugate mesh. The mesh has various applications in the surgical field, including hernioplasty and cranioplasty, and as an anti-adhesive, biocompatible barrier.

In the field of dentistry, BNC crystal-modified glass ionomer cement (GIC) is employed for reinforcement purposes with boosted physicochemical properties. The fluoro amino silicate family of glass is mixed with water-soluble, ionic polymers such as polyalkenoic acid to generate the glass ionomer cement, the anthropogenic dentin.

Bacterial nanocellulose-based glass ionomer cement is employed as a full restorative material for treating occlusal and proximal cavities of primary teeth, which are easily prone to decay. Other notable applications involve restoration of deciduous teeth, bonding agents for orthodontic brackets, the restoration of class III and class V carious lesions, tunnel restorations, fissure sealants, and atraumatic restorative materials in pediatric dentistry. In a nutshell, it is considered a suitable alternative to dental amalgams [19].

14.2.7 Dental Root Canal Treatment

The root canal refers to the nerve-rich, soft tissue region under the tough enamel and dentin coat. In cases of severe pulp necrosis progression due to bacterial infection, the dental pulp layer is harshly contaminated, and dental root canal treatment is needed to deal with such conditions of severe pain and suffering. This treatment's goal is to generate a contamination-free microenvironment for the affected tooth. Conventional decontamination treatments include mechanical instrumentation and chemical irrigation. Most of the contaminating endodontic bacteria remain deposited in the fluid phase of the dental root canal, and thick bacterial accumulation conjoins the dental root canal walls. Dental root canal treatment involves the removal of necrotic pulp leftovers, dental fillings, and the application of irrigant solutions. Under the mechanical instrumentation quota, cotton pellets and plant cellulose-based paper points are commonly employed to dry the dental root canal. The dried root canal is then plugged with chemical components like calcium hydroxide, aldehydes (e.g., formocresol), halides [e.g., iodine in potassium iodide (KI)], antibiotics, etc. as intracanal medication.

As a novel improvement, BNC has better biological characteristics in comparison to conventional plant-based cellulose. The paper points used in conventional treatments as a plug to remove contaminants and sterilize the infection site in the canal space often fail to showcase sufficient expansion property to attain close contact with the canal wall. The traditional paper points are also cursed in terms of lesser tensile strength under oral wet conditions, which eventually leads to outcomes like leftover residual fragments or lints, triggering a septic condition.

On the contrary, bacterial cellulose has a 10 times greater absorption rate and higher expansion capacity than plant cellulose-based paper points. Under wet conditions, bacterial cellulose possesses a greater value of tensile strength; i.e., it maintains physical integrity without being fragmented [20].

14.2.8 Chitosan-Based Hydrogel Reinforced with Oxidized Nanocellulose in the Field of Tooth Engineering

Bone tissue engineering is a multidisciplinary field that offers beneficial substitutes in biopharmaceutical applications. Chitosan-based bone reconstructions have high potential for tissue regeneration due to their low immunogenicity, biodegradability, and bioresorbable features [21]. The undulated physiological arrangement of teeth limits the accessibility and efficiency of regenerative treatments. Thus, the application of tools like injectable hydrogel serves a promising role in the regeneration of the dentin pulp complex [22]. Chitosan is a linear polysaccharide generated due to deacetylation of chitin, a natural biopolymer that is non-toxic, biocompatible, biodegradable, and easily processable [23]. Oxidized nanocellulose is produced easily by oxidizing the hydroxyl groups on the surface of cellulose nanocrystals [24]. The hydrogels were prepared with chitosan, mineral trioxide aggregate, and oxidized nanocellulose. To enhance the proliferation and differentiation of dental pulp stem cells, sol-gel hydrogels with higher accessibility are served by chitosan-based hydrogels. This type of hydrogel with a temperature/pH response is beneficial in terms of its high-water imbibing capacity, minimal invasiveness, porous networks, and capacity to mold perfectly in irregular regions, i.e., the teeth's physiological region [24]. Beta-glycerol-phosphate and genipin are used as chemical and ionic cross-linker agents alongside $NaHCO_3$ as co-gelling agents. The mineral trioxide aggregate is used as an endodontic filling material and differential media. The hydrogels are injected in liquid form, which then solidifies in the pulp layer. Nanocellulose acts as a scaffolding agent in the settling of the injectable hydrogel. The injectable hydrogels are known to promote the osteogenic differentiation of dental pulp stem cells [25].

14.2.9 Pickering Emulsion-Based Oral Drug Delivery

Pickering emulsions are a class of emulsions that are void of surfactants, and stabilized by solid-state components. The special structural features of Pickering emulsion couples have good stability, excellent biocompatibility, and biodegradability and ultimately possess a great potential to be used in oral drug delivery. As a drug delivery system, they multiply the bioavailability of weakly soluble drugs like curcumin, silybin, puerarin, rutin, etc. [26]. Nanocellulose-stabilized Pickering emulsions are of importance in the dental field. Nanocelluloses are bound to form oil/water emulsions, and the emulsions are generally manufactured by mechanical treatment of the mixture of oil and nanocellulose aqueous dispersion, in which nanocellulose is adsorbed at the oil–water interface, stabilizing the emulsion. At the oil–water interface, a hydrophobic edge plane appears due to the wetting ability of cellulose nanocrystals. Nanocellulose-stabilized Pickering emulsions showcase greater durability against coalescence, i.e., merging with additives. Nanocellulose-fixated emulsions are naturally thermostable in a wide range of temperatures by virtue of their structural integrity. When nanocelluloses lacking any surface charges are used as stabilizers, the generated emulsion shows good stability against salt and pH-based changes. Hence, the surface charge of nanocellulose plays a crucial role in the stability of the emulsions. Surface-carboxylated nanocelluloses, prepared by TEMPO-mediated oxidation, are often employed as stabilizers.

The nanocelluloses modified with surface additives can still be efficiently used to form emulsions at the oil–water interface. CNCs esterified by vinyl acetate form ethyl acetate in water, toluene in water, and cyclohexane in water emulsion systems, whereas CNCs esterified by vinyl cinnamate stabilize only cyclohexane in water emulsions.

These findings suggest that the phase behavior of the nanocellulose-stabilized Pickering emulsions can be manipulated by controlling the surface properties. Moreover, nanocellulose-stabilized Pickering emulsions show great potency as a drug delivery system in terms of chemical and physiological properties, dealing with less soluble drugs and boosting their bioavailability [27].

14.3 Conclusion

In the modern era of advanced technology and alarming rates of natural destruction, scientists are continuously looking for bioresources with qualitative properties that can be used as an alternative to hybrid components. Bacterial cellulose is a sustainable raw component that can be exploited in areas like industrial biotechnology and the biomedical sector for its qualitative aspects. The chapter focused on the advantages of nanocellulose, especially bacterial nanocellulose, as a biocompatible, non-toxic, easily available, and cost-effective biopolymer in the biomedical field, precisely under the dental science category [8]. The bacterial source-based nanocellulose, derived through static culture, is a premium raw material to be employed in different sectors under odontology as per its beneficial physicochemical features and surface chemistry. BNC is greatly employed as dressing material for healing wounds, natural or even surgical wounds concerning the oral mucosa. The static culture-based bacterial cellulose membrane is used for periodontal treatment, guided tissue regeneration treatment, and furcation treatment; as a drug delivery agent; in dental pulp tissue treatment and dental root canal treatment; in hydrogel preparation; and as dental scaffolds in surgery; it is also renowned in the field of tissue engineering [20, 24]. The chemical nature of native tissue impersonation allows bacterial cellulose to be used alongside affected tissues without having a second thought about side effects [13].

The biopolymer is often exploited in the formation of different additive complexes like glass ionomer cements, BC collagen, and BC-based hydrogel due to its 200 times greater water absorption and retention capacity over other sources [1, 6].

Moreover, the multifunctional scaffold material, bacterial nanocellulose, is being studied as a cartilage scaffold component under the dressing equipment genre, a vascular graft, and also as a DNA separation medium. As a substitute for autogenous material, it has an important role in guided tissue regeneration techniques. It is also required in the making of commercially accessible vibration membranes, used in the diaphragm of electroacoustic transducers. It is evident that nanocellulose, especially BNC, functions as a key substance in different applications of oral and dental care in the biomedical sector, and it has enormous potential to become an indispensable material required in rather less explored areas of dental science.

References

1. Gumrah Dumanli, A., Nanocellulose and its composites for biomedical applications. *Curr. Med. Chem.*, *24*, 5, 512–528, 2017.
2. Norrrahim, M.N.F., Nurazzi, N.M., Jenol, M.A., Farid, M.A.A., Janudin, N., Ujang, F.A., Yasim-Anuar, T.A.T., Najmuddin, S.U.F.S., Ilyas, R.A., Emerging development of nanocellulose as an antimicrobial material: An overview. *Mater. Adv.*, *2*, 11, 3538–3551, 2021.
3. Daicho, K., Saito, T., Fujisawa, S., Isogai, A., The crystallinity of nanocellulose: Dispersion-induced disordering of the grain boundary in biologically structured cellulose. *ACS Appl. Nano Mater.*, *1*, 10, 5774–5785, 2018.
4. Revin, V.V., Liyaskina, E.V., Parchaykina, M.V., Kuzmenko, T.P., Kurgaeva, I.V., Revin, V.D., Ullah, M.W., Bacterial cellulose-based polymer nanocomposites: A review. *Polymers*, *14*, 21, 4670, 2022.
5. Reshmy, R., Philip, E., Thomas, D., Madhavan, A., Sindhu, R., Binod, P., Varjani, S., Awasthi, M.K., Pandey, A., Bacterial nanocellulose: Engineering, production, and applications. *Bioengineered*, *12*, 2, 11463, 2021.

6. Mikkelsen, D., Flanagan, B.M., Dykes, G.A., Gidley, M.J., Influence of different carbon sources on bacterial cellulose production by Gluconacetobacter xylinus strain ATCC 53524. *J. Appl. Microbiol.*, *107*, 2, 576–583, 2009.
7. Abushammala, H., Krossing, I., Laborie, M.P., Ionic liquid-mediated technology to produce cellulose nanocrystals directly from wood. *Carbohydr. Polym.*, *134*, 609–616, 2015.
8. Tian, C., Yi, J., Wu, Y., Wu, Q., Qing, Y., Wang, L., Preparation of highly charged cellulose nanofibrils using high-pressure homogenization coupled with strong acid hydrolysis pretreatments. *Carbohydr. Polym.*, *136*, 485–492, 2016.
9. Guise, C. and Fangueiro, R., Biomedical applications of nanocellulose, in: *Natural Fibres: Advances in Science and Technology Towards Industrial Applications*, pp. 155–169, Springer, Dordrecht, 2016.
10. de Oliveira Barud, H.G., da Silva, R.R., Borges, M.A.C., Castro, G.R., Ribeiro, S.J.L., da Silva Barud, H., Bacterial nanocellulose in dentistry: Perspectives and challenges. *Molecules*, *26*, 1, 49, 2021.
11. Needleman, I., Worthington, H.V., Giedrys-Leeper, E., Tucker, R., Guided tissue regeneration for periodontal infra-bony defects. *Cochrane Database Syst. Rev.*, 2, 4, 1–35, 2006.
12. Jacob, S.A. and Amudha, D., Guided tissue regeneration: A review. *J. Dent. Health Oral. Disord. Ther.*, *6*, 3, 00197, 2017.
13. Kinane, D.F., Stathopoulou, P.G., Papapanou, P.N., Periodontal diseases. *Nat. Rev. Dis. Primers*, *3*, 1, 1–14, 2017.
14. Pilloni, A. and Rojas, M.A., Furcation involvement classification: A comprehensive review and a new system proposal. *Dent. J.*, *6*, 3, 34, 2018.
15. Amicarelli, R.G. and Alonso, C.A., Treatment of class II furcation lesions using an autogenous periosteal barrier. *Pract. Periodontics Aesthet. Dent.: PPAD*, *11*, 2, 237–44, 1999.
16. Scully, C. and Shotts, R., Mouth ulcers and other causes of orofacial soreness and pain. *BMJ*, *321*, 7254, 162–165, 2000.
17. Carvalho, J.P., Silva, A.C., Bastos, V., Oliveira, H., Pinto, R.J., Silvestre, A.J., Vilela, C., Freire, C.S., Nanocellulose-based patches loaded with hyaluronic acid and diclofenac towards aphthous stomatitis treatment. *Nanomaterials*, *10*, 4, 628, 2020.
18. Rossi, H.L., See, L.P., Foster, W., Pitake, S., Gibbs, J., Schmidt, B., Mitchell, C.H., Abdus-Saboor, I., Evoked and spontaneous pain assessment during tooth pulp injury. *Sci. Rep.*, *10*, 1, 1–12, 2020.
19. Anton-Sales, I., Roig-Sanchez, S., Traeger, K., Weis, C., Laromaine, A., Turon, P., Roig, A., *In vivo* soft tissue reinforcement with bacterial nanocellulose. *Biomater. Sci.*, *9*, 8, 3040–3050, 2021.
20. Yoshino, A., Tabuchi, M., Uo, M., Tatsumi, H., Hideshima, K., Kondo, S., Sekine, J., Applicability of bacterial cellulose as an alternative to paper points in endodontic treatment. *Actabiomaterialia*, *9*, 4, 6116–6122, 2013.

21. Tang, G., Tan, Z., Zeng, W., Wang, X., Shi, C., Liu, Y., He, H., Chen, R., Ye, X., Recent advances of chitosan-based injectable hydrogels for bone and dental tissue regeneration. *Front. Bioeng. Biotechnol.*, *8*, 587658, 2020.
22. DalirAbdolahinia, E., Alipour, M., Aghazadeh, M., Hassanpour, M., Ghorbani, M., Aghazadeh, Z., An injectable chitosan-based hydrogel reinforced by oxidized nanocrystalline cellulose and mineral trioxide aggregate designed for tooth engineering applications. *Cellulose*, *29*, 6, 3453–3465, 2022.
23. Zhang, Z., Zhang, L., Li, C., Xie, X., Li, G., Hu, Z., Li, S., Research progress of chitosan-based biomimetic materials. *Mar. Drugs*, *19*, 7, 372, 2021.
24. Ding, C., Yue, C., Su, J., Wang, H., Yang, N., Cheng, B., Effects of oxidized cellulose nanocrystals on the structure and mechanical properties of regenerated collagen fibers. *Cellulose*, *29*, 14, 7677–7690, 2022.
25. Samiei, M., Abdolahinia, E.D., Fathi, M., Barar, J., Omidi, Y., Chitosan-based bioactive hydrogels for osteogenic differentiation of dental pulp stem cells. *J. Drug Delivery Sci. Technol.*, *73*, 103478, 2022.
26. Tai, Z., Huang, Y., Zhu, Q., Wu, W., Yi, T., Chen, Z., Lu, Y., Utility of Pickering emulsions in improved oral drug delivery. *Drug Discovery Today*, *25*, 11, 2038–2045, 2020.
27. Fujisawa, S., Togawa, E., Kuroda, K., Nanocellulose-stabilized Pickering emulsions and their applications. *Sci. Technol. Adv. Mater.*, *18*, 1, 959–971, 2017.

15

Future Prospects of Bacterial Nanocellulose and Its Composites

Anbazhagan Thirumalai[1], Mahashweta Mitra Ghosh[2], Agnishwar Girigoswami[1] and Koyeli Girigoswami[1]*

[1]*Medical Bionanotechnology, Faculty of Allied Health Sciences, Chettinad Hospital and Research Institute, Chettinad Academy of Research and Education, Chettinad Health City, Rajiv Gandhi Salai, Kelambakkam, Chengalpet Dist., Tamil Nadu, India*

[2]*Department of Microbiology, St. Xavier's College (Kolkata), Kolkata, West Bengal, India*

Abstract

The environment-friendly properties of bacterial nanocellulose (BNC) and their applications are gaining attention day by day. The cellulose produced by microbes has many benefits that led researchers to investigate the scope of microbial hosts for BNC production. New techniques have been developed for large-scale production of BNC. BNC has a 3D network of micro- and nanofibrils and has several unique properties, such as increased tensile strength, crystallinity, purity, water holding capacity, and high modulus, which make them useful in the reinforcement of novel nanostructured composites. In the future, the production techniques of BNC will improve, including static culture, agitated/shaking culture, cell-free production, various receptor-based fermentation, and low-cost substrates as carbon sources to overcome the barriers incorporated with the industrial-scale production of BNC. Currently, *in situ* and *ex situ* fermentation processes are used for the production of BNC-based polymers, and they are widely used with metal or metal oxide nanoparticles to develop nanocomposites in diagnostics and regenerative medicine. Therefore, in this chapter, we have discussed the future prospects of

**Corresponding author*: koyelig@gmail.com
Anbazhagan Thirumalai: ORCID: https://orcid.org/0000-0002-2659-5191
Koyeli Girigoswami: ORCID: https://orcid.org/0000-0003-1554-5241
Agnishwar Girigoswami: ORCID: https://orcid.org/0000-0003-0475-2544

Mainak Mukhopadhyay and Debalina Bhattacharya (eds.) Nanocellulose: A Biopolymer for Biomedical Applications, (373–390) © 2024 Scrivener Publishing LLC

BNC and their several functional forms in biomedical applications like antimicrobial activity, wound healing activity, drug delivery systems, biosensing, tissue engineering, etc.

Keywords: Bacterial nanocellulose, biomedical applications, tissue engineering, biosensing

Abbreviations

BNC	Bacterial nanocellulose
BC	Bacterial cellulose
PC	Plant cellulose
PVA	Poly(vinyl alcohol)
PEG	Poly(ethylene glycol)
PLA	Polylactic acid
Ag	Silver
Au	Gold
Cu	Copper
N-CD	Nitrogen-doped carbon dots
Fe^{3+}	Iron (III)
AgNP	Silver nanoparticle
UV method	Ultraviolet method
PET	Polyethylene terephthalate
ePTFE	Expanded polyethylene tetrafluoroethylene

15.1 Introduction

Many different bacterial strains naturally produce bacterial nanocellulose (BNC), which is a biodegradable material. It comprises thin, microfibrillar strands of cellulose that are interconnected to form a highly ordered three-dimensional network. BNC has a lot of unique properties like high tensile strength, high modulus, and high toughness, which make it a suitable material for a wide range of applications [1]. BNC can be used as a standalone material or as a reinforcement in composite materials. BNC composite materials from a matrix (such as a polymer) are reinforced with BNC fibers. These composites can have improved mechanical properties compared to the matrix material alone. There are several methods for BNC production, such as the traditional fermentation method and the more recent electrospinning method. In the traditional fermentation method, the cultivation of bacteria is generally performed in a liquid medium, whereas in the electrospinning method, an electric field is used for the production of BNC fibers. BNC can be produced in many forms, including sheets, films, and fibers, and can be functionalized with various molecules to tailor its properties for certain applications [2].

One of the major advantages of BNC is its high tensile strength and high modulus, which make it ideal for use in structural applications. BNC also has high toughness, which can withstand a large amount of deformation without breaking. Another advantage of BNC is its biocompatibility and biodegradability [3] and that it is recognized as non-toxic, which means it can be safely used in medical and healthcare applications. As it is biodegradable in nature, it can be degraded into natural substances by microorganisms, making it a potentially sustainable alternative to synthetic materials.

Moreover, BNC has good barrier properties, which make it suitable for packaging applications. It has a low moisture content and can extend the shelf life of food products by preventing the penetration of gasses [2, 4]. In addition to its physical and mechanical properties, BNC also has a high surface area and porosity, making it ideal for use in filtration and separation applications as well. It has been explored as a filter material for water treatment and a membrane material for desalination. Overall, BNC's unique combination of properties makes it a promising material with many potential applications [5].

While BNC has many potential advantages, there are also some challenges that need to be addressed. One major challenge is the high cost of production, which has limited the widespread application of BNC. Another challenge is the limited availability of BNC [6]. Currently, BNC is produced on a small scale, and there is a need to develop large-scale production methods in order to meet industrial demand.

There is also a need for further research and development to fully understand the properties and behavior of BNC and optimize its use in various applications. This includes understanding how to functionalize BNC with various additives to tailor its properties for specific applications and how to process BNC into different forms such as sheets, films, and fibers.

15.2 Cellulose

The most common natural polymer used worldwide is cellulose, which is readily available, inexpensive, and the most abundant biopolymer on earth compared to other biopolymers. Each and every year, 10^{14} tons of cellulose pulp are produced worldwide; this production shows the economic importance of polymers and how they are used in our daily lives [7]. Almost 70 years ago, Ranby first reported the production of colloidal cellulose by sulfuric acid hydrolysis [8]. The primary intermolecular and intramolecular hydrogen bonds between cellulose molecules prevent it from being soluble in water and most common solvents despite its high hydrogen bonding. A wide range of applications are possible due to cellulose's poor solubility. Chemical modifications are performed on cellulose to improve its processing ability. These chemical modifications yield cellulosic (cellulose derivatives), which are generally strong, biocompatible, reproducible, and recyclable and are utilized in different industrial applications as well as the biomedical field of research [9]. Pure cellulose is obtained by various chemical techniques that require highly polluting chemicals, including carbon monoxide, carbon disulfide, carbon dioxide, chlorine gas, and caustic soda. This makes it harmful to nature.

Multiple sources of cellulose production are found, such as naturally occurring chemical reactions, microbes, plants, and enzymatic synthesis. The production of cellulose synthesis varies with the variety of plants, like cotton fibers, which have 90%, wood, which has 70%, and dried hemp, which has 40–50%. Multiple microorganisms, such as algae, fungi, and bacteria, are used to produce cellulose. Donini *et al.* compared the cellulose production from microbes and plants to list out the advantages of cellulose production from microbes [10]. Unlike plant cellulose, bacterial cellulose is produced in its purest form without pectin, hemicelluloses, lignin, or other compounds present in plant pulp. Compared to plant cellulose, bacterial cellulose has higher mechanical properties [11].

15.3 Bacterial Nanocellulose (BNC)

In 1886, Brown first reported the production of cellulose by using *Acetobacter xylinum*. After that, multiple bacterial strains were used to produce cellulose [12]. BNC are highly active materials and are used in drug delivery systems, the fabrication of various polymer nanocomposites, food applications, tissue engineering, etc.

15.3.1 BC/Natural Polymer Composites

BC has been successfully combined with various natural and synthetic polymeric matrices to produce novel nanocomposite materials for various applications [13]. BC-hemicellulose/pectin nanocomposites are the first BC polymer nanocomposites that were made by incubating *Gluconacetobacter xylinum* with pectin or hemicelluloses like xylans, glucomannans, and xyloglucans in either a moving or a still environment. N-deacetylated chitin produces chitosan, which is a polysaccharide with unique physicochemical and biological properties that include biocompatibility, biodegradability, film formation, and antimicrobial activity. Transparent nanocomposite films were greenly fabricated by casting a water-based preparation of BC nanofibrils and chitosan. The BC-chitosan nanocomposite films were used as antimicrobial and biodegradable packaging because they showed superior mechanical properties and were flexible, as well as more transparent than the unfilled chitosan films [14].

Among the most abundant natural polymers, starch is considered a raw material for bio-based composites and innovative materials. The BC mechanical property was increased when mixed with thermoplastic starch, showing the increasing mechanical reinforcement of BC and better biocompatibility between two polysaccharide chains [15]. Hot pressing, solvent casting, melting mixing, impregnation, and drying are the prominent techniques used to prepare BC/thermoplastic starch nanocomposites, which are mostly used for applications in food packaging. Thermoplastic starch is the most stable polysaccharide, and it plays a major role in the films' thermal stability, influences the films transparency, and influences their mechanical properties. Adding chitosan in place of starch will increase the films' mechanical performance, transparency, and antimicrobial properties, but it will decrease their thermal stability [16]. BC and pullulan nanocomposite films were developed by casting a water-based suspension entirely on microbial polysaccharides. Compared to pullulan films, these nanocomposite films showed higher mechanical properties and thermal stability (degradation temperature increased up to 40°C) [17]. Novel protein/BC biocomposites for biomedical applications have also been developed from collagen, gelatin, and fibrin. The BC membrane incorporated with type I collagen during *in situ* formation is one of the first works reported for increasing the features of BC for wound dressing [18]. Glutaraldehyde-treated BC/fibrin nanocomposites were crosslinked with two polymers, showing better mechanical properties with the native blood vessels.

15.3.2 BC/Water-Soluble Synthetic Polymer Nanocomposites

PVA [Poly(vinyl alcohol)] is known as a linear synthetic polymer, and researchers had an interest in developing cellulose-based nanocomposites because of their better biocompatibility, water solubility, and improved mechanical properties. A BC culture medium containing PVA results in nanocomposites with better ion absorption properties and water retention. BC/PVA nanocomposites were chemically crosslinking PVA with glutaraldehyde after the drying, and BC impregnation with PVA would increase the thermal stability and mechanical properties [19]. Traditional, non-biodegradable plastics are replaced by these composites. There are a number of applications in biotechnology and medicine for poly(ethylene glycol) (PEG), and it has novel properties. In a 1% PEG solution, the BC membrane was immersed by the freeze-drying technique to fabricate the BC/PEG biocompatible porous nanocomposites [20]. The strong interaction between the PEG chain and BC showed increasing thermal stability from 263°C to 293°C. The BC-PEG nanocomposite had excellent biocompatibility by showing better cell adhesion than pure BC [21]. Because of the highly interconnected structure of BC/PEG nanocomposites, it showed increasing mechanical, thermal, cell adhesion, and proliferation properties.

15.3.3 BC/Thermoplastic Nanocomposites

The high performance and simplicity of processing BC nanofibrils with synthetic thermoplastics (thermosetting matrices) have been extensively investigated by several researchers. Thermoplastic polymers are a type of polymer that can be melted and reshaped when heated and solidified when they are cooled. Bacterial cellulose/thermoplastic nanocomposites have been explored for various applications because of their exclusive combination of the high strength and stiffness of bacterial cellulose with the flexibility and processability of thermoplastic polymers. These materials have been studied for use in food packaging, biodegradable packaging, and drug delivery systems [22]. They have also been explored for use in tissue engineering and medical devices because of their biocompatibility and the potential for controlled drug release. These BC/thermosetting nanocomposites lack adhesion between the non-polar synthetic polymer and polar BC nanofibers. PLA/BC transparent nanocomposites were prepared by drying the BC pellicle with added chloroform. In comparison to the PLA matrix, BC/PLA nanocomposites showed better Young's modulus, and their tensile strength was elevated by 146% and 203% [23]. The use of organic solvent limited the BC nanofibrils and aliphatic polyester matrix adhesion, and it also exhibited a lack of applications in food packaging and biomedical devices. To overcome these limitations, new transparent nanocomposites were synthesized using the comprehensible mechanism of acetylated BC nanofibrils and PLA. This BC/PLA transparent nanocomposite showed improved mechanical properties, thermal stability, and water resistance [23].

15.3.4 BNC-Based Electroconductive Polymer Nanocomposites

The BC/electroconductive polymer (like polypyrrole and polydimethylsiloxane) nanocomposites were prepared by the high conductivity of the polymer matrix combined with the increased surface area, and the nanostructured nature of the fibers of bacterial cellulose resulted in increased electrical and thermal conductivity. Additionally, the use of bacterial cellulose as a reinforcing agent could improve the mechanical strength and durability of the nanocomposite material. Several approaches can be used to prepare BC/electroconductive polymer nanocomposites, including solution casting, *in situ* polymerization, and melt blending. *In situ* polymerization involves the synthesis of the electroconductive polymer within the network of bacterial cellulose fibers, resulting in a highly homogeneous and well-interconnected material [24].

One of the challenges in preparing BC/electroconductive polymer nanocomposites was the compatibility between the two materials. The high crystallinity and hydrophilic nature of bacterial cellulose could make it difficult to disperse in certain polymers, leading to poor dispersion and reduced conductivity in the final nanocomposite material [25]. To overcome this challenge, various compatible agents and surface modification techniques can be used to improve the interaction between the bacterial cellulose and the polymer matrix. BC/electroconductive polymer nanocomposites possess a versatile range of potential applications because of their unique combination of mechanical strength, electrical conductivity, and polymer flexibility.

15.4 Bacterial Nanocellulose-Based Hybrid Nanocomposites

15.4.1 BNC/Metal and Metal Oxide Nanocomposites

The new class of nanomaterials is called inorganic metal nanoparticles, such as Ag, Au, Cu, and their respective oxides. Because of the symbiotic effect of combined chemical and physical interaction with natural cellulosic materials, plant-based fibrous cellulosic materials are improving their antimicrobial, optical, electrical, mechanical, and thermal properties. Biomedicine can benefit from bacterial cellulose–metal nanocomposites by using them as scaffolds to be implanted in tissue or as drug delivery systems [26]. A step forward, bacterial cellulose–metal oxide nanocomposites may have applications in energy storage, sensors, and for treatment of wastewater.

A controlled release of drugs can be achieved by using bacterial cellulose because of its high surface area and porous structure. The fabrication of metal or metal oxide nanoparticles into the bacterial cellulose matrix can further enhance the drug-release properties of the nanocomposites [27]. For example, silver nanoparticles are known to possess antimicrobial properties, which could be useful in the development of antimicrobial dressings. Similarly, zinc oxide nanoparticles have been incorporated into the drug delivery systems based on bacterial cellulose for their ability to enhance drug release and improve the stability of the system [27].

There has been a growing interest in using bacterial cellulose composites with metal nanocomposites as scaffolds in tissue engineering to facilitate the proliferation and growth of tissue cells. The mechanical properties of the nanocomposite scaffold can be tuned by adjusting the size and distribution of the nanoparticles, which can influence cell differentiation and behavior. Due to gold nanoparticles' ability to promote neural stem cell differentiation into neurons, BNC-gold nanocomposites have been explored as scaffolds for neural tissue engineering [28].

15.4.2 Bacterial Cellulose Carbon Nanocomposites

BC/carbon nanocomposites are materials that are made by incorporating carbon nanomaterials into a matrix of bacterial cellulose. One of the interesting properties of these materials is that they are both strong, electrically conductive, and have improved mechanical and thermal properties as well [29]. A wide variety of applications can be made using these materials, including electronic applications, energy storage, and structural applications. Several methods are used to make BC/carbon nanocomposites. A commonly used method is to incorporate carbon nanomaterials inside the bacterial cellulose during the synthesis process by adding them to the culture medium in which the bacteria are grown. Another method is to mechanically mix the carbon nanomaterials with the bacterial cellulose after it has been produced [30].

In addition to the methods used to incorporate carbon nanomaterials into bacterial cellulose, several different types of carbon nanomaterials can be used. These include graphene, carbon nanotubes, and carbon nanofibers. Each of these materials has its own unique properties, and the choice of which one to use will depend on the particular application for which the bacterial cellulose carbon nanocomposite is being designed. Carbon nanotubes are extremely strong and conductive, and when they are combined with the elevated strength and water retention possessed by bacterial cellulose, the resulting material exhibits a number of interesting properties. One of the main advantages of bacterial cellulose carbon nanotube nanocomposites is their high strength and toughness. In addition to their mechanical and electrical properties, BC/carbon nanotube nanocomposites may also have potential in the field of biomedicine. Bacterial cellulose is biocompatible and has been used in medical applications such as tissue engineering and drug delivery [31]. The addition of carbon nanotubes may further enhance the physicochemical properties of these materials for use in the biomedical field.

Graphene is a two-dimensional form of carbon known for its huge surface area, high electrical conductivity, and extremely high strength. When it is combined with bacterial cellulose, the resulting material has a number of interesting properties, like strength and toughness. Due to their unique combinational properties, BC/Graphene nanocomposites have potential applications in the biomedical field like wound dressing, drug delivery, biosensing, and tissue engineering [32]. The different types of BNC prepared by incorporating the various nanoparticles are shown in Figure 15.1.

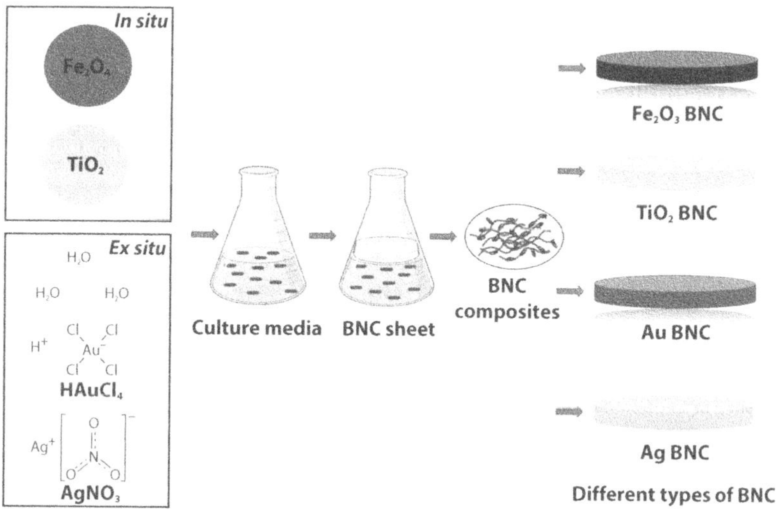

Figure 15.1 The different types of BNC nanocomposites prepared by incorporation of different nanoparticles.

15.5 Biomedical Applications of BNC

15.5.1 Drug Delivery

A drug delivery system is defined as a technology that facilitates the delivery of therapeutic drugs with a targeted or controlled release [33, 34]. The drug-release mechanism and behavior of BNC are based on the composite materials. It has a versatile application due to its high surface area, biocompatibility, biodegradability, and adaptable surface chemistry. Generally, the BC membranes are loaded with drugs through immersion in an aqueous drug solution after lyophilization to maximize absorption. Antimicrobial drugs and anti-inflammatory drugs (diclofenac and ibuprofen) are the most commonly used drugs to be incorporated into BC. Cacicedo *et al.* demonstrated how local drug delivery can enhance cancer treatment by combining nanostructured lipid carriers and BC hydrogel. They used doxorubicin as a model drug to test their implant [35].

15.5.2 Biosensing

Biosensors are emerging tools for detecting biomarkers in various fields like biomedicine and environmental monitoring. Several features must be considered important while developing biosensors, including high sensitivity, biocatalytic activity, selectivity, cost-effectiveness, and environmental friendliness. The BNC is the most favorable material for engineering biosensors due to its large surface area, mechanical strength, high crystallinity, biodegradability, 3D structures, and ability to be easily modified and functionalized with conductive materials, nanoparticles, and biomolecules. The surface optimization of cellulose will increase the sensitivity of biosensors by adding conductive materials such as graphene oxides, gold nanoparticles, carbon nanotubes, and conductive polymers to the surface. The magnetic and electrical conductivity of the BNC were increased by using gold, palladium, titanium dioxide, zinc oxide, ferrous oxide, and platinum nanoparticles. A BC-based biosensor is prepared by altering the culture medium or by incorporating the material in the second phase after obtaining the BC network [36, 37]. If the BC network is destroyed first, different techniques are used to fabricate the BC-based biosensors by combining them with other materials. For the detection of Fe^{3+} ions in aqueous solutions, Lv *et al.* constructed a fluorescent-based BC biosensor composed of BC and nitrogen-doped carbon dots (BC/N-CDs) [38].

15.5.3 Wound Dressing Material

Multiple wound dressing methods are used to protect wounds from injury or keep them out of the external environment. The importance of wound dressing is that the material should allow gaseous exchange, maintain a moist environment, absorb exudates, prevent microbial infections, facilitate healing, support angiogenesis, and be affordable. Also, it should be non-allergenic, non-toxic, non-adherent, and easy to remove. As a result of the microfibrillar structures in BC, a flexible 3D scaffold serves as a physical barrier against pathogens, thereby promoting cell attachment and tissue granulation, which are both necessary for an accelerated regeneration process. BC has favorable characteristics related to wound healing like biocompatibility, decent flexibility, non-toxicity, robust water-holding capacity, elasticity, and vapor permeability [39].

15.5.4 Antimicrobial Activity

If the nanocellulose is not an antimicrobial agent, it cannot protect humans from developing wound infections. BC has a deficiency of antibacterial activity, which confines its wound dressing applications. To overcome these limitations, BC was functionalized by adding antimicrobial agents. In recent years, a huge amount of research has been reported on the use of bacterial nanocellulose incorporated with multiple antimicrobial agents. AgNP/PVA/BNC films were developed through the UV method and reduction methods. In both medium and raw beef broth, the BNC films inhibited the growth of *E. coli* O157:H7. Compared to reduction method films, UV films have higher antimicrobial properties in all antimicrobial experiments [40].

15.5.5 Artificial Skin

Bacterial cellulose is used not only as a treatment for wound healing but also to induce the recovery of epithelial cells (after epithelial damage) and semipermanent artificial skins because of its high water content, unique network structure, and mechanical characteristics. An alternative to autografting cell-seeded BC skin substitutes is used to improve their treatment efficacy for patients with severe damage to their skin, like Stevens–Johnson syndrome and burn victims [41]. A macroporous BC hydrogel was fabricated via physical destruction, growth factors, and sequential modifications with elastin, collagen, and hyaluronan that were used to enhance its biocompatibility and biodegradability [42].

15.5.6 Tissue Engineering

Nanocellulose materials have emerged to advance biomedical engineering, such as regenerative medicine and tissue engineering [43]. BNC is a natural polymer; it is considered a unique scaffold material because of its unique material properties. BNC-based scaffolds are mostly classified into three classes: BNC/matrix biocomposite, BNC membrane, and surface-modified BNC pellicle scaffolds [44].

15.5.7 Vascular Tissue Engineering

The increasing number of coronary heart diseases and their emergence in healthcare require new blood vessel replacement. Synthetic blood vessels like expanded polytetrafluorethylene and polyethylene terephthalates are commonly utilized vascular materials for grafting the lower extremity bypass surgery. Although these synthetic materials yield a satisfactory rate of success with large-diameter blood vessels only, small-diameter vessels having <6 mm diameter show very poor performance because of the early thrombosis. BC is a favorable material in biomedical applications compared to synthetic polymers, and it has several advantages. BC can be engineered into tube shapes with any particular diameter and length because of its unique production techniques, and BC tubes increase blood compatibility and show better compliance and mechanical properties [45].

15.5.8 Bone Tissue Engineering

In order to produce bone tissue, collagen is used as the matrix, and hydroxyapatite is the mineralizing material. BC can act as an alternative to collagen fibers because it has numerous advantages, such as high crystallinity, biocompatibility, moldability, huge tensile strength in wet and dry states, and fine meshy networks of its fibrils. The morphological and nanostructure similarities between BC and collagen have led to its application as a potential bone tissue scaffold, which could provide maximum amalgamation with the cells and fluids of our body. Using 3D printing, Cakmak *et al.* engineered a composite scaffold made from BC, gelatin, polycaprolactone, and hydroxyapatite for bone and tissue engineering [46].

15.5.9 Cartilage Tissue Engineering

Articular cartilage has a limited ability to regenerate, so the regeneration problem is important. Chondrocytes must be able to penetrate and proliferate within the porous material in order to produce the extracellular matrix required for cartilage repair. For cartilage tissue engineering, BC scaffolds are suitable due to their biocompatibility and mechanical strength. BC cannot penetrate into the cartilage cells because of their small pore size of <10 μm. To overcome this small pore size, Xun *et al.* developed macroporous BC scaffolds with nearly 200 μm pore size by using the freeze-drying technique [47].

15.5.10 Anticoagulant Character

Artificial organs and blood-contacting biomaterials, such as artificial hearts, blood vessels, and pumps, are needed to improve their biocompatibility for clinical trials. Decron, ePTFE, and polyurethane are vascular substitutes used in our daily lives, but they have some limitations. Intimal fibrous hyperplasia and thromboembolism/thrombosis are the major causes of the failure of vascular grafts. Fink *et al.* showed that BNC does not induce plasma coagulation compared with commercial vascular graft tubes made of ePTFE (expanded polyethylene tetrafluoroethylene) and PET (polyethylene terephthalate). BC had excellent performance and induced the least and slowest coagulation cascade activation when compared to PET and ePTFE [48]. The different applications of BNC are shown in Figure 15.2.

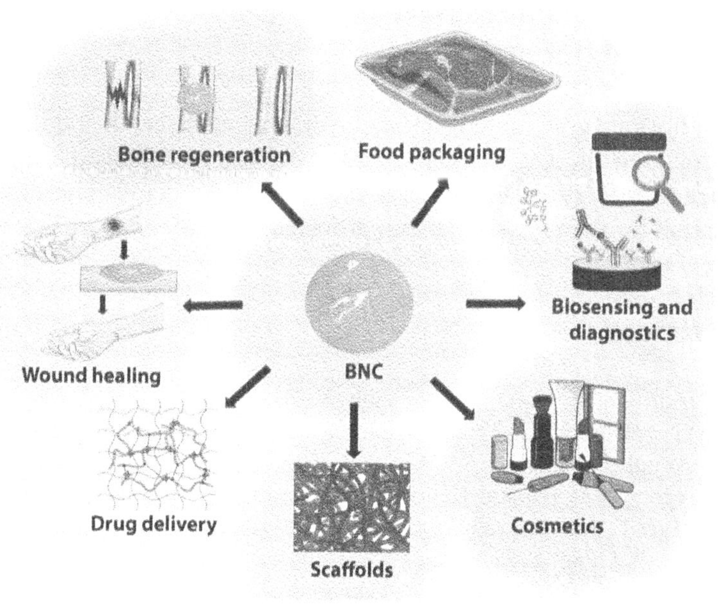

Figure 15.2 The application of bacterial nanocellulose in different fields.

15.6 Conclusion

Bacterial nanocellulose is a flexible material used in various applications in multiple fields. Because of its excellent mechanical properties, 3D network structure, availability, and ability to host augmentation to achieve superior and synergistic properties, it has been used more frequently and extensively. It is widely known that BNC is of high industrial concern and has an enormous range of applications; additional impetus is required for making this environmentally friendly material a competitive and economical product. BNC has also contributed to biomedical fields like drug delivery, biosensing, tissue engineering, wound dressing, and medical implants. As a result, continuous efforts are made to make the procurement of BNC feasible through yield enhancement, cost reduction, and a variety of physical, mechanical, and biological factors tailored to optimize its performance. To increase production yield and reduce production costs, low-cost substrates are used in fermentation processes. The addition of extraneous materials like micro- or nanoparticles during the *in situ* or *ex situ* fermentation process will significantly increase the BC properties.

Overall, the future of BNC and its composites looks bright, with many potential applications in various industries. However, further research and development are needed to fully realize the potential of this material.

Acknowledgments

The authors are grateful to Chettinad Academy of Research and Education for providing the CARE Seed Grant (004/Regr./AR-Research/2022-06). Anbazhagan Thirumalai is grateful to Chettinad Academy of Research and Education for the PhD scholarship provided.

References

1. Randhawa, A., Dutta, S.D., Ganguly, K., Patil, T.V., Patel, D.K., Lim, K.-T., A review of properties of nanocellulose, its synthesis, and potential in biomedical applications. *Appl. Sci.*, *12*, 7090, 2022.
2. Gregory, D.A., Tripathi, L., Fricker, A.T., Asare, E., Orlando, I., Raghavendran, V., Roy, I., Bacterial cellulose: A smart biomaterial with diverse applications. *Mater. Sci. Eng.: R: Rep.*, *145*, 100623, 2021.
3. Nascimento, D.M., Nunes, Y.L., Figueirêdo, M.C., de Azeredo, H.M., Aouada, F.A., Feitosa, J.P., Rosa, M.F., Dufresne, A., Nanocellulose nanocomposite

hydrogels: Technological and environmental issues. *Green Chem.*, *20*, 2428–2448, 2018.

4. Bangar, S.P. and Whiteside, W.S., Nano-cellulose reinforced starch bio composite films-A review on green composites. *Int. J. Biol. Macromol.*, *185*, 849–860, 2021.
5. Liu, Y., Liu, H., Shen, Z., Nanocellulose based filtration membrane in industrial waste water treatment: A review. *Materials*, *14*, 5398, 2021.
6. Reshmy, R., Philip, E., Thomas, D., Madhavan, A., Sindhu, R., Binod, P., Varjani, S., Awasthi, M.K., Pandey, A., Bacterial nanocellulose: Engineering, production, and applications. *Bioengineered*, *12*, 11463, 2021.
7. Felgueiras, C., Azoia, N.G., Gonçalves, C., Gama, M., Dourado, F., Trends on the cellulose-based textiles: Raw materials and technologies. *Front. Bioeng. Biotechnol.*, *9*, 608826, 2021.
8. Rånby, B.G., Fibrous macromolecular systems. Cellulose and muscle. The colloidal properties of cellulose micelles. *Discuss. Faraday Soc.*, *11*, 158–164, 1951.
9. Gopi, S., Balakrishnan, P., Chandradhara, D., Poovathankandy, D., Thomas, S., General scenarios of cellulose and its use in the biomedical field. *Mater. Today Chem.*, *13*, 59–78, 2019.
10. Donini, Í., De Salvi, D., Fukumoto, F., Lustri, W., Barud, H., Marchetto, R., Messaddeq, Y., Ribeiro, S., Biosynthesis and recent advances in bacterial cellulose production. *Eclet. Quím.*, *35*, 165–178, 2010.
11. Fu, L., Zhang, J., Yang, G., Present status and applications of bacterial cellulose-based materials for skin tissue repair. *Carbohydr. Polym.*, *92*, 1432–1442, 2013.
12. Jang, W.D., Hwang, J.H., Kim, H.U., Ryu, J.Y., Lee, S.Y., Bacterial cellulose as an example product for sustainable production and consumption. *Microb. Biotechnol.*, *10*, 1181, 2017.
13. Shah, N., Ul-Islam, M., Khattak, W.A., Park, J.K., Overview of bacterial cellulose composites: A multipurpose advanced material. *Carbohydr. Polym.*, *98*, 1585–1598, 2013.
14. Fernandes, S.C., Oliveira, L., Freire, C.S., Silvestre, A.J., Neto, C.P., Gandini, A., Desbriéres, J., Novel transparent nanocomposite films based on chitosan and bacterial cellulose. *Green Chem.*, *11*, 2023–2029, 2009.
15. Benito-González, I., López-Rubio, A., Martínez-Sanz, M., Potential of lignocellulosic fractions from Posidonia oceanica to improve barrier and mechanical properties of bio-based packaging materials. *Int. J. Biol. Macromol.*, *118*, 542–551, 2018.
16. Thakur, V.K., *Nanocellulose polymer nanocomposites: Fundamentals and applications*, John Wiley & Sons, USA, 2014.
17. Shashikumar, C., Mitra, S., Singha, S., Microbial polysaccharides (MPs) in food packaging, in: *Biopolymer-Based Food Packaging: Innovations and Technology Applications*, pp. 225–263, 2022.
18. Wiegand, C., Elsner, P., Hipler, U.-C., Klemm, D., Protease and ROS activities influenced by a composite of bacterial cellulose and collagen type I *in vitro*. *Cellulose*, *13*, 689–696, 2006.

19. Qiu, K. and Netravali, A.N., Bacterial cellulose-based membrane-like biodegradable composites using cross-linked and noncross-linked polyvinyl alcohol. *J. Mater. Sci.*, *47*, 6066–6075, 2012.
20. Sun, Y., Meng, C., Zheng, Y., Xie, Y., He, W., Wang, Y., Qiao, K., Yue, L., The effects of two biocompatible plasticizers on the performance of dry bacterial cellulose membrane: A comparative study. *Cellulose*, *25*, 5893–5908, 2018.
21. Cai, Z. and Kim, J., Bacterial cellulose/poly (ethylene glycol) composite: Characterization and first evaluation of biocompatibility. *Cellulose*, *17*, 83–91, 2010.
22. Wang, L., Gardner, D.J., Wang, J., Yang, Y., Tekinalp, H.L., Tajvidi, M., Li, K., Zhao, X., Neivandt, D.J., Han, Y., Towards industrial-scale production of cellulose nanocomposites using melt processing: A critical review on structure-processing-property relationships. *Compos. Part B: Eng.*, *201*, 108297, 2020.
23. Ávila Ramírez, J.A., Cerrutti, P., Bernal, C., Errea, M.I., Foresti, M.L., Nanocomposites based on poly (lactic acid) and bacterial cellulose acetylated by an α-hydroxyacid catalyzed route. *J. Polym. Environ.*, *27*, 510–520, 2019.
24. Liu, W., Du, H., Zheng, T., Si, C., Biomedical applications of bacterial cellulose based composite hydrogels. *Curr. Med. Chem.*, *28*, 8319–8332, 2021.
25. Ambrosio-Martín, J., Fabra, M., Lopez-Rubio, A., Lagaron, J., Melt polycondensation to improve the dispersion of bacterial cellulose into polylactide via melt compounding: Enhanced barrier and mechanical properties. *Cellulose*, *22*, 1201–1226, 2015.
26. Badshah, M., Ullah, H., Wahid, F., Khan, T., Bacterial cellulose-based metallic green nanocomposites for biomedical and pharmaceutical applications. *Curr. Pharm. Des.*, *26*, 5866–5880, 2020.
27. Swingler, S., Gupta, A., Gibson, H., Kowalczuk, M., Heaselgrave, W., Radecka, I., Recent advances and applications of bacterial cellulose in biomedicine. *Polymers*, *13*, 412, 2021.
28. Jabbari, F., Babaeipour, V., Bakhtiari, S., Bacterial cellulose-based composites for nerve tissue engineering. *Int. J. Biol. Macromol.*, 217, 120–130, 2022.
29. Hu, W., Chen, S., Yang, J., Li, Z., Wang, H., Functionalized bacterial cellulose derivatives and nanocomposites. *Carbohydr. Polym.*, *101*, 1043–1060, 2014.
30. Miyashiro, D., Hamano, R., Umemura, K., A review of applications using mixed materials of cellulose, nanocellulose and carbon nanotubes. *Nanomaterials*, *10*, 186, 2020.
31. Liu, W., Du, H., Zhang, M., Liu, K., Liu, H., Xie, H., Zhang, X., Si, C., Bacterial cellulose-based composite scaffolds for biomedical applications: A review. *ACS Sustain. Chem. Eng.*, *8*, 7536–7562, 2020.
32. Jawaid, M. and Mohammad, F., *Nanocellulose and nanohydrogel matrices: Biotechnological and biomedical applications*, John Wiley & Sons, USA, 2017.
33. Deepika, R., Girigoswami, K., Murugesan, R., Girigoswami, A., Influence of divalent cation on morphology and drug delivery efficiency of mixed polymer nanoparticles. *Curr. Drug Delivery*, *15*, 652–657, 2018.

34. Girigoswami, K., Pallavi, P., Girigoswami, A., Targeting cancer stem cells by nanoenabled drug delivery, in: *Cancer Stem Cells: New Horizons in Cancer Therapies*, pp. 313–337, 2020.
35. Cacicedo, M.L., Islan, G.A., León, I.E., Alvarez, V.A., Chourpa, I., Allard-Vannier, E., García-Aranda, N., Díaz-Riascos, Z., Fernández, Y., Schwartz Jr., S., Bacterial cellulose hydrogel loaded with lipid nanoparticles for localized cancer treatment. *Colloids Surf. B: Biointerfaces*, *170*, 596–608, 2018.
36. Li, X., Li, D., Zhang, Y., Lv, P., Feng, Q., Wei, Q., Encapsulation of enzyme by metal-organic framework for single-enzymatic biofuel cell-based self-powered biosensor. *Nano Energy*, *68*, 104308, 2020.
37. Sanchis, M., Carsí, M., Gómez, C., Culebras, M., Gonzales, K., Torres, F., Monitoring molecular dynamics of bacterial cellulose composites reinforced with graphene oxide by carboxymethyl cellulose addition. *Carbohydr. Polym.*, *157*, 353–360, 2017.
38. Lv, P., Yao, Y., Li, D., Zhou, H., Naeem, M.A., Feng, Q., Huang, J., Cai, Y., Wei, Q., Self-assembly of nitrogen-doped carbon dots anchored on bacterial cellulose and their application in iron ion detection. *Carbohydr. Polym.*, *172*, 93–101, 2017.
39. Naz, S., Ali, J.S., Zia, M., Nanocellulose isolation characterization and applications: A journey from non-remedial to biomedical claims. *Bio-Des. Manuf.*, *2*, 187–212, 2019.
40. Wang, W., Yu, Z., Alsammarraie, F.K., Kong, F., Lin, M., Mustapha, A., Properties and antimicrobial activity of polyvinyl alcohol-modified bacterial nanocellulose packaging films incorporated with silver nanoparticles. *Food Hydrocolloids*, *100*, 105411, 2020.
41. Petersen, N. and Gatenholm, P., Bacterial cellulose-based materials and medical devices: Current state and perspectives. *Appl. Microbiol. Biotechnol.*, *91*, 1277–1286, 2011.
42. Lin, Y.-K., Chen, K.-H., Ou, K.-L., Liu, M., Effects of different extracellular matrices and growth factor immobilization on biodegradability and biocompatibility of macroporous bacterial cellulose. *J. Bioact. Compat. Polym.*, *26*, 508–518, 2011.
43. Sathyaraj, W.V., Prabakaran, L., Bhoopathy, J., Dharmalingam, S., Karthikeyan, R., Atchudan, R., Therapeutic efficacy of polymeric biomaterials in treating diabetic wounds—An upcoming wound healing technology. *Polymers*, *15*, 1205, 2023.
44. Lin, D., Lopez-Sanchez, P., Li, R., Li, Z., Production of bacterial cellulose by Gluconacetobacter hansenii CGMCC 3917 using only waste beer yeast as nutrient source. *Bioresour. Technol.*, *151*, 113–119, 2014.
45. Lee, S.E. and Park, Y.S., The role of bacterial cellulose in artificial blood vessels. *Mol. Cell. Toxicol.*, *13*, 257–261, 2017.
46. Cakmak, A.M., Unal, S., Sahin, A., Oktar, F.N., Sengor, M., Ekren, N., Gunduz, O., Kalaskar, D.M., 3D printed polycaprolactone/gelatin/bacterial

cellulose/hydroxyapatite composite scaffold for bone tissue engineering. *Polymers*, *12*, 1962, 2020.

47. Xun, X., Li, Y., Zhu, X., Zhang, Q., Lu, Y., Yang, Z., Wan, Y., Yao, F., Deng, X., Luo, H., Fabrication of robust, shape recoverable, macroporous bacterial cellulose scaffolds for cartilage tissue engineering. *Macromol. Biosci.*, *21*, 2100167, 2021.
48. Fink, H., Faxälv, L., Molnar, G.F., Drotz, K., Risberg, B., Lindahl, T.L., Sellborn, A., Real-time measurements of coagulation on bacterial cellulose and conventional vascular graft materials. *Acta Biomater.*, *6*, 1125–1130, 2010.

Index

2,2,6,6-tetra-methylpiperidine-1-oxyl radical (TEMPO) oxidation, 28
3D printing, 207

Acetobacter, 231, 234–235
Acetylation, 142–143, 149, 154, 230
Achromobacter, 231
Acid hydrolysis, 230–231, 233
Acid treatment/acid hydrolysis, 32
Active targeting, 224, 225
Acylation/alkanoylation, 142, 144
  dispersivity, 145, 150
  zeta potential, 145, 150
Aerogel, 209
Agrobacterium, 231, 234
Alkali treatment, 31
Amidation, 12
Angiogenesis, 195, 200–201
Anhydroglucose unit (AGUs), 3
Antibacterial, 267, 268, 273, 279, 285–287, 289, 291–293
Antibacterial activity, 197–200, 202–206, 208
Applications of electrospun fibers,
  biomedical applications, 101–105
Atomic force microscopy (AFM), 82
ATP, 147
Azotobacter, 231

Bacterial cellulose (BC), 162, 163, 165, 166, 168–169, 171–174, 176, 178, 224, 234–242, 244, 250, 361–362, 366, 368–369
Bacterial nanocellulose (BNC), 64, 65, 224, 232, 234, 235, 242, 244, 247, 248
  BC/natural polymer composites, 376–377
  BC/thermoplastic nanocomposites, 378–379
  BC/water-soluble synthetic polymer nanocomposites, 377–378
  BNC-based electroconductive polymer nanocomposites, 379
Bacterial nanocellulose-based hybrid nanocomposites,
  bacterial cellulose carbon nanocomposites, 380–381
  BNC/metal and metal oxide nanocomposites, 379–380
Baker's yeast, 146–147
Ball milling, 34
BCS operon, 66–72
BcsA and BcsB, 70–71
BcsC, 71
BcsD, 71
BglAx, 72
Biocompatibility, 160, 161, 164, 165, 169, 171–174, 178, 180–181, 193–194, 196, 204
Biodegradability, 165, 166, 180, 181, 193, 196, 210
Biomass, 147, 268, 270
Biomaterial, 277, 283, 287
Biomedical, 267, 268, 271, 277–279, 292

Biomedical applications of BNC,
  anticoagulant character, 385
  antimicrobial activity, 383
  artificial skin, 383
  biosensing, 382
  bone tissue engineering, 384
  cartilage tissue engineering, 384–385
  drug delivery, 382
  tissue engineering, 384
  vascular tissue engineering, 384
  wound dressing material, 383
Bionanomaterials, 160
Bleaching, 32
Blood clotting, 199, 202
Blood coagulation, 198
Bone regeneration, 147

Canker sore, 359, 364
Carboxymethylation, 230
Carboxymethylcellulose, 246, 251
Casting, 274–276, 278, 292
CcpAx, 71–72
Cellulose, 147–148, 154, 267–293, 375–376
  extraction, 340
  origin, 339
  sources, 340
Cellulose fiber synthesis sources, 4
Cellulose nanocrystals (CNC), 28, 162, 164–170, 176, 178, 224, 233, 236, 240, 247
Cellulose nanofiber, 28
Cellulose nanofibrils (CNF), 162, 166–169, 171, 178, 224, 232, 233, 238, 240–242, 244, 246–248
Centrifugation, 37
Chemical characterization, 47
Chemical modification, 141
Chronic wounds, 193, 195–197, 201
CmxAx, 72
Coating, 276, 278, 279, 281
Collagen, 196–198, 200–202, 208, 211
Composite, 169–173, 175, 196, 208
Cryogel, 210
Crystalline domain, 198
Crystalline nanocellulose, 360
Crystallinity index, 360
Current challenges,
  functionalization, 20
  mass production, 20

Dermatitis, 204
Dermis, 209–210
Desalination/desalting, 30
Detection of polymer nanocomposites, 15
Dewaxing, 31
Dialysis, 37
Differential scanning calorimetry (DSC), 15, 83
Dispersion of nanocellulose, 14
Docetaxel (DTX), 226, 246
Drug carriers, 226, 235, 237–238
Drying, 31
Dynamic modulus analyzer (DMA), 15

ECMs, 238, 241
Electrophoretic light scattering microscopy (ELS), 82
Electrospinning, 89, 201–202
Electrospinning variations,
  emulsion electrospinning, 100
  melt electrospinning, 100
  needle-based electrospinning, 97–100
  needleless electrospinning, 100
  solution electrospinning, 101
Elemental analysis, 48
Endotransglucosylase, 148
Enzymatic hydrolysis, 28
Epidermis, 195, 209–210
Escherichia, 231
Esterification, 11, 142–143, 146–147, 149
  degree of substitution (DS), 142, 149

homogeneous route, 142
one-pot reaction, 143
Experiment setup, 92
Extraction, 28
Extraction of nanocellulose,
chemical extraction, 9
mechanical extraction, 6, 9
Exuding wounds, 203–204
Eyes of biomaterial, 18, 19

Factors affecting the process of electrospinning,
environmental aspects, 96–97
material parameters, 94–96
operating parameters, 94
properties of solvents, 96
Fibroblasts, 195, 197, 201, 206–207
Films, 204–205, 211
Flame resistance, 147
Fourier transform infrared spectroscopy (FTIR), 15, 47, 81
Furcation lesion,
class I, 362–364
class II, 362–364

Grafting, 146, 148
Granulation tissue, 195–197, 200, 208
GRAS, 361
Grinding, 34

Hemostasis, 195, 198–199
Hexokinases, 146
History of electrospinning, 90
Homogenization, 28
HS media, 78
Hydrocolloids, 194, 203, 211
Hydrogels, 16, 17, 194, 196–199, 206, 211, 224, 227, 238–239, 242–243, 252, 367
Hydrolases, 147
Hyperpigmentation, 196

Inflammation, 195–196, 199, 201, 208
Inflammatory cells, 195, 197
Interfacial polyelectrolyte complexation, 17
Ionic liquid extraction, 31

Kidney-targeted drug delivery by nanocellulose, 309–312
Komagataeibacter xylinus, 66–76

Laccases, 146–147
Lactoferrin, 208
Leukocytes, 195
Lignin, 147
Lignocellulosic nanofibers, 16
Lipases, 146–147

Macrophages, 196
Magnetic field, 236, 243
Make up of cellulose fibers,
chemical makeup, 337
structural makeup, 338–339
MCC, 244, 252
Mechanical characterization, 48
Mechanical properties, 162, 163, 168, 171–172, 174, 178, 180
Membranes, 204–206, 211
Methotrexate, 226, 243
MFC, 233–235, 243
Microfluidization, 34
Microwave-assisted hydrolysis, 33
miRNA, 227
Modification of function for nanocelluloses, 10
Morphology, 43

Nanocellulose, 159–164, 166, 167, 169–181, 267–293
hydrogel, 329
isolation, 322, 324
properties, 326, 327
source, 323
sources and preparation, 119–125
structure, 321
types and nomenclature, 118–119

Nanocellulose-based nanocomposites,
  nanocellulose properties, 125–126
  nanocomposite fabrication, 126–130
Nanocellulose—concept and formulation, 304–305
  bacterial cellulose (BC), 307–308
  cellulose nanocrystals (CNCs), 307
  cellulose nanofibers (CNFs), 306
  cellulose nanofibrils (CNFib), 308
  methods of extraction,
    chemical extraction, 306
    mechanical extraction, 305–306
  nanocellulose aerogels, 309
Nanocellulose-reinforced polymer composites, 13, 19
Nanocomposites, 114–116, 170–171, 175–176, 267, 268, 273–293
Nanocrystal, 194, 200, 202, 209
Nanofibers, 196, 199–203, 210–211
Nanofibrils, 194, 203, 205, 207
Nanofilaments, 17
Nanofillers, 116–117
Nanomaterials, 160, 170, 176, 180–181
Nanoparticles, 199, 201, 207–208
Natural biopolymer, 2
Neovascularization, 196, 206
Neutrophils, 195, 196
Nomenclature, 5
Nuclear magnetic resonance (NMR), 81

Ophthalmic hydrogel, 330
Oral drug delivery, 367–368
Oxidation, 146

Particle size analysis, 83
Periodontal disease, 362–364
Physical characterization, 43
Physical techniques, 148
  corona, 148, 153–154
  flame, 148, 153–154
  ion-beam, 148, 153–154
  plasma, 148, 152
Physiochemical performance, 14
Pickering emulsion, 359, 367–368
Poly(e-caprolactone), 16
Polylactic acid, 16
Polymer grafting, 142–143, 149
  grafting from, 143
  grafting onto, 143, 145
Polymerization, 270, 271, 276, 278
Post-treatment, 28
Pre-treatment, 28
Principle of the process, 93
Processing, 274, 276, 278
Proliferation, 193, 195–196, 198–199, 201–207, 210
Properties of nanocellulose-based nanocomposites,
  barrier property, 132–133
  mechanical property, 131
  optical property, 133
  other properties, 133
  thermal property, 132

Re-epithelialization, 196, 202, 205, 208
Reinforcement, 274, 275
Renal disease and its implications,
  causes of renal disease, 302–303
  diagnosis of renal disease, 303–304
  implications of renal disease,
    accumulation of waste products, 301
    anemia, 301
    bone disease, 302
    electrolyte imbalances, 301
    hypertension, 301
  structure and function of the kidneys, 300
  symptoms of renal disease, 303
  types of renal disease,
    acute kidney injury, 300
    chronic kidney disease, 300
    glomerulonephritis, 300
    polycystic kidney disease, 301
Retting, 30

Rhizobium, 231, 234
Root canal treatment, 359, 366

Salmonella, 231
Sarcina, 231, 234
Scanning electron microscopy (SEM), 15, 43, 81–82
Scanning probe microscopy (SPM), 15
Scanning tunneling microscope (STM), 15
Silylation, 11, 142, 144, 150, 154
  silane coupling, 144
  solvent exchange, 144, 150
siRNA, 227
Skin regeneration, 193, 198, 207, 211
Small angle x-ray diffraction (SAXS), 15
Sonication, 34
Sponges, 196, 199–201
Static culture, 77–78
Steam explosion, 28
Stirred culture, 77–78
Sugar palm crystalline nanocellulose, 16
Sulfonation, 12, 142, 145, 151
  affinity, 144, 145
Surface chemistry, 163, 166
Surgical wound treatment, 362, 364

TEMPO, 142–143, 145–146, 151, 155
Tempo-mediated oxidation, 230, 233–234, 239
Thermal stability, 147
Thermo gravimetric analyzer (TGA), 15
Thermo mechanical analyzer (TMA), 15
Thermogravimetric analysis (TGA), 82–83
Thermogravimetry analysis, 46
Tissue engineering, 160–163, 165, 168, 172, 175, 176, 178–181
  blood vessel, 173–174
  bone, 176
  neural, 175
  skin, 172
Tissue grafting, 161, 172, 181
Tissue reinforcement, 359, 365
Transformation of cellulose,
  cellulose nanocrystals (CNC), 341
  chemical alterations, 342, 344–345
  microcrystalline cellulose (MCC), 344
  nanofibrillar cellulose (NFC), 341
Transmission electron microscopy (TEM), 15, 43, 82
Types of nanocellulose,
  bacterial nanocellulose (BNCs), 6–8
  nanocrystalline cellulose (NCCs), 6–8
  nanofibrillated cellulose (NFCs), 6–8

Ulcers, 196, 211

Wide angle x-ray diffraction (WAXED), 15
Wise treatment, 32

X-ray diffraction, 45
X-ray diffraction spectroscopy (XDR), 81
X-ray photoelectron spectroscopy, 48
Xylanases, 147
Xyloglucan, 147

Young's modulus, 194, 206, 208–209

Zeta potential, 43

Printed and bound by CPI Group (UK) Ltd, Croydon, CR0 4YY

17/09/2026

14964277-0003